Road Planning - Freeways and Country Roads

AF212464

Thomas Richter

Road Planning - Freeways and Country Roads

Springer

Thomas Richter
Hannover, Germany

ISBN 978-3-658-35188-5 ISBN 978-3-658-35189-2 (eBook)
https://doi.org/10.1007/978-3-658-35189-2

© Springer Fachmedien Wiesbaden GmbH, part of Springer Nature 2023
This book is a translation of the original German edition „Planung von Autobahnen und Landstraßen" by Richter, Thomas, published by Springer Fachmedien Wiesbaden GmbH in 2016. The translation was done with the help of artificial intelligence (machine translation by the service DeepL.com). A subsequent human revision was done primarily in terms of content, so that the book will read stylistically differently from a conventional translation. Springer Nature works continuously to further the development of tools for the production of books and on the related technologies to support the authors.
This work is subject to copyright. All rights are reserved by the Publisher, whether the whole or part of the material is concerned, specifically the rights of translation, reprinting, reuse of illustrations, recitation, broadcasting, reproduction on microfilms or in any other physical way, and transmission or information storage and retrieval, electronic adaptation, computer software, or by similar or dissimilar methodology now known or hereafter developed.
The use of general descriptive names, registered names, trademarks, service marks, etc. in this publication does not imply, even in the absence of a specific statement, that such names are exempt from the relevant protective laws and regulations and therefore free for general use.
The publisher, the authors, and the editors are safe to assume that the advice and information in this book are believed to be true and accurate at the date of publication. Neither the publisher nor the authors or the editors give a warranty, expressed or implied, with respect to the material contained herein or for any errors or omissions that may have been made. The publisher remains neutral with regard to jurisdictional claims in published maps and institutional affiliations.

This Springer imprint is published by the registered company Springer Fachmedien Wiesbaden GmbH, part of Springer Nature.
The registered company address is: Abraham-Lincoln-Str. 46, 65189 Wiesbaden, Germany

Preface

The planning of country roads and motorways is subject to a variety of requirements. The continuing growth in motor vehicle traffic, especially in the area of long-distance freight transport, poses great challenges to our long-distance road network. Financial resources for new construction and expansion are limited, and a lot of money is needed to maintain the infrastructure.

Furthermore, the objectives in road planning have changed. The safety and ease of long-distance transport continue to be important objectives, to which a number of other objectives have been added, which are becoming increasingly important. The creation and safeguarding of a healthy environment for the population and the preservation and promotion of undivided habitats for fauna and flora are also important objectives that must be taken into account in the planning of rural roads and motorways.

This book takes up these diverse requirements and first of all presents the objectives of road planning and the user needs. Then the planning process for motorways and rural roads is presented. After a presentation of the basic design and operational features, the design elements for motorways and rural roads are introduced and explained. Finally, the required roadside furniture is presented.

The book is based in particular on the FGSV rules and regulations, excerpts of which are reproduced here with the kind permission of the Forschungsgesellschaft für Straßen- und Verkehrswesen e.V. (Road and Transportation Research Association). Decisive for the application of the FGSV rules and regulations is their version with the latest issue date, which is available at FGSV Verlag, Wesselinger Str. 17, 50999 Köln, www.fgsv-verlag.de.

The book is aimed at students of transport and related subjects who are involved in the planning, design, and operation of motorways and rural roads. The book is also addressed to all practitioners in administrations, engineering offices, or construction companies who wish to have a summary of the FGSV regulations with additional explanations for the planning of motorways and rural roads.

Hannover, Germany Thomas Richter

Contents

Introduction

1

Summary

Roads outside built-up areas serve to connect nearby and far away regions and form the backbone of individual mobility and the exchange of goods. Germany's central location in Europe and the increasing national and international connected social life require safe and efficient roads, not least to support economic performance. This book takes up the aspects of planning, design and operation of rural roads and motorways and is aimed at students of transport and practitioners in engineering offices and administrations.

1.1 Historical Review of Country Roads and Motorways

The history of country roads goes back several thousand years. The first paved roads were found in Mesopotamia (around 4000 BC) and India (around 3000 BC). The Romans created an 80,000 km road network of high quality until 200 AD. The design strategy was long straight roads with tight curves for inevitable changes of direction. For strategic reasons, this was preferably laid on hills or other lookout points.

In the eighteenth century, road construction was intensified to improve trade opportunities. In the nineteenth century, railway construction complemented the expansion of transport routes and new design principles were developed. Larger radii were used and for the first time transition curves were proposed between straight lines and circular curves. At the beginning of the twentieth century, the emerging motor vehicles required a further adaptation of the design principles. The large speed differences between motorized and non-motorized traffic systems required overtaking opportunities and larger radii to allow safe driving.

© Springer Fachmedien Wiesbaden GmbH, part of Springer Nature 2023
T. Richter, *Road Planning - Freeways and Country Roads*,
https://doi.org/10.1007/978-3-658-35189-2_1

Until the beginning of the twentieth century, the streets were similarly constructed. They were given a curvature and a clamping between curbs to improve water drainage and durability. They were usually only partially paved with adjoining earth paths for pedestrians and horsemen.

In 1924 the Studiengesellschaft für Automobilstraßenbau (STUFA) was founded, which presented the first program for a German highway network of 15,000 km. In 1926 the "HAFRABA"-Verein was founded, whose aim was to prepare and build a road from Hamburg via Frankfurt/Main to Basel reserved for motor traffic only. The first motorway from Cologne to Bonn was finally opened in 1932. By 1945, it had been expanded to a network of about 3800 km, with up to 120,000 workers being employed in motorway construction at peak times.

After 1945, the road design was adapted to the driving dynamics of the vehicles. Demand planning was based on the rapidly growing motorization and the forecasts derived from it. These forecasts were often exceeded by reality during this period. At the beginning of the 1970s, there was increasing resistance to uncontrolled road construction and its negative consequences, such as the destruction of nature, pollution from noise and exhaust gases or the growing number of road deaths, which peaked in 1970 with 19,193 (in former Western Germany). Due to extensive innovations in the active and passive safety of vehicles, improvements in rescue services, including medical progress, and the improvement of the infrastructure, it has been possible, despite an almost threefold increase in mileage, to reduce the number of road deaths continuously to well below 4000.

1.2 Development of Transport

In recent years, the rapid increase in transport performance has calmed down considerably and in some regions, car transport performance has stagnated. The situation is different for road freight transport. Here, disproportionately high growth rates have been recorded in recent years and forecasts indicate a further increase in transport performance. However, it should be noted that the forecasts have been significantly reduced. Whereas the interdependence forecast 2025 from 2007 still predicted an increase in annual transport performance of 84.6% from 2004 to 2025 (20 years), the current interdependence forecast 2030 from 2014 only predicts an increase in annual transport performance of 38.9% from 2010 to 2030 (19 years). In passenger transport, too, the forecasts have been reduced from +16.0% from 2004 to 2025 to +12.9% from 2010 to 2030 (Table 1.1).

Since a decrease in the population is expected in Germany as a whole, but growth and shrinkage rates of up to 30% are possible in the individual regions (PTV/TCI 2009), the development of transport performance will vary greatly from region to region, especially in local transport. For this reason, regional model forecasts are more used to estimate mileage and less global trend forecasts (FGSV 2015a, b).

Table 1.1 Development of road transport performance. (Sources: Analysis [Bundesanstalt für Straßenwesen 2012 (Federal Highway Research Institute 2012)] Forecast 2025 [ITP Munich and BVU Freiburg 2007] Forecast 2030 [BVU Intraplan and IVV Planco 2014])

	Annual mileage passenger car	Annual mileage of trucks	Annual transport performance Truck
Year	[Billion kilometres]	[Billion kilometres]	[Billion t · km]
1970	212.9	28.4	
1990	431.5	38.9	
2000	559.5	71.6	346.3
2004/2005	578.2 (2005)		392.5 (2004)
2010	599.0		434.0
2025	671.4		704.3
2030	657.0		607.4

1.3 Road Planning Objectives

The general objective of road planning is to transport people and goods with the highest degree of traffic safety and appropriate traffic quality. The negative effects of road traffic on people and nature must be kept as low as possible or excluded from the outset. Due to the integration into the general social development, the long-term nature of the decision to build new roads and the need for investment and maintenance, the necessity of building new roads must be critically examined, also with regard to the preservation of nature.

The strong growth in passenger and freight transport services over the last 40 years has also led to a significant increase in traffic congestion on the federal trunk roads, as new road construction lags far behind the growth in transport services. From 1980 to 2010, the length of the federal motorways increased by 76% while the transport performance increased by 185%. The federal motorways are the backbone of the federal trunk roads where 5% of the long-distance road network handles about 31% of the traffic performance (Bundesanstalt für Straßenwesen 2012).

Road Network Design 2

Summary

The road network design is based on federal, state and regional planning. The functional structure of the road network is based on the system of central cities. With greater central-local function, the demands on the quality of connections increase. In addition to spatial planning requirements, the demands of the environment, the landscape and the settlement area must be taken into account.

Within the framework of road network design, a distinction is made between traffic (connection and development) and non-traffic functions (stay or functions resulting from the use of the peripheral and surrounding area in addition to pure development). Outside built-up areas, the connection functions dominate, whereby the development functions can overlap in many ways, especially in the secondary network and on many existing routes.

The functional structure of the road network is regulated in the FGSV's guidelines in the RIN (FGSV, *Richtlinien für integrierte Netzgestaltung (RIN),* FGSV Verlag, Köln, 2008e).

2.1 Functions of the Roads

In modern economies, roads serve to transport people and goods. National and international social life, the central location of Germany within Europe and the maintenance of economic performance will continue to require safe and efficient road networks in the future. In addition to maintaining a healthy environment, good accessibility to infrastructure facilities (jobs, training facilities, supply facilities and recreational facilities) is a key feature of the quality of life for the population. The daily commuter and shopper traffic, as well as recreational traffic, are highly dependent on sufficient road networks in the area. In

© Springer Fachmedien Wiesbaden GmbH, part of Springer Nature 2023

T. Richter, *Road Planning - Freeways and Country Roads*,
https://doi.org/10.1007/978-3-658-35189-2_2

addition to private motorized and non-motorized transport, good road connections are also of great importance for public road transport.

Road network design aims to plan the individual connections in such a way that, in accordance with the regional planning objective of good accessibility for the individual road sections, safe traffic handling and appropriate traffic quality for all road users are ensured. When setting quality levels, however, the aim of saving travel time and transport costs and ensuring adequate traffic safety must be carefully weighed against the objectives of environmental protection and cost-efficiency. Therefore, the dimensioning and design of roads or road sections shall be based on appropriately agreed targets for average car speeds, depending on the function of the link. Because of the variety of concerns to be considered, this is done by defining a range of target driving speeds from which, depending on competing objectives and distance, the necessary specifications for the design and traffic engineering dimensioning can be determined flexibly and appropriately.

Development and residence are essential functions for roads within built-up areas. However, country roads with a low connection function often serve to develop the adjacent agricultural and forestry areas as well as any existing residential and commercial areas.

2.2 Category Groups

The classification of a road section into a category group results from its location and the different usage requirements. In the case of built-up roads, in particular, these usage claims are influenced to a very large extent by the adjacent buildings and the type and intensity of their use. Outside built-up areas, a distinction is made between motorways and country roads. The RIN (FGSV 2008e) distinguishes between five-category groups:

- AS motorways,
- LS country roads,
- VS non-built-up main roads,
- HS built-up main roads and
- ES development roads.

The category group AS and LS comprises non-build-up roads outside cities, which mainly serve to connect municipalities or parts of municipalities. The quality requirements of the connecting function are decisive for the design of these road sections.

The category group VS comprises non-build-up main roads in cities and their adjoining areas with a primarily connecting function. The fact, that the roads are non-build-up ensures that this connecting function overlaps with the development function only minor. The quality requirements of the connecting function are decisive for the design of these road sections. However, due to their location within built-up areas, these requirements are less stringent than for roads in category groups AS and LS. Main roads without built-up areas are single carriageway or two-lane roads, and their junctions are usually signalled intersections or roundabouts and their permissible maximum speed is 50 or 70 km/h, depending on the surroundings.

The HS category group includes built-up roads that serve both connecting and access and recreational functions. The design of these road sections is often determined by the quality requirements arising from the connecting function, which can, however, often be limited by the type and size of adjacent buildings. The junctions are usually at-grade and signalled or roundabouts. The permissible maximum speed is usually 50 km/h. The category group ES comprises built-up roads that are primarily used for the development of properties and buildings. In order to achieve maximum compatibility with the development and recreational function, the maximum speed should be limited to 30 km/h.

2.3 Connection Function

The importance of a connection depends on the central-local function of the two cities connected by the road. The RIN (FGSV 2008e) differentiates between the following levels:

- MR: metropolitan regions with international or national scope,
- RC: regional centres as administrative, supply, cultural and economic centres,
- MC: Medium-sized centres as centres for the coverage of higher demand and as focal points for trade, industry and services,
- BC: basic centres, sub-centres and small centres as centres of basic supply to cover daily needs for the respective local area,
- M: Municipalities without a central local function.

Further information on the system of central locations can be found on the homepage of the Federal Institute for Construction, Urban and Regional Research (Bundesinstitut für Bau-. Stadt- und Raumforschung, BBSR; www.bbsr.de).

The RIN (FGSV 2008e) define the target figures for accessibility on the basis of the provisions of the Spatial Planning Act (Raumordnungsgesetzes, ROG) and the Conference of Ministers for Spatial Planning (Ministerkonferenz für Raumordnung, MKRO). Table 2.1 shows the target values for the accessibility of the central cities from the residential locations, while Table 2.2 shows the desired accessibility of the central cities from neighbouring central cities.

Table 2.3 shows the connection function levels of the planned or existing connections derived from the centrality of the cities to be connected. From the desired maximum travel

Table 2.1 Target value for the accessibility of central cities from the residential locations (FGSV 2008e)

	Travel time in minutes	
Central city	By car	In public transport
Basic centre	≤ 20	≤ 30
Medium-sized centre	≤ 30	≤ 45
Regional centre	≤ 60	≤ 90

Table 2.2 Target values for the accessibility of central cities from neighbouring central cities with the same centrality function (FGSV 2008e)

	Travel time in minutes	
Central city	By car	In public transport
Basic centre	≤ 25	≤ 40
Medium-sized centre	≤ 45	≤ 65
Regional centre	≤ 120	≤ 90
Metropolitan region	≤ 180	≤ 180

Table 2.3 Connection function levels for connections and links (FGSV 2008e)

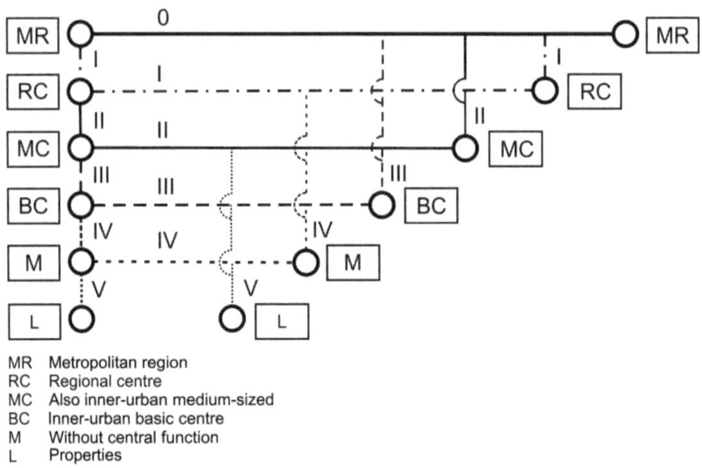

time and the average distances of the central cities, the desired average car speeds can be derived, which are shown in Table 2.4. However, these are not strictly converted over the distance, rather a reasonable bandwidth is given.

2.4 Road Networks in Europe and Germany

After the Second World War, the Western European states increasingly oriented their economic policies towards common principles. For this reason, transport policy in the individual states and the regulations applicable there were unified. In the fifties, road signs were aligned. In addition, multilateral agreements were reached on safety and vehicle regulations. The European Economic Commission of the UNO was the primary institution involved in this process: ECE (Economic Commission for Europe). From the very beginning, the transport policy of the Federal Republic of Germany was oriented towards the following three developments and maxims:

- Integration of Germany into the Western community of states.
- Germany is a densely populated industrial country, an export nation and a transit country, which means that Europe is becoming increasingly important for Germany.

Table 2.4 Categories of traffic routes for motor vehicle traffic and target average car speeds (FGSV 2008e)

Category group		Category		Standard distance range [km]	Desired car driving speeds [km/h]
AS	Motorways	AS 0/1	Long distance motorway	40-500	100-120
		AS II	urban motorway	10-70	70-80
LS	Country roads	LS I	Trunk road	40-160	80-90
		LS II	Supra-regional road	10-70	70-80
		LS III	Regional road	5-35	60-70
		LS IV	Local area road	0-15	50-60
		LS V	Connecting road	-	none
VS	Non-built-up main roads	VS II	Non-built-up main road	-	40-60
		VS III	Non-built-up main road	-	30-50
HS	Built-up main roads	HS III	Cross-town link, inner-urban main road	-	20-30
		HS IV	Cross-town link, inner-urban main road	-	15-25
ES	Development roads	ES IV	Collecting road	-	none
		ES V	Residential road	-	none

- Easier west-east travel to and from Berlin should be facilitated, and the opportunity of reunification should be kept open (now obsolete).

After the Second World War, the idea of the Western European states to build a road network developed and marked according to uniform criteria was born. The ECE then submitted a plan with route layouts and place names with the corresponding E- (European) road numbers. The design features and the marking for E-roads can also be found here.

The European road is not a separate uniform road type like the motorway. This concerns both constructional and legal features. The above-mentioned roads are made up of roads that can belong to different national road types. Thus, an E-road can be routed on the territory of the Federal Republic of Germany via a federal motorway or a federal road, or alternately via both road types. The course of an E-road is not fixed and its route can also be changed if, for example, a newly built trunk road results in improved conditions in the destination of an E-road.

In the Maastricht Treaty of 1992—which primarily agreed on the foundation of the European Union (EU) -, the heads of government of the European states agreed (in the Treaty of Amsterdam of 1999 in Articles 154–156) on an EU contribution to the establishment and development of trans-European networks. These are to be established in the fields of transport, telecommunications and energy infrastructure. The following acronyms were created in linguistic usage:

- TENs for all European networks (Trans-European Networks),
- TEN-T for the transport networks (Trans-European Network—Transport) and
- TERN for the Trans-European Road Network.

The Trans-European Road Network (TERN) is part of the Trans-European Transport Networks (TEN-T). It is made up of different sized parts of the national trunk road networks of the then 15, now 28 EU member states. The governments of the member states decide which roads in a given country are classified here according to

- Long-distance function,
- Connection of agglomerations,
- Connecting peripheral regions of the Union and
- Connection of hubs of other modes of transport (inland waterways, maritime transport, airports, KLV terminals, railway stations, etc.).

Between 2000 and 2006, approx. EUR 860 billion were invested in the EU's entire transport infrastructure, of which about 24% were invested in the road network (source: www.europa.eu). To satisfy current and future demand, the EU estimates that between 2010 and 2030, approx. EUR 1500 billion will have to be invested further in infrastructure. Of this, approx. EUR 550 billion will be spent on the expansion of the TEN-T by 2020, with the road network accounting for about 20%. Added to this is the maintenance of a total of about 4.5 million km of roads in the EU.

Targets with Measurement and Description Variables

3

Summary

In principle, motorways and rural roads should fulfil their spatial planning function with adequate traffic quality and a high level of traffic safety. In doing so, they should protect the surrounding area and use only a small amount of resources. To this end, the following requirements must be taken into account in planning:

- they should connect the origin-destination traffic as directly as possible,
- they should allow an average car speed appropriate to regional planning,
- they should have sufficient traffic quality on all sections of the route
- they should be safe to use for all modes of transport,
- they should be produced and operated at the lowest possible costs,
- wherever possible, they should have separate traffic areas for motor vehicle traffic and non-motorised and agricultural traffic,
- they should have consistent route characteristics, i.e. alignment and cross-section should be coordinated and not change fundamentally on longer lines,
- they should adapt well to the terrain and local constraints,
- if possible, they should not take up valuable land,
- they should not impair settlement and spatial development,
- they should be kept as far away as possible from environmentally sensitive areas and
- they should minimize noise and pollutant emissions and their propagation by appropriate alignment.

The multi-layered effects of rural roads require a balance that takes into account not only the objectives of safe and functional traffic flow but also the protection of natural resources, the limited resources and the effects on settlement areas and urban design. The basis for the consideration on different stages of planning is usually formed by

© Springer Fachmedien Wiesbaden GmbH, part of Springer Nature 2023

T. Richter, *Road Planning - Freeways and Country Roads*,
https://doi.org/10.1007/978-3-658-35189-2_3

several design variants. The most advantageous variant can be worked out by weighing up the target areas of traffic quality, traffic safety, environment and costs. In addition to line planning, further technical contributions are required for consideration. These are, in particular, traffic studies, which present the basic requirements and the effects on traffic, and environmental studies such as environmental impact studies or FFH impact assessments (FFH = Flora-Fauna-Habitat). In addition, urban development, calculations of immissions or economic feasibility studies are often required.

3.1 Target Field Traffic Quality

In the target field of traffic quality, two proofs are required for motorways and country roads. On the one hand, it must be demonstrated that the average car speeds aimed for in the RIN (FGSV 2008e) for the respective road categories are achieved. On the other hand, proof must be provided in accordance with the HBS (FGSV 2015a) that the desired quality level of traffic flow is achieved (Table 3.1).

The measures for the quality of traffic flow are different for motorways and country roads. For motorways, the degree of utilization is used, whereas for country roads the traffic density is the measure of traffic quality. Traffic density is more suitable for country roads since the curvature and gradients mean that very different average car speeds can be reached even with low traffic volumes. In contrast, traffic density is a measure that describes the individual's ability to drive at his or her desired speed.

In cycle traffic, tracks with few detours, flat, comfortable and, as far as possible, minimally affected by motor vehicle traffic are essential prerequisites for good traffic quality.

Other modes of transport to be considered in the target field of traffic quality with an appropriate quality are pedestrian traffic, public transport and agricultural traffic.

3.2 Target Field Road Safety

The design and operating characteristics of motorways and rural roads determine not only the traffic load but also their traffic safety. They are safe, especially for road users who are not familiar with the area, and guarantee a good quality of traffic flow if they are recognizable (in good time) as a whole or in larger sections, especially in the intersection areas, are neatly arranged, comprehensible and (sufficiently) passable.

The RAA (FGSV 2008d) and the RAL (FGSV 2012a) therefore pursue the goal of designing roads in one category as uniformly as possible (standardization) and making roads in different categories as distinguishable as possible (recognizability). The design elements to be used within a category are (pre-)selected in such a way that they can be expected to provide the highest possible degree of traffic safety, taking into account appropriate traffic quality and economic efficiency (Table 3.2).

Table 3.1 Relationship between the assessment of the traffic quality of individual facilities (QSV) and the assessment of the quality of supply of network sections (SAQ_N) for motor vehicle traffic according to the HBS (FGSV 2015a) and the assessment of the quality of supply of connections (SAQ_V) according to the RIN (FGSV 2008e)

HBS Single facility Road section Intersection	Determination of the degree of utilization or density on road section and waiting times for traffic streams in intersections
	Comparison with the capacity of the road sections and the capacity of traffic streams in intersections
	Evaluation of traffic quality on road sections and of traffic streams in intersections (QSV)
	Summarized assessment of traffic quality on roads and at intersections
Network section	Determination of the travel speed on road sections and the waiting times in intersections
	Determination of the achievable average car driving speed on the network segments
	Comparison with the average car travel speed aimed for on network sections according to the RIN
	Evaluation of the quality level of network sections (SAQ_N)
RIN Connection	Summary of the car travel speeds on the network sections of a connection
	Comparison with the beeline speed for this connection
	Evaluation of the quality level of this connection (SAQ_V)

Table 3.3 shows some key data on the occurrence of accidents in 2013. It becomes clear that rural roads are the core problem of accidents involving fatalities. On the country roads, accidents in longitudinal traffic, which are often related to overtaking, and driving accidents, which are often related to collisions with trees, dominate. In addition, there is a large number of accidents at junctions, which often occur at non-signaled, single-level crossings. Of all accidents with fatalities on country roads, 30% can be assigned to the type of accident oncoming vehicle, 30% to the type of accident stray off the road to the right or left (often with subsequent obstacle collision) and 16% to the type of accident turning/intersecting vehicle (data from 2009).

The RAL (FGSV 2012a) starts with these three main categories and tries to secure overtaking, to create obstacle-free side areas and to offer safe types of junctions (Table 3.4).

Table 3.2 Target field traffic quality (Based on FGSV 2012a)

Goals	Evaluation criteria	Possible measurement and description variables	Possible influence
Appropriate quality of traffic flow in motor vehicle traffic	Travel quality	• Utilization rate of the junctions and sections [%]. • Average car driving speed [km/h][a] • Traffic density [car/km][a] • Average waiting time during rush hours [s]	• Cross-section design • Alignment • Separation of slow road users • Secured overtaking opportunities • Number of junctions • Junction types • Avoid long waiting times • Control and, if necessary, coordinate light signal systems depending on traffic • Provide sufficient space for maintenance service
Adequate quality of the traffic flow in bicycle traffic	Travel quality	• Average driving speed [km/h] • Detour length [km] • Height differences • Number and type of crossing possibilities	• Separation from motor vehicle traffic • Low gradients • Direct route guidance • Safe crossing possibilities

[a] To be determined according to the HBS (FGSV 2015a)

Table 3.3 Road traffic accidents with personal injury 2013 (Bundesanstalt für Straßenwesen 2012)

	Motorways	Country roads	City roads
Accidents with personal injuries	18.452	73.003	199.650
Deaths	428	1934	977
Injured	29.202	103.419	241.521

3.3 Target Field Environment

The target field Environment describes the effects of the roads on the immediately adjacent and wider surroundings. The areas of impact outside and inside built-up areas are to be considered equally, since, for example, bypasses can have positive effects in traffic reduced cross-town links.

Traffic immissions (noise, air pollution, vibrations) depend not only on speed, driving style, road surface, gradient and distance between the route and the immission points, but also to a large extent on traffic volume and traffic composition. However, immissions can be reduced with accompanying environmental measures (e.g. active and passive noise protection or immission reducing plant strips). Subjective impairments can also be reduced by a design adapted to the natural environment and adapted planting that does not endanger traffic safety.

Table 3.4 Target field traffic safety (Based on FGSV 2012a)

Goals	Evaluation criteria	Possible measurement and description variables	Possible influence
High traffic safety for all types of traffic	Accidents	• Accident density [U/km · a] • Accident cost density [€/km · a] • Accident rate [U/1000Kfzkm] • Accident cost rate [€/1000Kfzkm]	• Promote a steady speed appropriate to the network function through adapted design elements • Sufficient stopping sight distance • Good radius ratio • Avoid discontinuities or, if unavoidable, clarify them • Avoid zones with weak water drainage • Coordinate turning points in the layout and elevation plan • Avoid overtaking in oncoming traffic • Design junctions in such a way that they are recognizable, comprehensible, clearly arranged and passable • Separate road network for cyclists, pedestrians and agricultural traffic • Safe, clear guidance of bicycle traffic at junctions • Sight distance at junctions • Clarify the course of the road through appropriate street planting • Provide obstacle-free side space • Provide for planting unattractive for the game in sufficient distance
	Speed behavior in motor vehicle traffic	• Local speeds [km/h] • Speed profile over a distance [km/h]	
	Situational safety parameters	• Interactions • Encounters • Conflicts	

Low land use by motorways and rural roads should be aimed for. In doing so, it is always important to consider the quality of the original use of the respective areas in comparison to the intended traffic use. As a rule, it can be assumed that agriculturally used areas are of a significantly lower quality than essentially natural areas.

The fragmentation of areas with important land functions is, in addition to the direct land use, an important assessment variable. The protection of coherent areas with important area functions for the water balance and humans, plants and animals should be particularly aimed at by means of a low level of fragmentation. The water balance can be impaired by disturbances of the underground or surface water.

Local recreation areas are of essential importance as rest areas for the population. It must therefore be ensured that local recreation areas if they are intersected by motorways or country roads, can still perform this function. On the other hand, the construction of bypasses can strengthen inner-city recreation areas by shifting traffic.

The landscape-compatible road design is to be ensured by minimizing the impact on the landscape. Especially in hilly terrain, the integration of country roads into the landscape often contradicts the required size of the design elements. The landscape is usually less impaired if the routing is based on the natural terrain relief and embankments are avoided. The integration of country roads into the landscape can be promoted by visually effective protective plantings. On the other hand, the integration of structural noise protection facilities into the landscape is problematic (Table 3.5).

Effects on social usability are to be expected both in the built-up areas adjacent to the rural roads and in built-up road spaces within built-up areas if these are relieved of motor vehicle traffic by new rural roads. Evaluation criteria for the local road design and other effects within the town are, in particular, the separating effect of the road and the quality of life and stay.

3.4 Target Field Costs

The total costs of a road construction measure should be kept as low as possible. The total costs include not only the investment costs but also the running costs for the maintenance and conservation of the roads. However, when assessing the profitability of road construction measures, not only the total costs must be considered. All benefit components must also be included, regardless of whether they are quantifiable or monetizable. These include in particular the road user costs, which are made up of operating costs and driving time savings. The basis for an economic assessment of road construction investments with benefit-cost factors is EWS (FGSV 1997). An important component of the EMS is the monetarization and evaluation of these benefits. With benefit-specific calculation methods, an attempt is made to calculate the changes to be expected from a design variant

- the operating costs,
- the driving times,
- of the accident,
- the noise pollution,
- of the pollutant load,
- the climate impact,
- the separating effect of roads and
- the availability of space for pedestrians and cyclists to assess.

To determine total expenses, all cost and benefit components (insofar as they can be monetized) are reported as annual costs. The results of economic feasibility studies are

Table 3.5 Target field environment (Based on FGSV 2012a)

Goals	Evaluation criteria	Possible measurement and description variables	Possible influence
Low immission loads • Biotopes • Local recreation areas • Nature and landscape conservation areas • FFH and Natura 2000 areas • Water protection areas • Agricultural and forestry areas • Settlement areas • Cold or fresh air lanes	• Noise immissions • Air pollution: CO, CO_2, NOX, HC, lead, particulates, dust, ...	• Rating level [dB(A)]. • Concentration [mg/m^3]	• Take up little valuable space • Little change to the natural environment • Keeping a distance from those who are too adversely affected • Limiting negative effects to a small number of affected persons • Keep dams and cuts low • Provide lateral protective planting • Check crossing facilities for animals • Put on the noise barrier
Low land consumption • Areas see above.	• Type of land use • Space quality • Soil quality	• Traffic area size [m^2] • Additional area size [m^2] • Remaining area size [m^2] • Ground score	

Goals	Evaluation criteria	Possible measurement and description variables	Possible influence
Low surface cutting • Areas see above	• Functional separating effect • Reduction of the possibility of use • Islanding, chambering • Animal migration • Networking of habitats (biotope network) • Biodiversity • Underground and surface water • Soil quality	• Area size [m^2] • Distance [m] • Length [m] • Traffic volumes [vehicle/h] • Width of carriageway/additional facilities [m] • Number of wildlife accidents [n] • Land use in the surroundings • Insulation of partial areas [m^2] • Dam height [m] • Ground score	
Good microclimate	• Surface sealing • Biomass	• Degree of sealing [%]. • Percentage of infiltration [%]. • Green volume [m^3, m^2]	
Landscape-compatible road design	• Adaptation to the landscape • Optical guidance	• Dam and cutting heights [m] • Degree of coordination of the design elements with the topography • Recognition of direction and distance	

decision-making assistance, but they are usually not sufficient to comprehensively assess the advantages of a design variant. Multi-criteria impact analyses therefore also contain non-quantifiable evaluation criteria and, in contrast to cost-benefit analyses, avoid the general monetarization of all evaluation criteria (Table 3.6).

Table 3.6 Target field costs (Based on FGSV 2012a)

Goals	Evaluation criteria	Possible measurement and description variables	Possible influence
High economic efficiency	Planning effort	• Planning and coordination costs [€]	• Open planning process • Avoid conflict-prone areas
	Investment expenditure	• Land acquisition costs [€] • Building costs [€]	• Alignment adapted to the terrain • Avoid areas worthy of protection • Minimize Bridges and tunnels • Avoid active noise protection measures • Open drainage
	Maintenance and operating expenses	• Structural maintenance costs [€/a]. • Operating costs [€/a]	• Consider later necessary maintenance and service work in the design • Ensure maintenance with as little disruption to traffic as possible
	Road user costs	• Accident costs [€/1000 · Kfz · km] • Fuel consumption costs [€]. • Basic operating costs [€] • Time costs [€]	• Use traffic safe design elements • Choose low and flat alignment • Ensure adequate average car speeds even during peak hours
	Total expenditure	Benefit-cost factor	

Usage Requirements 4

Summary

When designing freeways and rural roads, a wide range of qualitative and quantitative usage requirements must be taken into account, and these must be determined as the basis for the design. In addition to the traffic-related requirements, non-transport-related requirements for use must also be taken into account appropriately.

Basic dimensions for the traffic spaces of motor vehicle traffic can be derived from the dimensions of the design vehicle, the movement spaces and the safety spaces (Fig. 4.1). The design vehicle for rural roads is the largest vehicle generally approved according to the Road Traffic Licensing Regulations (Straßenverkehrs-Zulassungs-Ordnung (StVZO)) (BMVBS 2012). It is 2.55 m wide, 4.00 m high and 18.75 m long. Larger vehicle dimensions are possible but are not relevant for design. Refrigerated vehicles can be up to 2.60 m wide and special long trucks can be up to 25.25 m long. Further details on vehicle dimensions are included (FGSV 2001a).

In addition to the dimensions of the vehicle, room for manoeuvre and safety spaces must be taken into account. The traffic spaces for motor vehicle traffic are composed of the space occupied by the design vehicle as well as the lateral and upper movement spaces. The height of the traffic spaces is 4.25 m.

The lateral movement clearances serve to compensate for driving and steering inaccuracies and as a safety clearance for protruding parts (e.g. mirrors). Furthermore, up to certain radii (R > 200 m), the areas additionally driven over by vehicles in curves are included. The ranges of movement also include the width allowances between vehicles required in case of encounters on undivided roads with oncoming traffic. The width of the lateral movement margins on motorways and country roads is primarily determined by the

© Springer Fachmedien Wiesbaden GmbH, part of Springer Nature 2023

T. Richter, *Road Planning - Freeways and Country Roads*,
https://doi.org/10.1007/978-3-658-35189-2_4

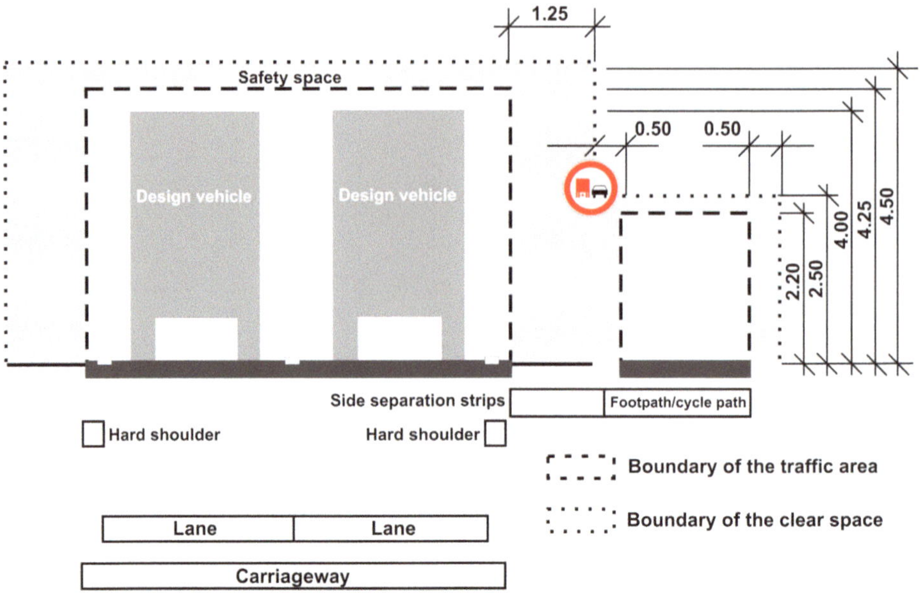

Fig. 4.1 Basic dimensions for the traffic area and the clear space (According to FGSV 2012a)

speeds to be expected and by the traffic load—in particular by heavy traffic—and, depending on the standard cross-section, is between 0.70 and 1.20 m, resulting in lane widths of between 3.25 and 3.75 m.

Basic dimensions for the traffic areas for bicycle traffic can be composed of the basic width of 0.60 m and the required movement margins of 0.20 m on both sides (minimum value 0.10 m). The lateral safety spaces for bicycle traffic are 0.25 m wide. The upper safety spaces for bicycle traffic have a height of 0.20 m.

Basic dimensions for pedestrian traffic areas can be made up of the basic width of pedestrians and the minimum required clearance of 0.20 m on both sides (minimum value 0.10 m). The traffic areas for pedestrian traffic are 2.20 m high.

The clear rooms for cyclists and pedestrians are usually 2.50 m high. Traffic signs on footpaths and cycle paths may project into the clear space up to the boundary of the traffic area.

The local public transport system has no special usage requirements compared to heavy goods traffic. The use of standards cross-section RQ 9, however, is not useful with frequent encounters of buses. (see Sect. 7.2).

Special usage requirements are placed on country roads in the military basic network for the passability of heavy military vehicles in accordance with the guidelines for the design and construction of roads for heavy military vehicles (Richtlinien für die Anlage und den Bau von Straßen (RABS). Military heavy vehicles reach widths of up to 3.75 m. The widths specified in the RABS are, in particular in the case of unrestricted encounter possibilities, significantly higher than the dimensions required for motor vehicle traffic and must therefore be carefully weighed up in each individual case, especially in relation to the target areas of traffic safety, environment and costs.

Civilian heavy and oversized transports do not constitute special usage requirements on country roads if they occur only occasionally. If they occur more frequently, the special space requirements must be taken into account. A problem here is the frequent use of roundabouts, as they cannot be passed through. Special solutions with drive-over roundabouts are then required.

Agricultural and forestry transport generally does not require larger traffic areas than general motor vehicle traffic. It should be noted, however, those special vehicles (e.g. combine harvesters) maybe 3.00 m wide according to the StVZO.

The maintenance and operation service contributes to ensuring traffic safety and securing the quality of rural roads through regular maintenance work. The leaflet by FGSV for the maintenance and operation service on roads describes the measures required for this. All drainage and road equipment, such as crash barriers, delineators and planting, must be accessible to the maintenance and operating service. In high dam areas, this may require special measures such as berms.

The usage requirement greenery results from ecological and design concerns of nature conservation and landscape design. The nature conservation laws of the federal and state governments require the preservation of the uniqueness, diversity and beauty of the landscape as well as of historical-cultural landscapes and oblige to restore or redesign the landscape in a way that is appropriate to the landscape in the event of constructing new roads. This obligation can generally only be fulfilled in the case of constructing new roads by planting trees and bushes in the road area in a manner appropriate to the character of the landscape. For reasons of nature conservation law, it is therefore not possible to do without planting trees and shrubs along roads. Furthermore, plantings serve as a guide element to clarify the course of the road.

The planting of trees at the roadside has to be critically weighed up against traffic safety, as accidents involving collisions with trees are particularly frequent and particularly serious on country roads. Trees should therefore be planted on fast-moving country roads away from the road space to be reached by vehicles coming off the road.

Planning Process

5

Summary

The planning process for motorways and rural roads is based on the general traffic planning process, which is described, for example, in the Guidelines for Traffic Planning (FGSV, *Leitfaden für Verkehrsplanungen*, FGSV Verlag, Köln, 2001b). Due to the special boundary conditions resulting from the legislative process (e.g. Federal Transport Infrastructure Plan (Bundesverkehrswegeplan)) and the close integration with environmental planning, the planning process for motorways and rural roads is divided into several planning stages and is much more formalized than other transport planning procedures.

5.1 General Transport Planning and Consideration Process

The task of transport planning is the foresighted systematic preparation and implementation of decision-making processes to enable or order transport processes in the sense of a given target concept by means of suitable measures. In this context, traffic planning should cover all modes of transport that are important for the planning area. It is a continuous task of the municipalities, the federal states and the federal government.

The planning area is the area for whose traffic development an action plan is to be drawn up. The area under investigation comprises the planning area itself and its area of traffic influence. The planning process according to Fig. 5.1 is also oriented towards the general traffic planning process when planning country roads and motorways.

For traffic plans a forecast period of between 15 and 25 years is useful. As a rule, traffic plans should be revised at intervals of about 5 years—but in any case when data, targets or problems have changed significantly. The development or updating of a traffic plan is a

© Springer Fachmedien Wiesbaden GmbH, part of Springer Nature 2023

T. Richter, *Road Planning - Freeways and Country Roads*,
https://doi.org/10.1007/978-3-658-35189-2_5

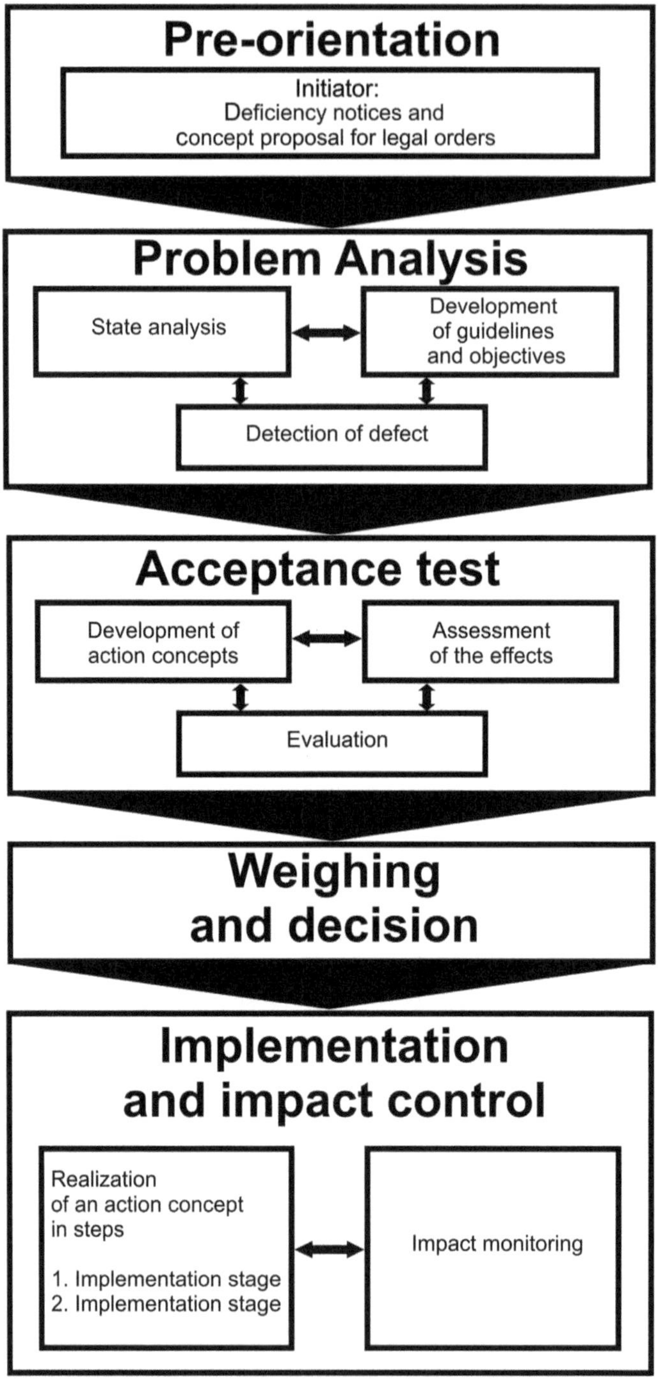

Fig. 5.1 Traffic planning process (FGSV 2010b)

process. This planning process is roughly divided into the phases of problem analysis, measures investigation and decision.

The target concept is developed from the target fields according to Chap. 3. Depending on the problem definition, some goals may be dropped or additional target fields and goals (e.g. urban development goals) may be added. The deficiencies arise as discrepancies between the goals or ideas of states and the actual states. The identification of deficiencies, therefore, requires not only the recognition of a state but also the existence of goals.

The analysis of the traffic, ecological and urban development conditions serves on the one hand to describe deficiencies and their effects. On the other hand, it is intended to show and quantify the connections between causes and effects. Finally, it should provide basic data for further work steps.

On the basis of the results of the problem analysis, measures can be designed which appear suitable to eliminate the deficiencies. These measures can be of a constructural or operational nature in the broadest sense. They can be combined into bundles of measures or plan cases. The effects of the individual measures must be estimated on the basis of the cause-and-effect relationships. Finally, an assessment is made against the background of the target concept.

When deciding on the implementation of traffic planning and design measures, the need to weigh up the effects of measures in a balanced way against each other arises from the constitutionally protected legal interests (e.g. right to physical integrity, right to free development of the personality, property rights).

Within the framework of the administrative court control of standards, guidelines and recommendations, it is therefore increasingly being examined whether there are deficits in the weighing up of these rights, that is, whether

- an assessment has taken place,
- the decisive influencing factors were included in the weighing and
- the influencing factors received the weight they deserved.

Balancing deficits are avoided if

- the evaluation criteria and values (weightings) used are identifiable,
- the weighing process is comprehensible and verifiable and
- all effects important for the decision-making process (e.g. traffic, economic, social, ecological, settlement structure, design) are included with a holistic target concept (evaluation concept).

The work steps must be separated:

- Assessment of the effects (quantitative or qualitative determination of the "quantity structure" of the effects),

- Evaluation of the effects (e.g. by subjective weightings based on the values of the evaluators) and
- Decision on the measures to be implemented (political task) within the transport planning or design process.

No formalized procedure makes it possible to work out an action plan for transport development directly on the basis of descriptions of objectives, deficiencies and state of art. It is, therefore, necessary to proceed iteratively and heuristically. Based on the current transport networks and their forms of operation, it should take into account

- the existing or specified
- boundary conditions,
- the forecast spatial development and
- the planning objectives

sensible planning cases are developed, whereby the zero-case forecast must always be taken into account.

At least the following work steps are required to assess the effects and evaluate measures and plans:

- Concretization of the target concept (from the problem analysis),
- Determination of the effects ("quantity structure"),
- Definition of eligibility and exclusion levels (reasonable limits) for the evaluation criteria,
- Scaling of evaluation criteria and indicators,
- Weighting of the evaluation criteria and indicators,
- Determination of the individual judgements (utility values) and, if necessary
- Merging of the individual judgements (utility values).

A number of environmentally relevant effects are operational. Thus, because the influencing variables (i.e. traffic load) are largely identical, there are very close interrelationships between the traffic assessment and the environmental assessment. The traffic parameters determined with traffic planning models are therefore often also an essential basis for the quantification of environmental and urban development criteria. An essential basis for the determination of effects for the quantifiable evaluation criteria (assessment criteria) and indicators are the model-based calculation results for the individual variants. Besides, the evaluation criteria (assessment criteria) and indicators that can only be described qualitatively must be determined.

The compulsive conversion of impacts into monetizable indicators (e.g. via avoidance cost approaches) does not usually fulfil the conditions of a balanced impact analysis. For this reason, multi-criteria impact analyses are increasingly used or cost-benefit analyses are supplemented by other procedures (e.g. spatial effectiveness analyses).

Against the background of the target concept and the defined target levels, the discernible effects of the measures or bundle of measures examined are to be assessed according to the same assessment criteria in each case. As a result of the evaluation (assessment) in particular the advantageousness, the urgency and the order of priority of measures are to be pointed out comprehensibly. Further information on evaluation procedures can be found in the EWS (FGSV 1997), the FAR (FGSV 2002a) or in the notes on areas of application of procedures for decision-making in traffic planning (FGSV 2010b).

In the decision phase, the decision-makers have to decide what measures are to be taken to influence further traffic developments. The result is a technically justified and politically secured action concept for the transport development of a planning area, which should be jointly supported by citizens, politicians and planners.

5.2 Planning Process for Motorways and Rural Roads

The planning process is based on the general planning process but is divided into several stages due to the required approval procedures. First of all, demand planning is carried out. On the basis of notifications from the federal states, an evaluation procedure is carried out for motorways and federal roads by the Federal Ministry of Transport and Digital Infrastructure (Bundesministerium für Verkehr und digitale Infrastruktur (BMVI)). Extensive investigations and planning are already required for the notifications of the federal states. Figure 5.2 shows an example of the notification of demand for a federal road in Lower Saxony. The site plan contains the line preferred in a variant investigation as well as an assessment of the spatial resistances. Figure 5.3 shows a corresponding layout proposal for a possible alignment.

All measures reported are subjected to an impact analysis by the BMVI and ranked on the basis of an economic evaluation. There are similar, but much simpler procedures and specifications for the state, district and municipal roads in state development plans, regional spatial planning programmes and transport development plans. An essential basic determination is a traffic examination. Here the traffic effects of changes in the road network are determined.

Once the measure has been included in the priority or the second measures list with planning permission, the preliminary planning stage follows. The measures of the Federal Transport Infrastructure Plan are legally secured in the Law on the Extension of the Federal Highways (Gesetz über den Ausbau der Bundesfernstraßen). This means that the need is basically established, but it is still possible to take legal action against the advantageousness of the measure and the chosen variant or the encroachment on the property or personal rights. It can increasingly be observed that ultimately the Federal Administrative Court (Bundesverwaltungsgericht) is the final instance in Germany to decide on the correctness of planning approval decisions.

Within the framework of the line determination or regional planning procedure, all possible variants are shown and evaluated. Figures 5.4 and 5.5 show the example of the

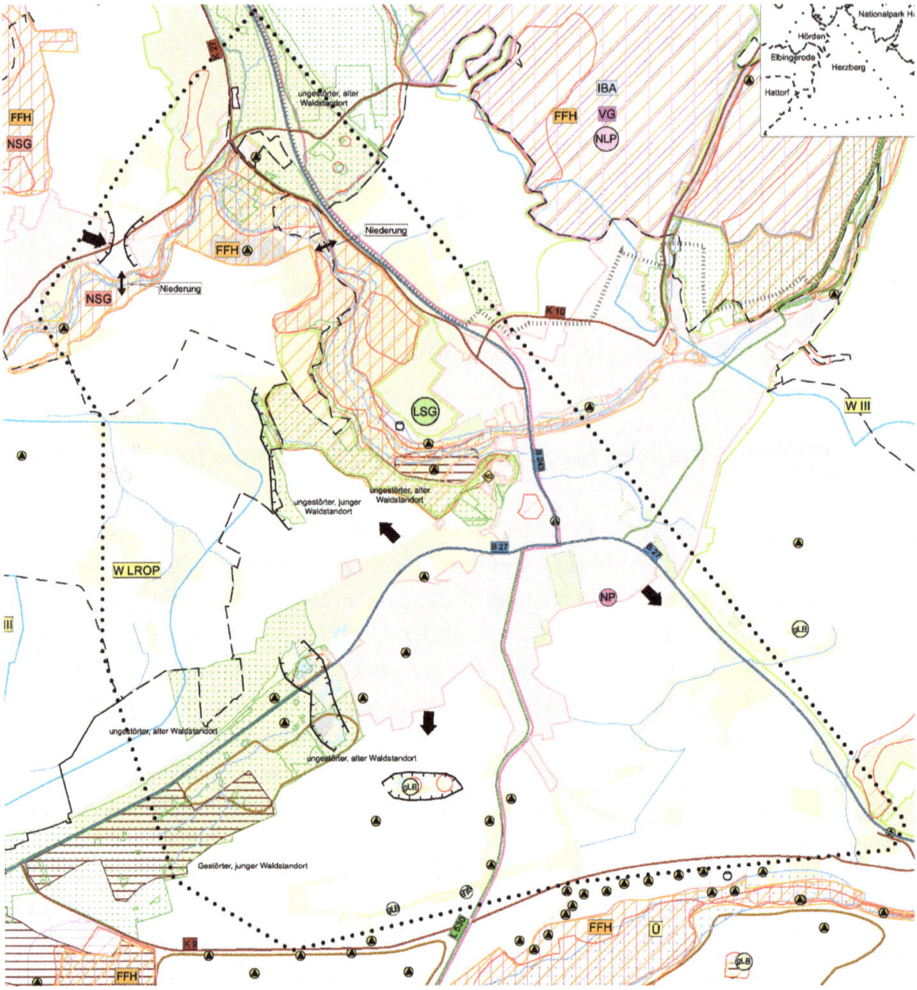

Fig. 5.2 Example of the demand notification of a federal road in Lower Saxony, here: a spatial analysis (Source: NLStbV zGB)

continued construction of the A33 in Osnabrück. The evaluation of the environment and the identification of conflicting areas that are particularly suitable for a route are carried out according to the "Guidelines for the preparation of environmental impact studies in road construction (Richtlinien für die Erstellung von Umweltverträglichkeitsstudien im Straßenbau (RUVS))" (FGSV 2011c). The environmental impact study (EIS) records, describes and evaluates the effects of a project on the environment in terms of space and variants. In addition, the compatibility with the conservation objectives of any existing sites of community importance according to the Habitats Directive (FFH-Richtlinie) or European bird sanctuaries according to the Birds Directive (Vogelschutzrichtlinie) (Natura 2000 sites) must be examined in all planning. In addition to imposing new construction

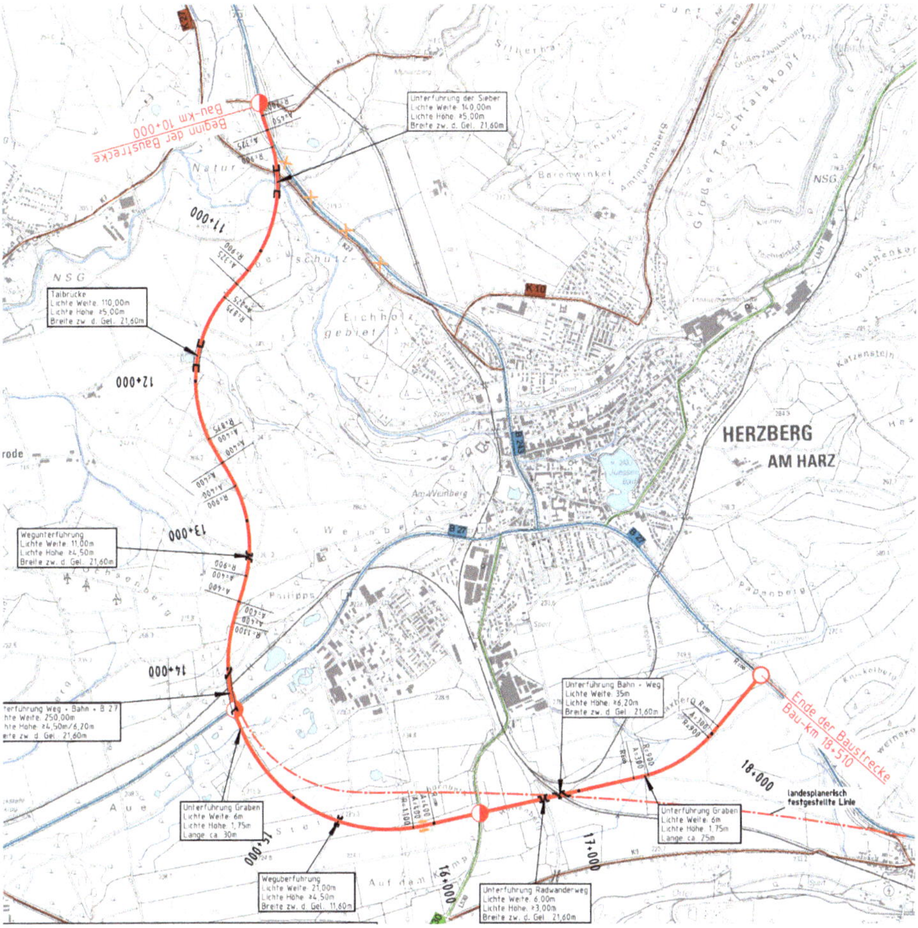

Fig. 5.3 Example of the demand notification of a federal road in Lower Saxony, here: route proposal (Source: NLStbV zGB)

variants, expansion variants and the zero variant (no new construction or expansion) must also be evaluated.

Depending on the design class, the basic design and operational characteristics must be pre-selected as early as the stages of the application by the federal states and preliminary planning in order to be able to develop the layout variants in the horizontal and vertical alignment. The junctions shall be defined and displayed according to location, type and operational form. The same applies to the bridges, tunnels and noise barriers which shall be presented at least with their basic dimensions.

Subsequently, all variants of the preliminary planning are to be evaluated with regard to the achievable traffic quality and the achievable average car driving speed using the methods according to the HBS (FGSV 2015a) and compared to the desired average car driving speed according to the RIN (FGSV 2008e).

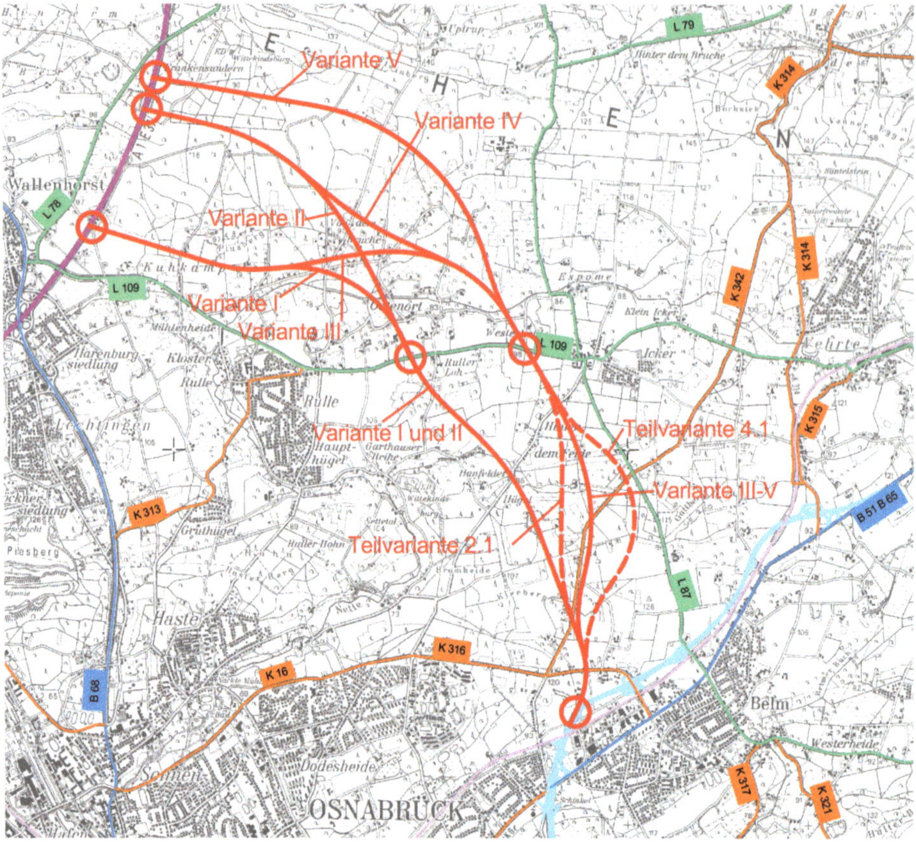

Fig. 5.4 Example of preliminary planning/line determination (Source: NLStbV rGB OS)

In the design planning, which is usually only carried out for the preferred variant determined in the preliminary planning, the design is concretized on a larger scale in the site plan, the longitudinal section and the cross-section and a cost calculation is carried out. The results of expert opinions, for example on the subsoil, the required immissions protection or drainage conditions, are integrated. The result is the preliminary design, which is prepared in accordance with the guidelines for the design of uniform design documents in road construction (RE) (FGSV 2012b). Figure 5.6 shows the example of the A33/B51n Belm bypass. In this planning stage, the environmental and nature conservation issues are dealt with in greater depth. For this purpose, the accompanying landscape conservation plan will be developed according to the Guidelines for the accompanying landscape conservation planning in road construction (Richtlinien für die landschaftspflegerische Begleitplanung im Straßenbau) (RLBP) (FGSV 2012c). The preliminary draft is used for internal administrative and technical examination and approval of the road construction measure.

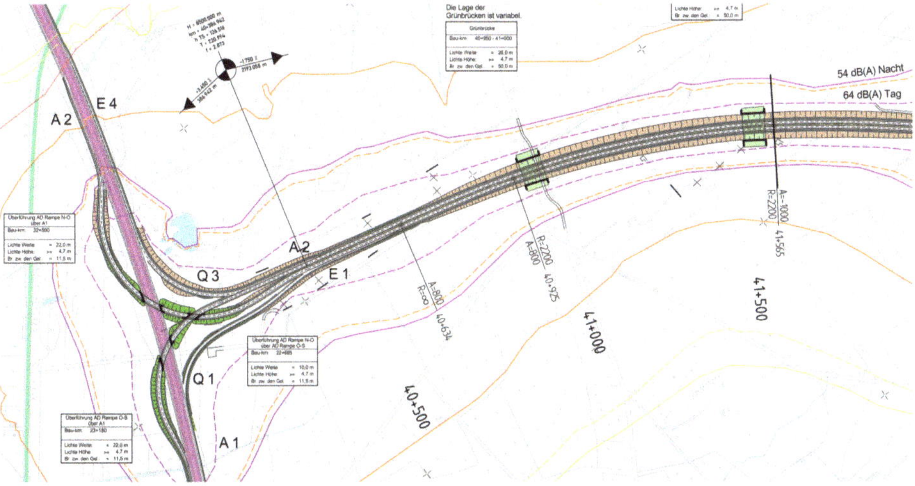

Fig. 5.5 Example of a preliminary planning/display of the individual lines (Source: NLStbV rGB OS)

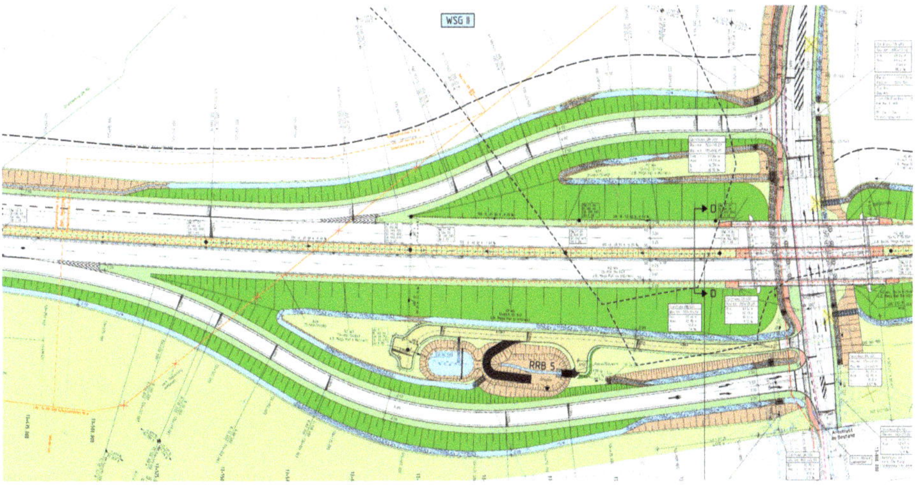

Fig. 5.6 Example of a draft design (detail) (Source: NLStbV rGB OS)

In the approval planning, design planning is further developed and supplemented. In this process, all legally relevant details are presented with sufficient accuracy—usually on the same scale as the design planning. The approval planning should make it possible for all parties involved in the planning approval procedure to recognize the type and extent of the impact. It is the basis for the overall weighing of all public and private interests in the planning approval procedure and the regulation of land acquisition and compensation issues. As a rule, the approval planning is drawn up on the basis of the guidelines for planning approval in accordance with the Federal Highway Law (Richtlinien für die Planfeststellung nach dem Bundesfernstraßengesetz) (PlafeR) (BMVBS 2008) and also includes a list of structures and a land acquisition register (example in Fig. 5.7) with land acquisition plan.

The implementation planning (detailed design, construction design) serves as preparation for the execution of the construction work. The regulations from the planning approval procedure are adopted and incorporated into the design with their effects on the project. In addition, a further level of detail is required (e.g. drainage plans, cross-sections, ceiling height plans for junctions; for example in Fig. 5.8). Many documents such as stake-out

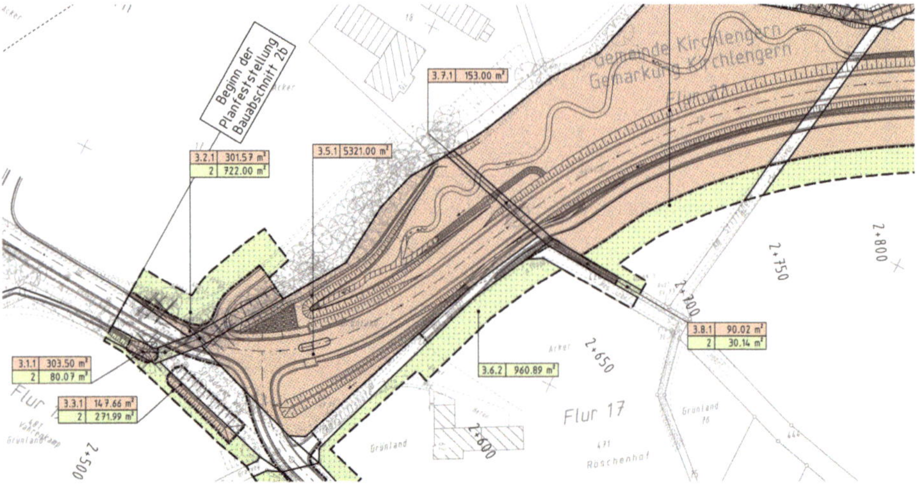

Fig. 5.7 Example of a land acquisition plan for the planning approval documents (extract) (Source: Gemeinde Kirchlengern)

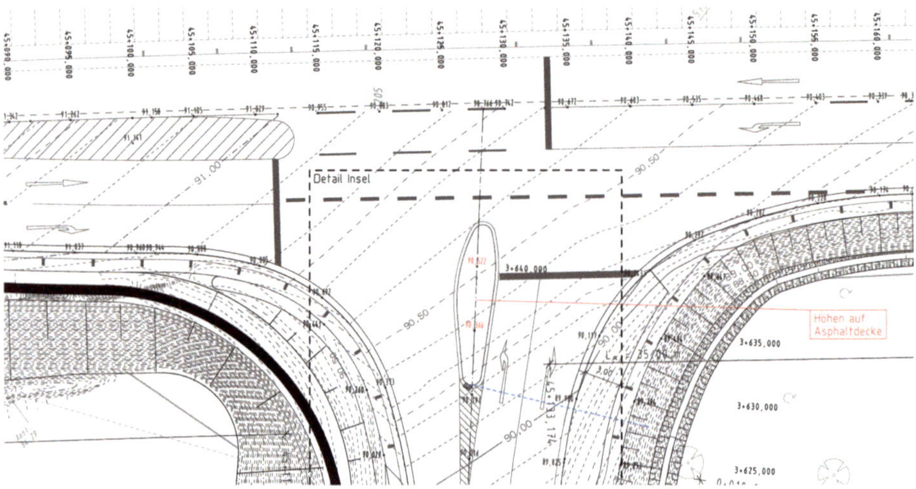

Fig. 5.8 Example of a ceiling height plan at a node (section) (Source: Gemeinde Kirchlengern)

lists, planum and surface books are also transferred digitally. The design must be concretized to such an extent that it can serve as a reliable basis for the invitation to tender and the award of contracts for the construction work and the execution of the construction work. Further components of the implementation planning are signposting and marking plans, signal location plans, line plans, landscape conservation implementation plans/plant plans, structural designs, and, if necessary, other specialist plans.

The scale in the design stages are becoming increasingly larger. The demand planning is usually displayed on a scale of 1:25,000. Preliminary planning as a basis for regional planning procedures and the determination of lines is developed at a scale of 1:10,000 or 1:5000. The preliminary draft is usually prepared on a scale of 1:1000, whereby the documents for the approval procedures are often handed over on a scale of 1:5000. The documents for the planning approval procedure are presented on a scale of 1:1000. The implementation planning is often prepared in even greater detail. The site plans are prepared on a scale of 1:500 or 1000, depending on the complexity of the measure. Even larger scales may be required for individual areas such as intersections.

In the "Guidelines for the planning process and the uniform design of design documents in road construction (Richtlinien zum Planungsprozess und für die einheitliche Gestaltung von Entwurfsunterlagen im Straßenbau (RE))" (FGSV 2012b), the required design documents for the planning stages preliminary planning to approval planning and, by analogy, also to implementation planning are presented.

6 Road Categories and Design Classes

Summary

The road categories are determined according to Chap. 2 of the RIN (FGSV, *Richtlinien für integrierte Netzgestaltung (RIN)*, FGSV Verlag, Köln, 2008e). According to the RIN, there are already possibilities to upgrade road categories in the case of major traffic links and a lot of transit traffic or to downgrade them in the case of weak traffic links or a good offer from other transport systems. Furthermore, for rural roads according to the RAL (FGSV, *Richtlinien für die Anlage von Landstraßen (RAL)*, FGSV Verlag, Köln, 2012a), there is the possibility of upgrading or downgrading design categories depending on traffic volume. However, both possibilities should not be used and double upgrading and downgrading should be avoided.

6.1 Motorways

Motorways are assigned to the category group AS according to the RIN (FGSV 2008e). The connection function levels 0, I and II come into question. Motorways are usually located outside built-up areas, but can also pass through built-up areas or be urban motorways. The road categories are assigned design classes according to Fig. 6.1.

Motorways according to the RAA (FGSV 2008a) can be federal motorways (blue signs) or motorway-like dual-carriageway (federal) roads (yellow signs). The roads, which are dual carriageways but not federal motorways, are usually intended for shorter connections and large volumes of traffic. Therefore, the demands on the design elements and the achievable average car speed are lower and the design class EKA 2 is planned. The same applies to urban motorways, which are assigned to design class EKA 3.

The design class of a road section between two centres or connections of the same or higher rank should remain the same, even if there are different traffic volumes on the

© Springer Fachmedien Wiesbaden GmbH, part of Springer Nature 2023

T. Richter, *Road Planning - Freeways and Country Roads*,
https://doi.org/10.1007/978-3-658-35189-2_6

Road category	AS 0 / AS I		AS II		
Location to built-up areas	outside or inside		outside or inside	outside	within
Road dedication	Motorway	not Motorway	Motorway	not Motorway	all
Designation	Long distance Motorway	Motorway-like road	Supra-regional motorway	Motorway-like road	Urban motorway
Design class	EKA 1 A	EKA 2	EKA 1 B	EKA 2	EKA 3

Fig. 6.1 Design classes for roads of category AS (FGSV 2008d)

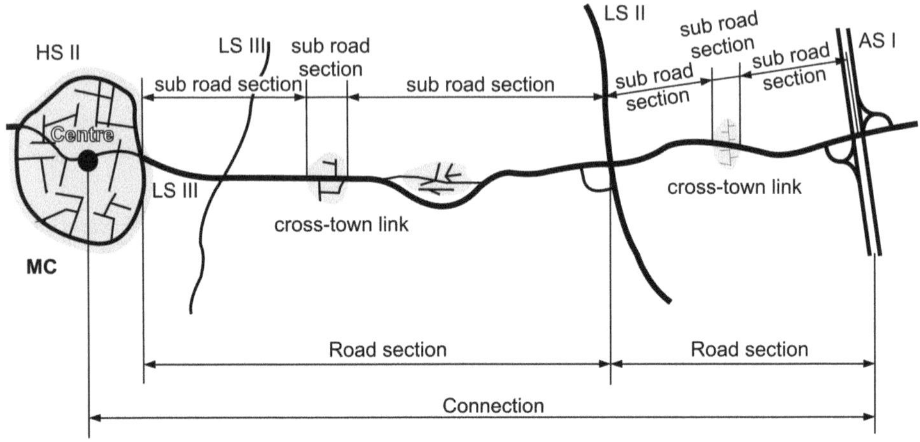

Fig. 6.2 Example of an EKL 3 road section with subsections (Based on FGSV 2012a)

individual sections. However, changes from EKA 2 to EKA 3 may occur when regional motorways merge into urban motorways within built-up areas. Changes must always be made at motorway junctions.

6.2 Country Roads

According to the RIN (FGSV 2008e), rural roads are assigned to category group LS. The connection function levels I to IV come into question. There are currently no guidelines for country roads of the road category LS V. They can be planned according to EKL 4, if necessary even more adapted. If there is a very low probability of encounter cases of large vehicles, the width of the carriageway could be reduced to the encounter case of a large vehicle/car at about 50 km/h (about 5,50 m).

As in the case of motorways, the design class should remain the same for lines between centres or links of the same or higher rank. Figure 6.2 shows the example of EKL 3 road section in a LS III connection. In contrast to the motorways, the country roads of different design classes should also differ significantly from each other in order to make clear to the driver the design class and the associated expectations of his driving behaviour. For reasons of traffic safety, it is particularly important that the driver can maintain his driving behaviour and in particular his speed behaviour as constantly as possible and that any breaks in the continuity that may be necessary due to constraint points are clearly indicated.

The road categories according to the RIN are first transferred 1 to 1 to the design classes. This means that the design classes EKL 1 to EKL 4 are assigned to LS I to IV. In principle, the designation of the country road (federal, state or district road) has no influence on the design class, but in the case of a network-wide classification, many federal roads would be assigned neither to EKL 1 nor EKL 2, which would be problematic due to the different funding. If many federal and state roads were to be classified as state, district or municipal roads, the new owners would incur high costs for maintenance and possibly renewal.

In the case of very high traffic demand, it may be advisable to choose a higher design class for safety reasons alone, since the accident costs saved by a cross-section with more lanes and, where the appropriate, separation of carriageways exceed the additional investment costs (see Sect. 9.1). In addition, a higher traffic quality can usually be achieved in the higher design class.

In the case of very low demand for transport, it might make sense for economic reasons to choose a lower design class as well. The resulting potentially higher accident costs are compensated by the lower construction costs. In addition, from the author's point of view, it is still an open question whether accident cost rates for certain standard cross-sections are the same across the entire spectrum of traffic loads or whether lower accident cost rates are to be expected for very low traffic volumes, and whether accident cost rates increase disproportionately for loads close to the capacity.

A gradation of a LS 3 road into EKL 4 is not provided for in the RAL. Upgrading or downgrading should only be carried out if it extends over several network sections and thus results in a reasonable route.

As there are many aspects to be taken into account when considering the question of upgrading and downgrading, the RAL (FGSV 2012a) provides guidance as to where an examination of upgrading or downgrading could be useful. All relevant objectives (traffic safety, traffic quality, environmental compatibility and construction costs) must then be included in the review process. With regard to traffic quality, it must be demonstrated, particularly in the case of a gradation, that the desired average car speed in accordance with the RIN (FGSV 2008e) is achieved (Fig. 6.3).

	Traffic demand on the road section ADT Cross-section [Veh./24h]	
Road category	**Check a lower-level EKL**	**Check a higher-level EKL**
LS I	< 12,000	
LS II	< 8,000	> 15,000
LS III		> 13,000
LS IV		> 3,000*

*) higher-level EKL mandatory (also for SV > 150 Veh./24h)

Fig. 6.3 Reverence values for upgrades and downgrades of design classes depending on road category and traffic demand (FGSV 2012a)

Design Characteristics of the Design Classes 7

Summary

The design characteristics for motorways and country roads are based on planning speeds that differ for the individual design classes in analogy to the achievable average car speeds required by the RIN (FGSV, *Richtlinien für integrierte Netzgestaltung (RIN)*, FGSV Verlag, Köln, 2008e). In principle, it does not correspond to the speed limit, but the planning speed and the speed limit can be identical, especially on roads in categories EKA 2 and EKL 2.

7.1 Motorways

The design features of motorways are quite similar in the design classes and differ for the driver mainly in the width of the design elements (see Chap. 9). Further design features are shown in Table 7.1.

Motorways of EKA 1 are long-distance motorways of the road categories AS 0 and AS I. Here, the importance of the connection and the high average car speed to be reached are dominant. The design elements should be chosen accordingly spacious and the junction distance should be large in order to avoid instability in the traffic flow due to frequent turning. For the dynamic derivation of the design elements the speed 130 km/h is used as a basis. The cross-section should be wide enough to ensure a 4+0 work zone, i.e. two lanes per direction on a one-directional carriageway, even in the case of construction sites since alternative routes are usually not available and the reduction to one lane in one direction is usually not possible for a longer period of time for reasons of performance.

The motorways of EKA 1 B are inter-regional motorways of road category AS II and do not differ significantly from the motorways of EKA 1 A. Lower junction distances and

© Springer Fachmedien Wiesbaden GmbH, part of Springer Nature 2023

T. Richter, *Road Planning - Freeways and Country Roads*,
https://doi.org/10.1007/978-3-658-35189-2_7

Table 7.1 Design features for motorways (According to FGSV 2008a)

Design class	EKA 1 A	EKA 1 B	EKA 2	EKA 3
Designation	Long-distance motorway	Inter-regional motorway	Motorway-like road	Urban motorway
Signage	Z 330 StVO (motorway)		Z 331 StVO (motor road)	Z 330 or Z 331 StVO
Directional signing´	Blue		Yellow	Blue, yellow
Permissible maximum speed	None		None	≤ 100 km/h
Recommended distance between junctions	> 8000 m	> 5000 m	> 5000 m	None
Traffic management in work zones on four-lane roads	4+0 usually required		4+0 not mandatory	

lower limit values of the alignment parameters are permitted. The speed as the basis for the driving dynamics calculation of the design elements is 120 km/h.

Roads of EKA 2 are "motorway-like roads" and therefore not motorways with blue signs but also not urban motorways. They are often found in agglomeration areas for shorter or medium connections with high traffic volumes. The importance of the connection and the derived average car speeds to be achieved are significantly lower than in the case of EKA 1 and therefore lower routing parameters can be used, which allow a more adapted routing. A speed of 100 km/h is taken as a basis.

Roads of the EKA 3 are called urban highways and serve within built-up areas as tangents or radials to bundle large volumes of traffic. They can be motorways. The constraints from the surrounding area are often very high so that a very adapted alignment with partly special design and operational elements is required. The speed at which the design elements are derived from the driving dynamics is 80 km/h.

The guidance is the same for all motorways at the junctions on the superior road (the motorway). Only diverging and merging lanes are possible. Only the length of the diverging and merging lanes is different. The same applies to the marking, which is always similar to the design classes of the country roads.

7.2 Country Roads

Rural roads should allow a uniform average car driving speed appropriate for the network function according to the RIN (FGSV 2008e). For this reason, significantly different design and operating characteristics are specified for the four design classes and thus for the underlying planning speeds. This should also make clear the differences between the design classes and the expected driving behaviour. Distinctive features to distinguish the design classes are the longitudinal marking and, to a lesser extent, the overtaking rules and the types of junctions. The design speed is also used to determine the limits of individual design parameters based on driving dynamics (see Chap. 8). In contrast to motorways, the planning speeds for rural roads are not directly oriented to the desired achievable average car driving speeds, since the journeys on longer stretches of road through junctions or, in the case of design classes 1 and 2, through the constant alternation of one- and two-lane sections in the pattern of speed are very inhomogeneous. However, the choice of design elements must be based on the average car speeds that are driven on undisturbed road sections.

Table 7.2 shows the basic design features for rural roads in the four design classes.

7.2.1 Country Roads of EKL 1

Country roads of design class EKL 1 are long-distance roads of road category LS I and thus supra-regional connections with correspondingly high importance and long distances.

Table 7.2 Design features for rural roads (According to FGSV 2012a)

Design class		EKL 1	EKL 2	EKL 3	EKL 4
Planning speed [km/h]	Design/ operational characteristics	110	100	90	70
Kind of operation		Vehicles only road	General traffic	General traffic	General traffic
Cross-section		RQ 15.5	RQ 11.5+	RQ 11	RQ 9
Secured overtaking sections per direction		~ 40%	≥ 20%	None	None
Management of bicycle traffic		Street independent	Independent of the road or accompanying the road	Accompanying the road or on the road	On the road
Horizontal alignment	Alignment	Very elongated	Stretched	Adapted	Very adapted
Recommended radius range R [m]		≥ 500	400–900	300–600	200–400
Maximum longitudinal inclination max s [%].		4,5	5,5	6,5	8
Recommended vertical curvature H_K [m]		≥ 8000	≥ 6000	≥ 5000	≥ 3000
Junction control	Rule solution on the higher-level road	Only diverging and merging	Turn off/ crossing with traffic lights	Turn off/ crossing with/ without traffic lights	Turn off/ crossing without traffic lights

Fig. 7.1 Example of a country road of the EKL 1 (Photo: NLStbV zGB)

The average car speeds of 80 to 90 km/h to be achieved require a very stretched alignment. The recommended design elements are based on a planning speed of 110 km/h. The junctions are to be designed grade-separated or partially grade-separated so that the guidance in the junctions of the EKL 1 roads is always diverging and merging. The three-lane standard cross-section RQ 15.5 is applied, so that traffic-secured overtaking possibilities are available by the constant change of one- and two-lane sections per the direction of travel on approximately 40% of the route. In order to clearly separate the two directions of travel from each other and to illustrate the difference compared to the RQ 11.5+ in EKL 2, a 1.00 m wide central stipe with double lane boundaries and an additional green filling in between is used throughout (see also Fig. 7.1). Roads of EKL 1 are to be operated as motorways due to the desired driving speeds. For this reason, agricultural traffic and bicycle traffic must be carried on a separate road network.

7.2.2 Country Roads of the EKL 2

Country roads of level 2 are supra-regional roads of the road category LS II and differ clearly from country roads of level 1 in the design of the central separation, in the types of junctions and the constant alternation of two- and three-lane sections (see Fig. 7.2). The

Fig. 7.2 Example of a country road of the EKL 2 (Photo: TU Berlin)

average passenger car speeds to be achieved from 70 to 80 km/h require a stretched route. Nevertheless, in many cases, the desired average car travel speed is not reached due to the time lost at the signalized junctions. For this reason, the responsible committees of the FGSV are currently discussing the reduction of the average car speeds to be achieved. The recommended design elements are based on a planning speed of 100 km/h. The junctions are usually signaled in the course of the EKL 2 road. The highest traffic quality is provided by junctions with a partially grade-separated layout, as the crossing traffic is guided grade-separate. This has clear advantages, especially for frequent agricultural or non-motorized traffic, but must be weighed against the higher costs.

The three-lane standard cross-section RQ 11.5+ is used, which has a continuous change between two- and three-lane sections. In the three-lane sections, traffic-secured overtaking sections are alternately available for both directions of travel. It should be ensured that at least 20% of the route length in each direction of travel is equipped with overtaking lanes. This is usually sufficient to reduce the overtaking pressure. The two-lane sections should therefore always be separated from each other by two-lane boundaries (marking). In the RAL (FGSV 2012a), overtaking is nevertheless permitted (double broken line) if visibility conditions are sufficient and no particular risks are to be expected for other reasons. This regulation is owed to the StVO, which stipulates in § 45 (9) that "traffic signs and traffic facilities are only to be arranged where this is absolutely necessary due to the special circumstances". In principle, a section of road with permitted overtaking is not a particularly dangerous place, although the accident costs to be expected in such sections of the road are considerably higher than in three-lane sections with traffic-secured overtaking facilities.

Both directions of travel are separated by a 0.50 wide central reservation with a continuous double line. Agricultural traffic should be prevented and—if not possible—restricted to fast-moving vehicles (40 to 60 km/h). Cycling traffic should be excluded from the roadway. If agricultural traffic is excluded, the road can be operated as a vehicle only road.

7.2.3 Country Roads of the EKL 3

Country roads of EKL 3 are regional roads of the road category LS III, their characteristic feature is the regular continuous two-lane road with separation by a broken line. According to current practice, overtaking lanes are only arranged on gradients (climbing lanes) or in other areas where, due to poor traffic quality, a safe overtaking lane is indicated.

On sections where overtaking should not be permitted due to lack of visibility or other hazards such as access roads, it is advisable to prevent overtaking by means of a simple continuous lane marking in the middle of the road or by Z 276 StVO. These areas are determined according to Chap. 15.

On rural roads of the EKL 3, average car speeds of 60 to 70 km/h are to be reached. The planning speed on which the recommended design elements are based is 90 km/h. The alignment can be adapted accordingly to the design elements in comparison with design classes EKL 1 and EKL 2. Initially, signalized intersections and T- junctions as well as roundabouts are available as junction types. For safety reasons, the omission of a traffic light system is only possible at intersection traffic volumes under 5000 Veh./24 h, because at higher traffic volumes, the additional costs for the traffic light system are more than compensated by the saved accident costs. Further advice on the selection of the appropriate type of intersection is given in Chap. 10.

The two-lane standard cross-section RQ 11 is used. Figure 7.3 shows the example of an EKL 3 road. Agricultural traffic is usually permitted. The guidance of bicycle traffic depends on demand and traffic volume. In principle, it must first be proven with network planning for the bicycle traffic whether there is a regular demand at all. For example, there will often be no need for bypasses, because cyclists can continue to use the cross-town links, which are usually the more attractive connections.

If the need for a cycle connection is demonstrated, it must be examined whether guidance on the road is reasonable for safety reasons, based on the volume of a motor vehicle and cycle traffic, the proportion of heavy goods vehicles, the clarity of the route, the proportion of vulnerable cyclists and pedestrians and the speed of motor vehicle traffic.

Fig. 7.3 Example of a country road of the EKL 3 (Photo: Richter)

The practice of creating cycle paths varies greatly in the different regions of Germany. While in the Münsterland region, for example, cycle paths are generally laid out on connections of lesser importance and lower traffic volumes in new, converted or extended projects, in other regions cycle paths are rarely found even on highly congested connections, which significantly limits the attractiveness of cycle traffic.

In principle, therefore, regardless of the RAL reference values, the aim should always be to create a cycle traffic facility when the network importance is proven. Since the financial resources of the federal government, the federal states and the rural districts do not permit the construction of cycle traffic facilities on all country roads, Table 7.3 indicates when a cycle traffic facility is necessary for safety reasons. The basis of the table is speed limit V = 100 km/h, the share of heavy vehicle traffic = 10%, stretched lines and a low proportion of vulnerable cyclists and pedestrians. The author would recommend deleting the column with DTV 7000–10,000 Veh./24 h and from 7000 Veh./24 h onwards to always provide a bicycle traffic facility if necessary.

Cycle lanes along roads, detached cycle paths or the agricultural road network can be considered as cycle traffic facilities. In general, bicycle traffic is conducted in two-way traffic. Further information on cycle traffic is contained in the "Recommendations for cycle traffic facilities ERA (Empfehlungen für Radverkehrsanlagen ERA)" (FGSV 2010a).

Table 7.3 Indicative values for the suitability of a common footpath and cycle path on roads of EKL 3 according to RAL (FGSV 2012a)

Average daily motor vehicle traffic [Veh./24 h]	2500–4000	4000–7000	7000–10,000
Daily load in bicycle and pedestrian traffic [R and F/24 h].	> 200	> 100	> 50

Fig. 7.4 Example of an EKL 4 rural road, here different from the RAL with broken marking offset by 0.75 m (Photo: TU Berlin)

7.2.4 Country Roads of the EKL 4

Country roads of design class EKL 4 are local roads of road category LS IV. The characteristic feature is the single-lane standard cross-section RQ 9 with broken marking on both sides of the roadway offset from the edge of the roadway (see Fig. 7.4). The intended carriageway width also does not allow the marking of a broken marking in the middle. The aim is to use the road as a single lane to a large extent. Encounters between cars and cars with heavy goods vehicles are possible without any problems by sharing the roadway areas outside the broken marking. When two heavy goods vehicles meet, only small safety areas remain, so that full use of the lane width, a reduction in speed and increased attention are required. To limit these encounters, the EKL 4 should only be used up to a traffic volume of 3000 Veh./24 h or a heavy goods vehicle traffic volume of 150 HG Veh./24 h.

The average car speeds to be achieved are between 50 and 60 km per hour. The planning speed on which the recommended design elements are based is 70 km/h. The alignment should be very much adapted accordingly. It should be noted that a higher speed level than 70 km/h will be achieved if significantly higher than the recommended design elements are selected. However, as the encounters of motor vehicles at increasing speeds also require larger safety areas, which are not available with the RQ 9, it would make sense to consider limiting the maximum permissible speed to 70 km/h.

Interchanges are usually designed as intersections or T-junctions without traffic lights. Country roads of the EKL 4 are usually shared by agricultural traffic and bicycle traffic. If there is a particularly high network function or the route section has a special function in student traffic, pedestrian and bicycle paths accompanying the roadway can also be useful on roads of EKL 4.

7.2.5 Dual Two-lane Rural Roads

Nearby agglomerations or when connections are superimposed on one or more network sections, traffic volumes can increase significantly, so that the traffic quality required by the HBS (FGSV 2015a) or the average car speeds required by the RIN (FGSV 2008e) can no longer be achieved.

If this section is not very long (up to about 15 km) and the traffic volume does not exceed a value of 30.000 Veh./24 h, the dual two-lane RQ 21 (see Fig. 7.5) can be used in this section. The other design elements and the junction types do not change for design classes EKL 1 and EKL 2. On country roads of design class EKL 3, the elements of EKL 2 should be used. In the case of longer sections or higher traffic volumes, these sections should be designed according to the design class EKA 2 of RAA (FGSV 2008a).

Fig. 7.5 Example of a dual two-lane country road. (Photo: Richter)

Vehicle Dynamics

8

Summary

In vehicle dynamics, the interplay of forces of a moving vehicle is described. The vehicle is subjected to longitudinal and centrifugal forces which must always be safely dissipated by frictional forces between the tyres and the road surface. The aim of the driving dynamics design is to provide sufficient reserves against unintentional movements of the vehicle. A wide range of vehicles and road surfaces of very different quality must be taken into account.

8.1 State of the Art

The basic research on the vehicle dynamics of individual vehicles and the design variables derived from speeds was carried out between 1960 and 1980, e.g. (Durth 1974). The basic idea was that design elements were derived from vehicle dynamics on the basis of a set or estimated speeds, which could then be driven on supposedly safely. The problem of this methodology was that elements determined with some safety factors are driven through at significantly higher speeds than those set since the safety factors such as utilization factors of friction were often converted into speeds by the drivers. The RAS-L 1995 (FGSV 1995), like its predecessor guidelines, uses speed specifications as a basis for the vehicle dynamic design of design elements.

For the first time, the RAL (FGSV 2012a) is using a new methodology. For the design classes EKL 1 to EKL 4, design elements are specified which allow for the expectation of appropriate driving behaviour for the design class. The individual design parameters, such as the required stopping distance, can be traced back to assumed speeds in terms of driving dynamics. However, the design class specifies the element sizes and not the speeds that can be achieved on the basis of the alignment.

© Springer Fachmedien Wiesbaden GmbH, part of Springer Nature 2023

T. Richter, *Road Planning - Freeways and Country Roads*,
https://doi.org/10.1007/978-3-658-35189-2_8

8.2 Driver–Vehicle-Road Space: Control Loop

"Driving a car", that is, the driving process can be described with the help of a control engineering model. The driving process is a control loop in which the driver (controller) controls his vehicle (controlled system) according to his wish to reach a destination safely and quickly (controlled variable) under the influence of the road and traffic regulations (reference variables) as well as traffic and weather conditions (disturbance variables).

8.2.1 Driver

The driver takes in information about his sense of sight (optics), his sense of hearing (acoustics), his sense of touch (haptics) and his sense of balance (proprioceptors) that is important for his driving task.

Most of the stimuli perceived with the senses are subconsciously processed and not consciously perceived. The most important sensory perception for driving is seeing, as only the sense of sight offers the possibility of anticipatory perception. The other sensory organs merely exercise a momentary control function.

The driver's reaction and reaction time depend on the type and importance of the perceived information and individual influences. The reaction time in practical traffic situations is 1 to 2 seconds. RAL (FGSV 2012a) assumes a calculated reaction and impact time of $t_R = 2.0$ s. When driving attentively, real reaction times are usually much shorter.

8.2.2 Vehicle

The forces and torques occurring on the moving vehicle and the translational, rotational and oscillating movements caused by them are investigated in vehicle dynamics. It deals with the physical laws of vehicle movements as a function of vehicle and road conditions. Complex and therefore difficult to handle mathematical models are required for the complete calculation of the forces acting on a moving motor vehicle, as they are applied in the design and construction of vehicles.

On public roads, many different motor vehicles are moved on very differently designed routes and junctions by drivers who drive and react very differently. Consequently, an exact calculation fails due to the complexity of the input conditions, so that a point-mass model is used for simplification.

A vehicle can only be set in motion or held in motion if the forces of motion overcome the driving resistances. It is in an equilibrium of forces if the transmittable driving force is equal to the sum of all driving resistances acting on the vehicle. The maximum transmissible driving force is limited by the coefficient of friction between the tyre and the road and by the weight of the vehicle on the driving wheels.

8.2.3 Driving Resistances

The engine tractive force Z and the driving resistances W must be in equilibrium on the vehicle in every driving condition. The resistances can be divided into internal (mechanical) and external resistances. Internal resistances due to bearing, axle and gear friction, etc. and external resistances such as rolling resistance W_R, curvature resistance W_K, gradient resistance Ws, drag W_L and acceleration or deceleration resistance W_b must be overcome.

The rolling resistance W_R is caused by deformation work on the tyres and road surface, air friction in the tyres, lifting work due to unevenness of the road surface and suction between wheel and road. It varies according to the type of road surface:

Concrete	0.015–0.020,
Asphalt concrete	0.010–0.015,
Small pavement	0.015–0.020,
Earth path	0.040–0.080.

The curvature resistance W_K arises during driving along curves due to transverse slip. It is on average between 0.004 and 0.007 (Table 8.1).

The gradient resistance W_s results as a component of gravity and is

$$W_s \approx \pm G \cdot \frac{s}{100} [N]$$

with s = Longitudinal inclination [%].

The drag W_L results from the dynamic pressure of the air acting on the projection surface of the vehicle. It is generally the greatest driving resistance.

$$W_L = \frac{\rho}{2} \cdot c_w \cdot F \cdot v_R^2 [N]$$

with:

P = Air density = γ/g = 1226 [kg /m³],
c_W = Drag (air resistance) coefficient [−],
F = Cross-section of vehicle in direction of travel [m²],
V_R = Relative speed between vehicle and air [m/s].

Table 8.1 Drag coefficient and cross-section in the direction of travel

	Car	Truck	Motorbike
c_W [−]	0.25–0.40	0.80–1.20	0.50–1.50
F [m²]	2.0–2.4	5.0–8.0	1.0

When a vehicle is accelerated, not only the vehicle mass must be accelerated translationally. The rotating masses (engine, wheels) must also be accelerated. This resistance can be taken into account with a factor φ:

$$W_b = \varphi \cdot m \cdot b [N]$$

with

m = Vehicle mass [kg],
b = Acceleration [m/s^2],
φ = Coefficient.

8.2.4 Tractive and Braking Forces

To overcome the driving resistances, tractive forces are transmitted by the driven wheels of the vehicle and braking forces by the braked wheels.

$$Z - (W_i + W_R + W_K + W_s + W_L + W_B) = P$$

The residual force P causes an acceleration or deceleration of the vehicle. The tractive force Z is calculated:

$$Z \approx \frac{3.6 \cdot N_e}{V} \cdot \eta \cdot [KN]$$

with

N_e = Effective power [KW],
V = Speed [km/h],
η = Efficiency of the transmission truck: η ≈ 0.85 to 0.95, car: η ≈ 0.95 (automatic η ≈ 0.91).

8.2.5 The Street

The driving behaviour is strongly determined by the design of the driving space, that is, by the carriageway and the environment. The driver should not be overloaded by too much information, but also not become tired by monotony. Good visual guidance should relieve the driver in order to be able to process important information more easily.

The decisive structural feature of the road, in addition to the alignment, cross-sectional design, design of the driving area and junction design, is the condition of the road surface, since the forces must be transmitted as frictional forces between the road surface and the tyres. The force to be transmitted is the wheel normal force multiplied by the friction coefficient. At s = 0%, the wheel normal force corresponds to the weight force of the vehicle F_N. The coefficient of friction f is the measure of the force that can be transmitted from the tyre to the road surface. The maximum coefficient of friction is referred to as f_{max} and is reached at approximately 20% slip. This value is also known as the coefficient of static friction μ_{max}. The maximum transmittable forces are therefore

$$\max F = \mu_{max} \cdot F_N = f_{max} \cdot F_N.$$

A multitude of factors affect the maximum friction coefficients and modern vehicles can make better use of these by means of ABS and ESP and prevent the vehicles from entering larger brake slip ranges and thus unstable driving situations. In the model, a simplified point-mass model is used, since the various influences cannot be meaningfully considered individually.

Slip is the difference between the driving speed and the circumferential speed of the tyres. The peripheral speed is greater than the driving speed when accelerating and less than the driving speed when braking. During emergency braking with locked tyres, the slip is 100% and the coefficient of sliding friction μ_g is reached. This is well below μ_{max}. The optimum slip during braking is 10 to 30%. For this reason, ABS attempts not to exceed this value by "stutter braking".

When driving straight ahead, vertical or normal forces act as shown in Fig. 8.1 (forces perpendicular to the road surface and horizontal forces parallel to the road surface). In the dimensioning case of wet and clean road conditions, the maximum value of force transmission in tangential direction using the coefficient of friction up to the coefficient of static friction μ_{max}:

$$F_{T\,max} = F_N \cdot f_{T\,max}.$$

In addition to straight driving, driving around curves as shown in Fig. 8.2 is to be considered as a more important dimensioning parameter. This basically also applies to driving straight ahead, as the crossfall of the road requires constant counter-steering. During

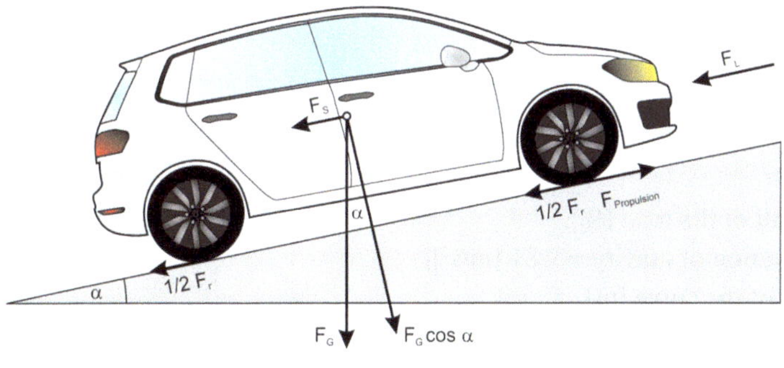

Fig. 8.1 Forces and driving resistances on the vehicle when driving straight ahead

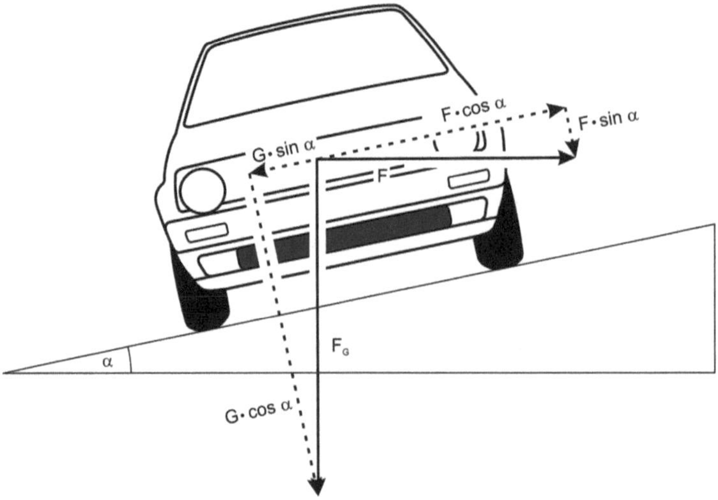

Fig. 8.2 Balance of forces in the radial direction

driving through a curve, centrifugal forces (see Fig. 8.2) occur, which must be absorbed in the radial direction by the frictional connection between the tyres and the road surface.

In order that no gliding occurs, the following must apply:

$$G \cdot \sin\alpha + f_R(v)\left[G \cdot \cos\alpha + F\ \sin\alpha\right] \geq F\ \cos\alpha$$

with

G = Weight force of the vehicle = $m \cdot g$,
$f_R(v)$ = Radial component of the friction coefficient,
F = Centrifugal force = $(m \cdot V^2)/R$.

After division by cos α and with tan α = q/100 [−] results in

$$\frac{v^2}{g \cdot R} \leq \frac{f_R(v) + \frac{q}{100}}{1 - f_R(v) \cdot \frac{q}{100}} \approx f_R(v) + \frac{q}{100}$$

with

Q = Crossfall of the road [%],
G = Acceleration of gravity = 9.81 [m/s²],
R = Radius of the curve [m],
V = Speed [m/s].

8.3 Coefficients of Friction

The coefficient of friction is a measure of the forces transmitted between the tyre and the vehicle. It is dependent on

- Road construction qualities of the surface structure,
- Climatic variables,
- Vehicle technical variables,
- Technical tyre variables and
- Driving performance.

Empirical investigations carried out in the 1980s with the Stuttgart friction meter (blocked towing wheel) provided the following results for the magnitude of the coefficient of sliding friction as a function of speed on wet roads (see Fig. 8.3).

The coefficient of sliding friction μ_g is speed-dependent. At high speeds, a low coefficient of sliding friction is initially available, which increases significantly during the

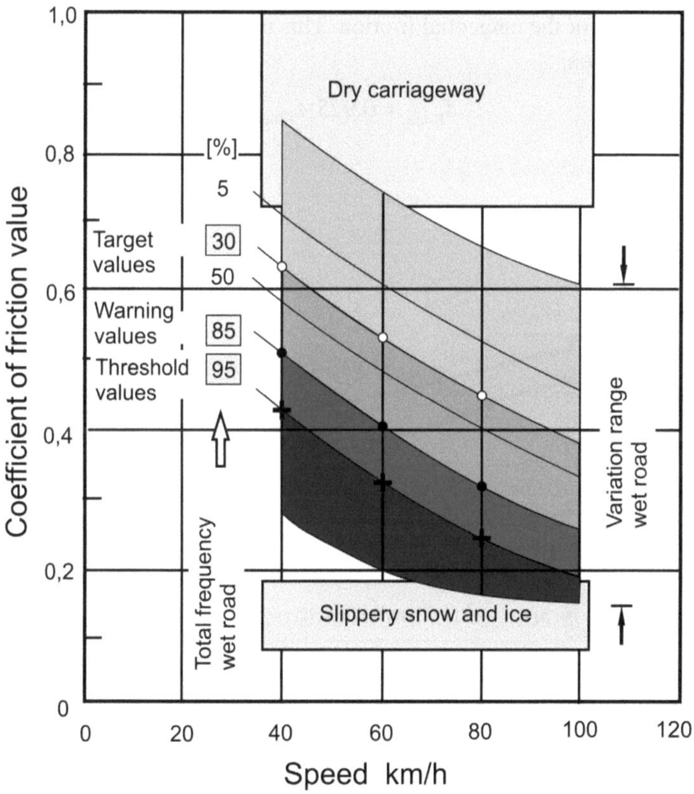

Fig. 8.3 Coefficient of sliding friction μ_g from older investigations with the Stuttgart friction meter

braking maneuver. For the design, for example, to determine minimum radii, speed-dependent coefficients of sliding friction must therefore be used. As a basis for the design, the coefficient of sliding friction was chosen in the RAS-L (FGSV 1995), which is complied with on 95% of the roads examined. The following applies:

$$\mu_g = f_{T\max} = 0.241 \cdot \left(\frac{V}{100}\right)^2 - 0.721 \cdot \left(\frac{V}{100}\right) + 0.708$$

with: V [km/h].

Due to the measuring method, f_{Tmax} results at a 100% slip (lagging of the blocked wheel). As Fig. 8.4 shows, there is a frictional reserve for normal driving conditions and thus a high level of safety.

The course of the coefficient of friction f_T depends on slip s. The maximum permissible tangential coefficient of friction f_{Tmax} for the design is defined as a proportion of the possible tangential coefficient of friction and corresponds to the sliding friction coefficient μ_g. The maximum possible value of the tangential friction is reached at approximately 10 to 30% slip.

Due to the construction of the car tyres, the value for the friction in the radial direction is different from that for the tangential friction. This is taken into account by the following empirical determination:

$$f_{R\max} = 0.925 \cdot f_{T\max}.$$

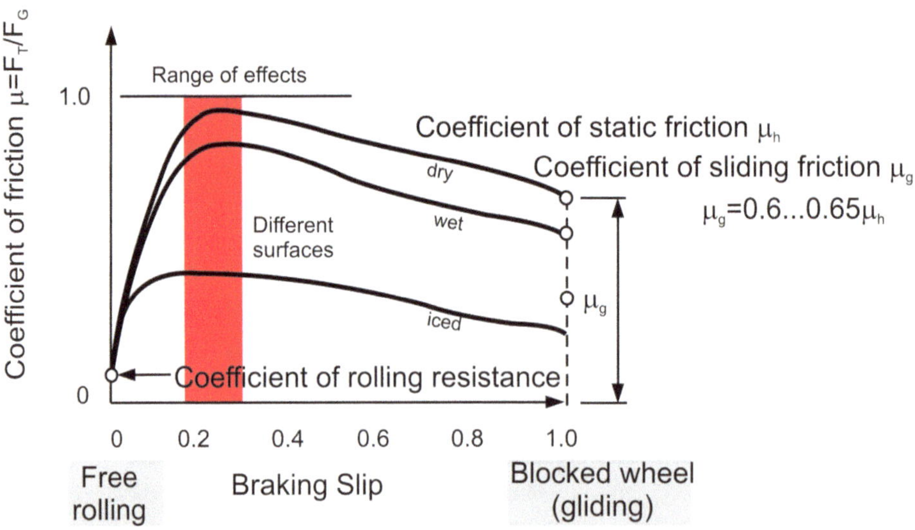

Fig. 8.4 Tangential coefficient of adhesion f_T from older investigations with the Stuttgart friction meter

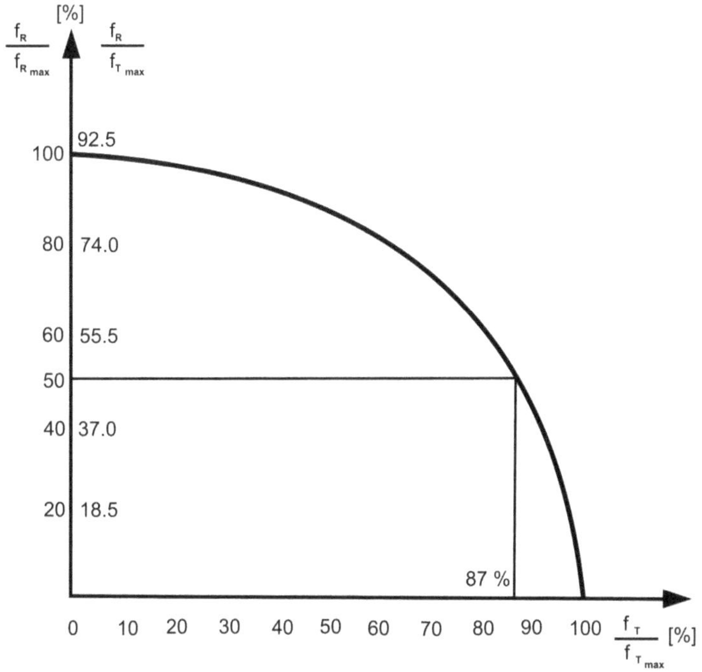

Fig. 8.5 Relationship between the tangential and radial coefficient of friction

The relationship between the used radial coefficient of friction and the remaining tangential friction coefficient is shown in Fig. 8.5.

Since the available total friction force represents the vector sum of radial and tangential friction, the following relationship applies:

$$\left(\frac{f_R}{f_{R\,max}}\right)^2 + \left(\frac{f_T}{f_{T\,max}}\right)^2 = 1.$$

For vehicle dynamics calculations, f_{Rmax} is reduced by a utilization factor. Therefore, applies:

$f_{Tzul} = f_{Tmax}$,
$f_{Rzul} = n \cdot f_{Rmax} = n \cdot 0.925 \cdot f_{Tmax}$,
$n = f$(road category, crossfall).

The technical test specifications for friction measurements in road construction currently use the SCRIM (Sidewayforce Coefficient Routine Investigation Machine) measuring method developed in Great Britain. The aim of this measuring method is to prove a friction value μ_{SCRIM} during acceptance of a newly built road and a (lower) value during warranty acceptance.

8.4 Driving Dynamics of the Individual Vehicle

The motion quantities can be derived from the basic quantities distance (s [m]) and time (t [s]):

Speed:	$v = ds / dt = \dot{s}$ [m/s],	$v = ds / dt = \dot{s}$ [m/s],
Acceleration:	$b = dv / dt = \dot{v} = \ddot{s}$ [m/s²],	$b = dv / dt = \dot{v} = \ddot{s}$ [m/s2],
Jerk:	$k = db / dt = \dot{b} = \ddot{v}$ [m/s³].	$k = db / dt = \dot{b} = \ddot{v}$ [m/s3].

8.4.1 Driving on a Straight, Horizontal Road

The following applies to a uniform movement:

$$s = v / t, \; b = 0.$$

For a uniformly accelerated (decelerated) movement, the following applies

$$b = \left(v_2 - v_1\right) / t = \left(v_2^2 - v_1^2\right) / 2s.$$

The following applies to a continuous and linearly variable acceleration

$$k = \left(b_2 - b_1\right) / t.$$

8.4.2 Driving on a Straight Line with a Gradient

The maximum longitudinal acceleration max b is limited by the engine power, the sum of the driving resistances, the friction between the tyres of the driven axle and the road surface and the gradient of the road.

$$\max\, b = \frac{g \cdot 3.6^2 \cdot N_b}{G \cdot v} \pm g \cdot \frac{s}{100}$$

with:

g: Acceleration due to gravity = 9.81 [m/s²],
N_b: Acceleration power [KW] = motor power—overcoming the driving resistances,
G: Vehicle weight [KN],
v: Speed [m/s],
s: Longitudinal inclination (gradient) [%].

The maximum braking deceleration max b is limited by the mechanical braking force in the brake system and the friction between the tyre and the road surface.

$$\max\ b = g \cdot \left(f_{Tvorh} \pm \frac{s}{100} \right)$$

with f_{Tvorh} = Tangential coefficient of friction.

For a passenger car, the max b = 5.0 to 6.0 m/s² is valid for wet roads and max b = 8.0 to 10.0 m/s² for dry roads. The Technical Inspection Agency requires at least 5.0 m/s² for the mandatory vehicle check every two years.

8.4.3 Driving through Curves

As long as the forces acting on the vehicle are in balance when driving through curves, the vehicle will not skid. It must apply (see Fig. 8.2):

$$\frac{v^2}{g \cdot R} \leq f_R(v) \pm \frac{q}{100}.$$

This means that the maximum speed at a given radius R must be calculated as follows:

$$\max\ v = \sqrt{g \cdot R \cdot \left(f_R(v) \pm \frac{q}{100} \right)}.$$

Or that the minimum radius at a given speed corresponds to the following term:

$$\min\ R = \frac{v^2}{g \cdot \left(f_R(v) \pm \frac{q}{100} \right)}.$$

The required coefficient of friction to prevent skidding is

$$f_R = \frac{v^2}{g \cdot R} \pm \frac{q}{100} \pm \frac{S_{v,h}}{G_{v,h}}$$

with

$S_{v,h}$ = Crosswind force on front or rear axle [N],
$G_{v,h}$ = Weight force on front or rear axle [N].

The freehand speed is the speed at which the entire centrifugal force is absorbed by the crossfall. It amounts to (crosswind neglected):

$$V_F = \sqrt{g \cdot R \cdot \frac{q}{100}}.$$

8.5 Driving Dynamics Derivation of Design Variables in the RAS-L 95

Since in the RAL (FGSV 2012a) the design parameters are set on the basis of planning speeds, the description of the driving dynamics derivation of the individual design elements is based on the RAL-L 95 (FGSV 1995), which is no longer valid.

8.5.1 Speeds

When processing the road design according to RAS-L 95 (FGSV 1995), the speeds were used with different effects as target value, design value and control value. In the RAL (FGSV 2012a) only the target average car driving speed according to the RIN (FGSV 2008e) and the planning speed are mentioned, but these do not have any direct influence on the design elements. These are determined by the design class.

In the RAS-L, different speed types were used, which will continue to be presented here, as they are used to derive individual design elements.

- Rated speed V_B,
- Design speed V_e,
- Speed V_{85},
- Intersection speed V_K.

The desired average car speeds were targets for dimensioning according to the HBS (FGSV 2015a), that is, mainly for the choice of a standard cross-section or the type of junction. The design speed V_B corresponded to the desired average car speeds.

According to the RAS-L (FGSV 1995), the design speed V_e represented an economic and technical key variable. It was determined under consideration of environmental concerns based on the intended network function of the road. The quality of the traffic flow intended for this function was taken into account. In practice, however, a different procedure was often applied. Instead of selecting a design speed and using correspondingly adapted design elements, the existing constraint points were used for alignment and the design speed was subsequently determined according to the smallest elements. This often led to unbalanced and unsteady solutions. Furthermore, the roads of the different categories were no longer clearly distinguishable.

The design speed was assigned limit and guide values of certain design elements as well as permissible ratio values for the combining of the design elements:

- Minimum curve radii,
- Clothoid parameters,
- Maximum longitudinal inclinations,
- Sag and summit minimum radius.

As a result, it had a decisive influence on the route characteristics and thus on the safety and quality of the traffic flow as well as on economic efficiency. The design speed V_e should remain constant over longer distances or at least over longer contiguous sections of road.

The speed V_{85} as a (pseudo) driving dynamics-based guide variable represented a target value of the actual driven speed level. It was used for the dimensioning of individual safety-relevant design elements and also allowed the design and its specifications to be checked primarily in the site plan. For country roads, it is based on the speed which 85% of unobstructed driving passenger cars on wet roads do not exceed. The speed V_{85} varies over a section of the road depending on the alignment. It was determined with:

- Crossfall in a circular curve,
- Minimum radii for negative crossfall,
- Required stop sight distance,
- Required overtaking sight distance.

On country roads, the speed V_{85} changes depending on the route geometry. It could be estimated for short sections in reconstruction and upgrading as a function of the individual radius according to Table 8.2. For the new construction, a relationship was established between the curve, road width and speed V_{85}. Curvature is the sum of the absolute changes of direction related to the length of the road. To determine the curvature, the road to be investigated had to be divided into sections of a similar curvature. This already contained inaccuracies due to different section sizes and boundaries.

The speed V_{85} was not allowed to exceed the design speed V_e by more than 20 km/h. If the difference $V_{85} - V_e$ was greater than 20 km/h, the design speed V_e of the road section

Table 8.2 Determination of V_{85} for individual curves (upgrading) and sections with similar curves (new construction) for road widths ≥ 6.50 m according to RAS-L (FGSV 1995)

Speed V_{85} [km/h]	Single radius [m]	Curvature [gon/km]
100	300	150
95	230	183
90	175	215
85	135	248
80	110	280
75	80	313
70	65	345

was to be increased or the expected speed V_{85} was to be reduced by means of suitable measures, e.g. by selecting smaller curve radii and increasing the curvature.

However, neither of these two approaches was effective and led to unbalanced solutions. Since in flat terrain the curvature was usually less than 150 gon/km, the V_{85} was at 100 km/h and the V_e had to be chosen at least 80 km/h, regardless of the road category. An increase of the curvature was usually not possible due to a large number of constraint points. However, this led to the fact that the country roads of different road categories were often equipped with similar design parameters due to the same V_{85}.

A further control step was to check the continuity of a route. For this purpose, the sums of the directional changes were plotted against the length of the road. If there were different gradients in sections, it had to be checked whether the speeds V_{85} determined for neighbouring sections did not differ by more than 10 km/h.

8.5.2 Minimum Radii

The minimum curve radius (min R) can be derived dynamically (see Sect. 8.4) as a function of the design speed V_e (see Sect. 8.5.1), the utilization factor n of the radial frictional (see Sect. 8.3) and the limits of the crossfall max q. The initial relationship for determining the relationship between the minimum radius of the curve, the crossfall of the curve and the design speed V_e is the clearance of forces on the vehicle in the curve (see Sect. 8.2). The size of the minimum radius is determined according to the following equation:

$$\min R = \frac{V_e^2}{3.6^2 \cdot g \cdot \left(\max f_R(V) \cdot n \pm \frac{q}{100} \right)}$$

with:

min R: Minimum curve radius [m],
V_e: Design speed [km/h],
g: Acceleration of gravity [m/s^2],
f_R: Max. radial coefficient of friction [−] = 0.925 · max f_T,
f_T: Max. tangential coefficient of friction [−],
n: Use of the radial coefficient of friction [−],
q: Crossfall [%] (use negative if facing the outside of the curve).

For circular curves with crossfall to the inner side of the curve, utilization factors n of the f_R of 50% were determined. Thus, an f_T of about 87% was still available for braking, for example. To avoid zones with weak drainage in the event of insufficient longitudinal gradient or to avoid central reservation drainage on dual two-lane roads, crossfalls to the outside of the curve were permitted in accordance with the RAS-L for larger radii. For safety

Table 8.3 Minimum radii according to the RAS-L (FGSV 1995)

V_e [km/h]	min R [m] q to the inside n = 50%	V_{85} [km/h]	min R [m] q to the outside n = 30%
50	80	70	600
60	120	80	950
70	180	90	1400
80	250	100	2100
90	340	110	3000
100	450	120	4100
120	720	130	5500

reasons, the utilization factor n of the f_R was set at 30% and the reference was the V_{85}, which was to approximate the real driving behaviour. Rounded with this formula, the minimum radii shown in Table 8.3 were obtained.

For example, for V = 90 km/h the minimum radii are calculated as follows:

$$\max\ f_T = 0.241 \cdot \left(\frac{V}{100}\right)^2 - 0.721 \cdot \left(\frac{V}{100}\right) + 0.708,$$

$$\max\ f_T = 0.241 \cdot \left(\frac{90}{100}\right)^2 - 0.721 \cdot \left(\frac{90}{100}\right) + 0.708 = 0.25431,$$

$$\min\ R = \frac{V^2}{3.6^2 \cdot g \cdot (\max\ f_R \cdot n + q)} = \frac{V^2}{127 \cdot (\max\ f_R \cdot n + q)}.$$

At q = 7.00% to the inside of the curve and n = 50%:

$$\min\ R = \frac{90^2}{127 \cdot (0.925 \cdot 0.25431 \cdot 0.5 + 0.07)} = 339.94\text{m}.$$

At q = 2.50% to the outside of the curve and n = 30%:

$$\min\ R = \frac{90^2}{127 \cdot (0.925 \cdot 0.25431 \cdot 0.3 - 0.025)} = 1399.56\text{m}.$$

8.5.3 Required Sight Distance

The German Road Traffic Regulations (Straßenverkehrsordnung) (BMVBS 2013) stipulate that the driver must choose his speed according to the visibility conditions. If this legal standard were to be observed, the existing sight distances would be a good way of influencing speed behaviour. However, measurements and observations of the relationship between visibility and speed show that existing visibility has only a limited effect on the choice of speed. The following reasons can be given for this:

- Distances and speeds are difficult to estimate, especially while driving.
- At lower speeds, the actual braking distances are underestimated. On fast-traffic federal roads and motorways, speeds are underestimated.
- Emergency braking before sudden and unexpectedly occurring obstacles are rare events. The driver has no feeling for the forces involved.

As a consequence, the road design is based on the assumption that sight distance has a significant influence on traffic safety, performance and driving comfort and that it must be considered as a design parameter. In order to implement this, sight distance analyses are required in the early design stages. By reducing the driving space to only one dimension, geometric sight distance is obtained. The minimum requirement here is to check the geometric sight distance in the driving space to ensure that it corresponds to the stopping sight distance required by driving dynamics.

The **required stopping sight distance S $_h$** is defined as the distance required by a driver travelling at a certain speed to stop in front of an unexpectedly occurring obstacle of defined height on the roadway. It is made up of the distance during the reaction and impact time (S_1) and the pure braking distance (S_2) (see also Chap. 15 and Fig. 15.1).

$$s_h = s_1 + s_2$$

$$s_1 = \frac{V_0}{3.6} \cdot t_R \text{ with } t_R = 2.0 s$$

$$s_2 = \frac{1}{3.6^2 \cdot g} \cdot \int_{V_1}^{V_2} \frac{V}{f_T(V) + \frac{s}{100} + \frac{W_L}{G}} \cdot dV$$

with:

s_h: Stop sight distance [m]
s_1: Distance traveled during the reaction and effect duration [m].
s_2: Pure braking distance [m]
V_0: Speed at the start of braking [km/h].
$V_{1,2}$: Speeds at the beginning and end of braking [km/h]
g: Acceleration of gravity [m/s^2]
f_T: Max. tangential coefficient of friction [−]

$$\max f_T = 0.241 \cdot \left(\frac{V}{100}\right)^2 - 0.721 \cdot \left(\frac{V}{100}\right) + 0.708$$

s: Longitudinal inclination (positive: slope, negative: descent) [%].
q: Crossfall [%] (use negative if facing the outside of the curve)

The required stopping sight distances were shown in the RAS-L (FGSV 1995) in a diagram as a function of the longitudinal inclination. In the RAL (FGSV 2012a), the required holding visibility was slightly reduced (see Chap. 15).

Standard Cross-sections 9

Summary

For the various design classes for motorways and rural roads, different standard cross-sections have been defined in the guidelines depending on the importance of the connection. These differ mainly in the number and width of the lanes. In the case of rural roads, they also differ in the number and regularity of overtaking lanes and in the markings.

9.1 State of the Art

9.1.1 Road Safety

For motorway cross-sections, results of investigations by Meewes are available (Meewes and Eckstein 2002). The somewhat wider standard cross-sections for long-distance motorways (RQ 43.5, RQ 36 and RQ 31) show a slightly lower accident cost rate compared to the standard cross-section for regional motorways (RQ 28). Dispensing with the hard shoulder further reduces traffic safety, especially when there is a high proportion of heavy traffic. Compensation is provided, on the other hand, by the speed restrictions often imposed and the overtaking bans for trucks. The accident cost rates are 18 to 20 EUR/1000 Kfz · km (price level 2000).

Within the framework of studies on the use of so-called intermediate cross-sections, Brannolte has determined accident cost rates for four cross-section types on the basis of accident data for the years 1984 to 1986 (Brannolte et al. 1993; Table 9.1).

On behalf of the Federal Highway Research Institute (Bundesanstalt für Straßenwesen, BASt), Palm and Schmidt investigated in 1999 the effects of different cross-sectional

© Springer Fachmedien Wiesbaden GmbH, part of Springer Nature 2023

T. Richter, *Road Planning - Freeways and Country Roads*,
https://doi.org/10.1007/978-3-658-35189-2_9

Table 9.1 Adjusted accident cost rates by cross-section type (Brannolte et al. 1993)

Standard cross-section	Roadway width	Accident cost rate UKR_a (PS + SS) [DM/1000 Kfz · km]
b2	>7.50–8.00	71
c2	>7.00–7.50	51
d2	>6.50–7.00	60
e2	6.00–6.50	72

Table 9.2 Adjusted accident cost rates by cross-section type. (Palm and Schmidt 1999)

Standard cross-section	Carriageway width	Accident cost rate UKR_a (PS + SS) [DM/1000 Kfz · km]
b2	>7.50–8.00	42.71
c2	>7.00–7.50	33.15
d2	>6.50–7.00	75.98
e2	6.00–6.50	65.25

Table 9.3 Adjusted accident cost rates by cross-section type. (Vieten et al. 2010, 2011)

Standard cross-section	Carriageway width	Accident cost rate UKR_a (PS + SS) [EUR/1000 Kfz · km]
RQ 11.5+ in the two-lane section[a]	8.50	23.1
RQ 11[a]	8.00	23.9
RQ 11 reduced at SV < 300 Veh./24h[a]	7.50	31.2
	7.00	36.0
	6.50	44.5
RQ 9[a]	6.00	57.9
	5.50	63.8
	5.00	55.8

[a] Due to different edge strip widths and marking shapes not directly comparable

widths on road safety on the basis of accident data from 1992 to 1995 and came to the following results (Palm and Schmidt 1999; Table 9.2).

Vieten presented two research reports in 2011, also commissioned by the BASt, which examined various elements of rural roads with regard to their effects on road safety (Vieten et al. 2010, 2011). For the different cross-sectional widths it comes to the following results (see Table 9.3).

The research work sometimes comes to different results. In general, the accident cost rate increases with narrow carriageway widths ≤ 7.00 m. In the work of Vieten, the accident cost rate continues to decrease for larger road widths, while in the investigations of Palm and Brannolte, a minimum is reached for road widths between 7.00 and 7.50 m. The RAL is mainly based on the more recent studies of Vieten (Vieten et al. 2010).

Various studies are also available on the standard cross-section RQ 15.5 or b2 + 1. In earlier studies, for example (Brannolte et al. 1993), a clear safety gain was always determined compared to two-lane standard cross-sections. Weber has determined an accident cost rate of 14.70 EUR/1000 Kfz · km (including accidents with slight material damage (LS)) on his eight investigated sections and is thus significantly below the values for two-lane sections (Weber and Löhe 2003). Vieten determined almost the same value in his investigation (14.80 EUR/1000 Kfz · km), although this value is somewhat less favourable, as no LS accidents (slight material damage) were considered. The increase in safety is usually based on the traffic-secured overtaking possibilities.

Dual two-lane standard cross-sections (RQ 21 or similar) were also investigated in Vieten (Vieten et al. 2010). An accident cost rate (PS + SS) of 9.80 EUR/1000 Kfz · km is given here. This cross-section is thus around 50% safer than the RQ 15.5. The reasoning behind this is essentially the centre separation, which largely prevents a coming off the road in oncoming traffic, and the lack of overtaking pressure, since overtaking can be carried out continuously. In a more recent study by Maier, the results were largely confirmed (Maier and Berger 2012). An accident cost rate (PS + SS) of 10.70 EUR/1000 Kfz · km was determined for the sections of the road without junctions. Comparisons with regard to the type of junction showed that a lower accident cost rate can be achieved on the sections without junctions when the route is managed grade-separated.

In summary, the evaluation of existing research shows that dual carriageway cross-sections are expected to provide the highest level of traffic safety. Single carriageway roads with three-lane cross-sections, on the other hand, are less safe but are clearly advantageous in terms of traffic safety compared to single carriageway with two-lane cross-sections. In the case of a single carriageway with two-lanes cross-sections, traffic safety is significantly reduced for lane widths ≤ 7.00 m The results are inconclusive for lane widths ≥ 7.00 m. For the RAL the most recent results from Vieten (Vieten et al. 2010) were used and 8.00 m wide carriageways were recommended.

9.1.2 Traffic Flow

In the run-up to the processing of the first HBS from 2001, extensive studies were carried out on the traffic flow and performance of motorways and country roads. These were partially supplemented and revised for the new edition of the HBS 2015.

Some of this research was carried out at the Ruhr University Bochum (Ruhr Universität Bochum) (Brilon and Breßler 2001; Brilon and Geistefeld 2010; Brilon et al. 2007). These deal in particular with the road sections without junctions. Richter, Steinauer and Zierke have also investigated the interaction of traffic safety and traffic flow of junctions and junction-free sections of road in their research (Richter 1993; Steinauer et al. 2008; Zierke 2010). According to this, the type of junction has an influence on the driving behaviour on the adjacent junction-free road sections, which leads to different traffic flows, especially in the case of the frequently existing short distances between junctions.

The junction-free sections of rural roads were allocated ever greater capacities in the generations of guidelines. Compared to the RAS-Q of 1982, the capacities have increased by 25 to 60%. Especially on regional motorways and urban motorways, real traffic volumes are often higher than the calculated capacities. This is mainly due to the risk-taking, non-compliant driving style, for example, the permanent falling below safety distances. For this reason, the question is increasingly being raised as to whether a rule-compliant behaviour should be assumed when designing a road traffic system or whether possible capacities should be used in disregard of the StVO.

9.1.3 Status of the Guidelines

For motorways, the RAA have a standard range of application for each standard cross-section depending on the traffic volume (FGSV 2008a). The final dimensioning is carried out according to the HBS (FGSV 2015a). Here the main factors influencing the capacity are taken into account. These are the number of lanes, the longitudinal slope, the proportion of heavy traffic, the location inside or outside metropolitan areas and any restrictions on the speed limit.

In the RAL, a single-carriageway standard cross-section is defined for each design class (FGSV 2012a). The final design is again based on the HBS (FGSV 2015a). If the verification according to the procedures of the HBS shows that the single-carriageway standard cross-section does not guarantee sufficient traffic quality, a dual carriageway standard cross-section can be used in the course of otherwise single carriageway roads on one or more successive network sections, for example, nearby agglomeration areas (up to a total length of approximately 15 km). Otherwise, it should be checked whether it would be better to use the next higher design class. The factors influencing the capacity of rural roads are longitudinal slope, curvature and the proportion of heavy traffic.

9.2 Standard Cross-sections for Motorways

The standard cross-sections for motorways are determined in accordance with the RAA (FGSV 2008a). The aim is to ensure that the design of the motorways is as uniform as possible.

9.2.1 Standard Cross-sections for Motorways of the EKA 1

The standard cross-sections of the motorways of EKA 1 are shown in Fig. 9.1. The somewhat more generous dimensions of the RQ 31 result mainly from the required 4+0 construction site routing and the possibility of temporarily opening the hard shoulder at a later date.

Table 9.4 shows the regular ranges and the lower and upper weighing ranges. Two standard cross-sections are generally possible in the weighing range. In this case, it must be

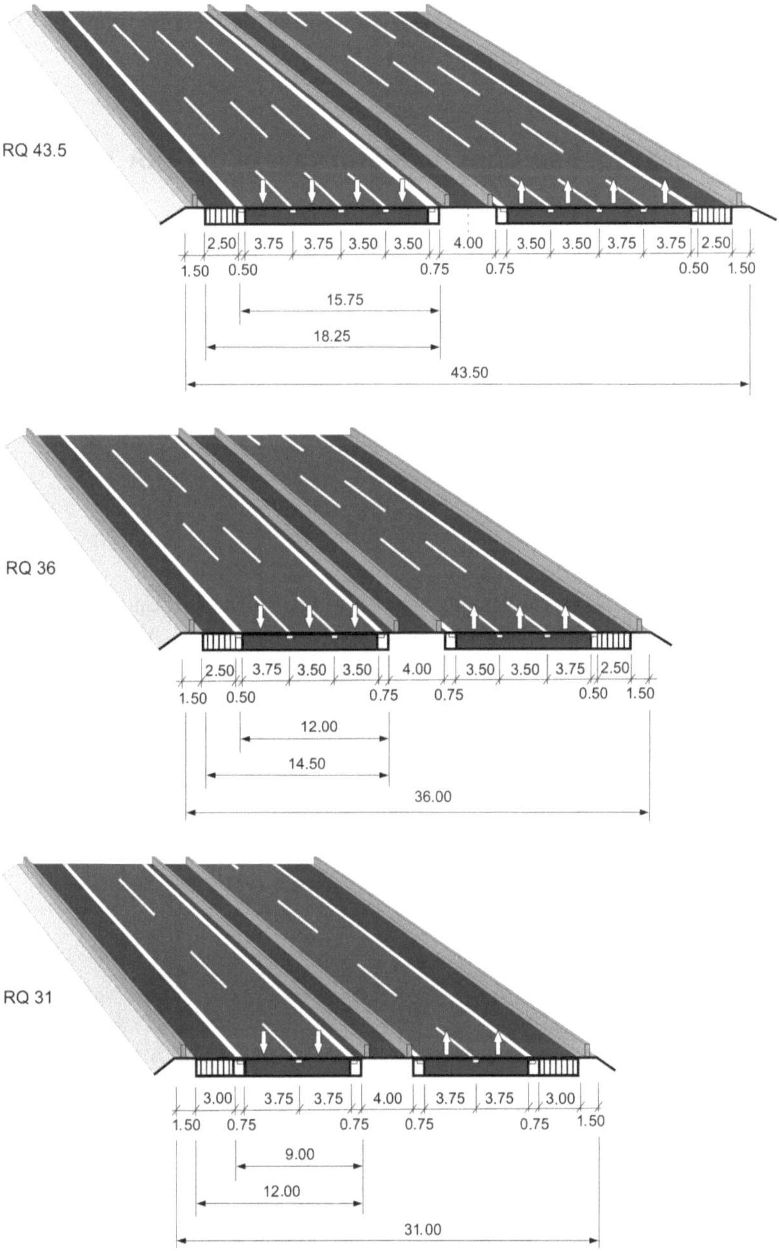

Fig. 9.1 Standard cross-sections for motorways of EKA 1, dimensions in [m] (According to FGSV 2008d)

checked which standard cross-section offers the most advantageous solution. Evaluation criteria for weighing can be found in Chap. 3.

9.2.2 Standard Cross-sections for Motorways of the EKA 2

In the EKA 2, only the RQ 28 is provided. If a 4+0 construction site routing is required due to the traffic volume and the lack of diversion possibilities on the neighboring road network, or if the traffic volume requires a six-lane cross-section, the standard cross-sections of EKA 1 must be used (Fig. 9.2).

The range of application is in the same level as the RQ 31, although due to the frequent location within metropolitan areas and the frequent limitation of the permissible maximum speed, larger capacities are usually available.

Table 9.4 Areas of application of the standard cross sections for motorways of the EKA 1 (According to FGSV 2008d)

Standard cross-section	Lower weighing range ADT [Veh./24 h]	Regular range ADT [Veh./24 h]	Upper weighing range ADT [Veh./24 h]
RQ 43.5	85.000–95.000	95.000–120.000	
RQ 36	58.000–62.000	62.000–102.000	102.000–115.000
RQ 31		18.000–68.000	68.000–73.000

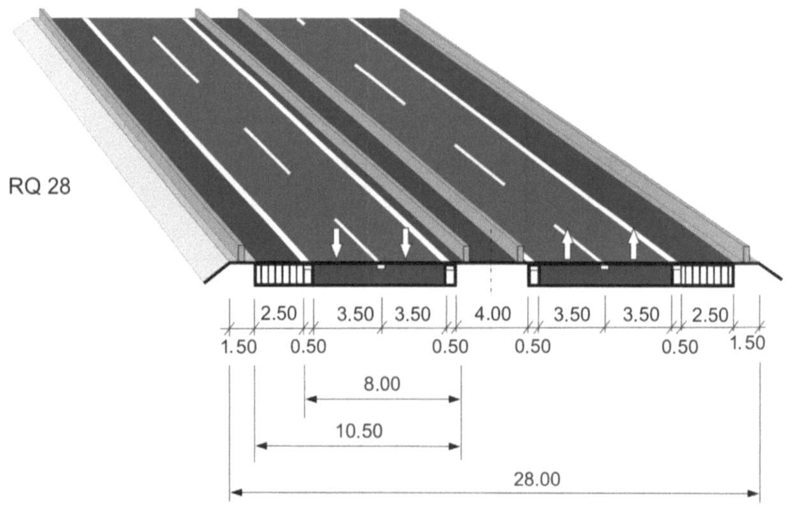

Fig. 9.2 Standard cross-section for EKA 2 motorways, dimensions in [m]. (According to FGSV 2008d)

9.2.3 Standard Cross-sections for Motorways of the EKA 3

The standard cross-sections for EKA 3 urban motorways are shown in Fig. 9.3. Compared to the standard cross-sections of EKA 1, these are significantly narrower. This is justified to the extent that significantly shorter distances are travelled at significantly lower speeds and the areas within built-up areas are limited. Furthermore, there is no need for complex construction site management, as alternative routes are usually available.

Table 9.5: Application areas of the standard cross-sections for motorways of the EKA 3 according to (FGSV 2008d) shows the standard application areas as well as the lower and upper consideration areas. Two standard cross-sections are generally possible in the weighing range. Here it must be checked which standard cross-section offers the most advantageous solution. Evaluation criteria for weighing can be found in Chap. 3.

In some sections of the motorway, the range of applications is clearly exceeded. On the six-lane, A 100 in Berlin, average daily traffic (ADT) of 186.000 Veh./24 h was counted in 2010 between three-way motorway interchange Funkturm and exit Kurfürstendamm. On the A3 near Cologne, the ADT was over 150.000 Veh./24 h in sections.

9.2.4 Definition of Standard Cross-sections for Motorways

The standard cross-sections usually offer sufficient traffic quality if they are operated within the standard range. For each planning, however, proof must be provided that the desired quality of traffic flow and the desired average car speeds are achieved. As a rule, the procedures of the HBS (FGSV 2015a) are used. For motorways it must be demonstrated that the expected traffic load can be handled with the desired quality of traffic flow on both directional lanes, that is, the degree of utilization is below the required level. It must also be demonstrated that the desired average car speed is reached.

For sections of motorways, the traffic volumes of one direction of travel are required in units of [vehicle/h] and the share of heavy goods vehicles b_{sv} in [%]. Both directions of the motorway must be edited separately.

The alignment in the site plan is not taken into account when assessing the quality of traffic flow on motorways. Larger longitudinal slopes have a particular effect on the travel speeds of heavy goods vehicles and thus also impede car traffic. Separate diagrams are therefore used for gradients with $s \geq 2\%$. Equivalent gradients are determined for short gradients (Fig. 9.4). The procedure does not take into account the summit and sag curves.

9.2.4.1 Quality Criteria

The RIN (FGSV 2008e) provide quality specifications for the target car driving speed (see Chap. 2). Since, for the same traffic volume, the targeted car travel speeds on sections of road without speed restrictions and major inclines are significantly higher than on sections with speed restrictions or major inclines, the targeted car travel speed is not sufficient as

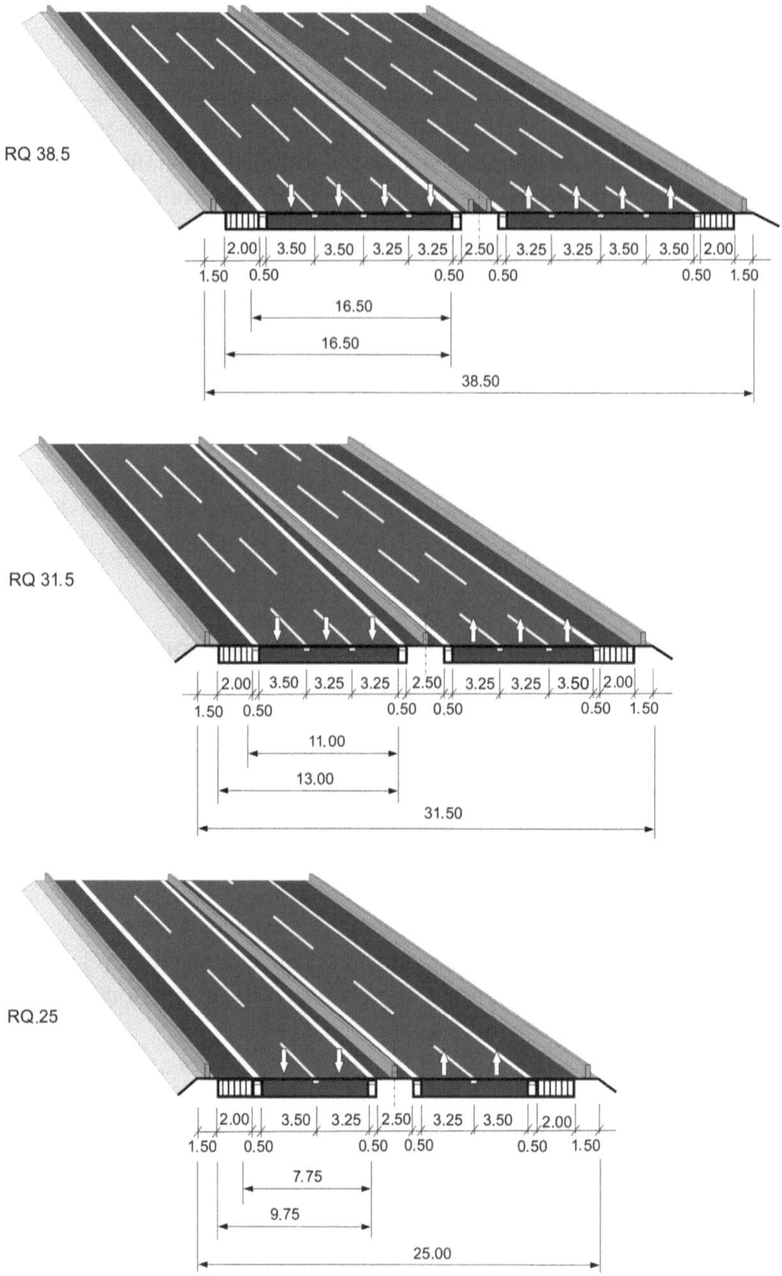

Fig. 9.3 Standard cross-sections for EKA 3 motorways, dimensions in [m] (According to FGSV 2008d)

Table 9.5 Areas of application of the standard cross sections for motorways of the EKA 3 (According to FGSV 2008d)

Standard cross-section	Lower weighing range ADT [Veh./24 h]	Regular range ADT [Veh./24 h]	Upper weighing range ADT [Veh./24 h]
RQ 38.5	85.000–95.000	95.000–120.000	
RQ 31.5	60.000–70.000	70.000–103.000	103.000–118.000
RQ 25	12.000–20.000	20.000–73.000	73.000–82.000

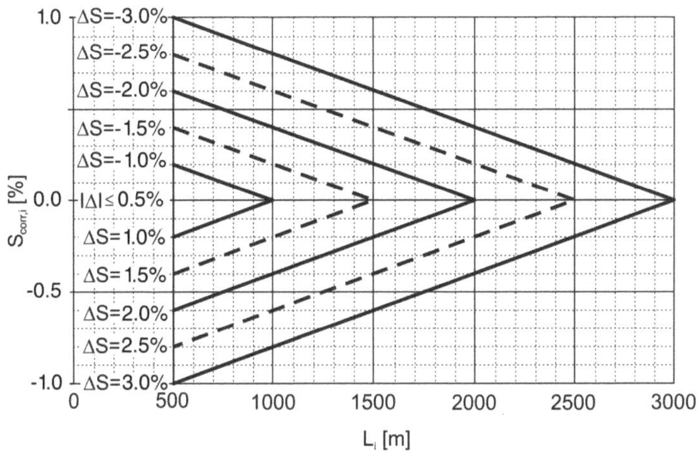

Fig. 9.4 Consideration of the longitudinal slope according to the HBS (FGSV 2015a)

the only quality criterion. The HBS (FGSV 2015a) therefore uses the degree of utilization as a quality criterion, which describes the degree of freedom of movement under traffic conditions.

$$a = \frac{q_B}{C}$$

with
a: Degree of utilization [–],
q_b: Rated traffic intensity [vehicle/h],
C: Capacity [vehicle/h].

9.2.4.2 Traffic Flow on Motorway Sections

The average driving speed V_F on a motorway section of length L is given by the following equation:

$$V_F = \frac{L}{\sum_{i=1}^{m} \frac{L_i}{V_{F,i}}}$$

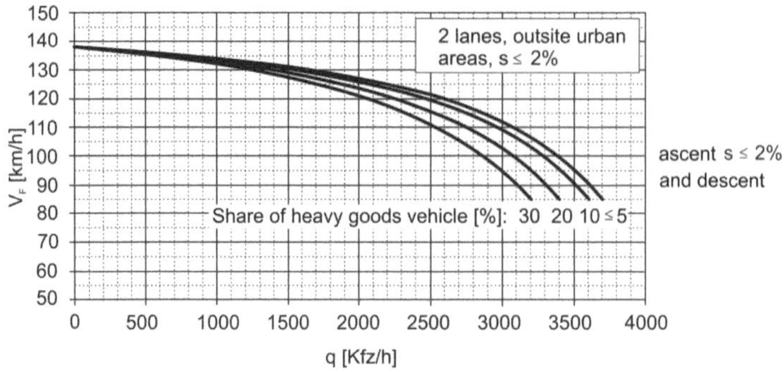

Fig. 9.5 Example for the determination of the average car speed for motorways according to the HBS (FGSV 2015a)

with

V_F: Average speed of passenger cars on the motorway section [km/h],
$V_{F,i}$: Average speed of passenger cars on subsection i [km/h],
L: Length of the motorway section [km],
L_i: Length of subsection i [km].

The average travel speeds of the passenger cars on the subsections can be taken from the HBS depending on the traffic load, the proportion of heavy goods vehicle, the mean equivalent gradient and other environmental and control conditions. Figure 9.5 shows the example of a two-lane carriageway outside conurbations with low longitudinal gradients.

The determined travel speed of the passenger cars $V_{F, Ges}$ for the motorway section is to be compared with the average passenger car travel speed required according to the RIN (FGSV 2008e). If the desired average car speed is not reached, check whether influencing factors can be changed or whether a lower average car speed can be accepted.

9.2.5 Standard Cross-sections for Bridges

The bridges should fit harmoniously into the alignment. This also includes that the lanes run through in unchanged width. Deviations occur in the hard shoulder due to drainage requirements or necessary construction site traffic routing. The width of the bridge edge construction also depends on the effective ranges of the selected safety barrier, so that different dimensions can be achieved here. In order to ensure a normal effective range W4 (W ≤ 1.30 m), the bridge edge construction must be 2.05 m wide. This also corresponds to the current RiZ-ING.[1] Figure 9.6 shows an example from the RAA.

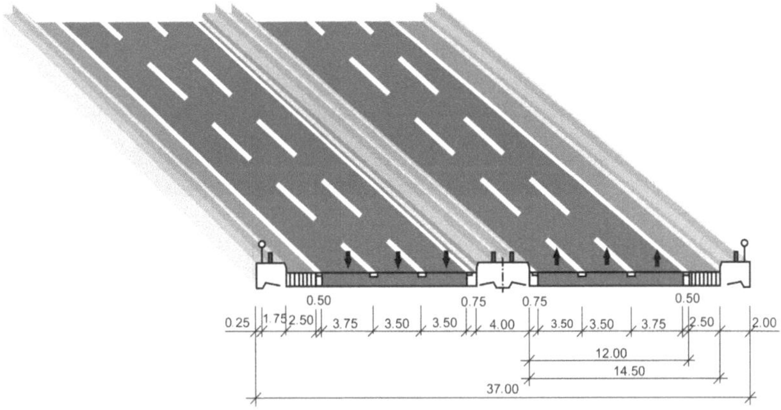

Fig. 9.6 RQ 36 B. (According to FGSV 2008d)

9.3 Standard Cross-sections for Rural Roads

The standard cross-sections for rural roads are determined according to the RAL (FGSV 2012a). The aim is to achieve the most uniform possible design of the country roads of one-design class and the possibility of distinguishing between country roads of different design classes to a large extent.

9.3.1 Standard Cross-sections for Roads of EKL 1

The standard cross-section for EKL 1 is RQ 15.5 (Fig. 9.7). The RQ 15.5 is a single-carriageway, three-lane cross-section, whereby the middle lane is alternately assigned to both directions of travel as an overtaking lane. Deducting the alternating sections results in approximately 40% traffic-secure overtaking possibilities for each direction of travel. The separation of the driving directions is made by a central traffic strip, which is 1.00 m wide and consists of two-lane boundaries (VZ according to StVO and therefore subject to regulation) and a green-coloured area in between (not subject to regulation). Due to the network function and the high speeds to be expected, the RQ 15.5 is to be operated as vehicles only road.

The length of the overtaking lanes must be a compromise between sufficient length of the two-lane section to satisfy overtaking needs and limitation of the length of the single-lane sections to avoid large platoons. The overtaking sections should be long between 1200 m (at least 1000 m) and 2000 m. Often, however, the intersections and topography determine the arrangement and length of the overtaking lanes. Therefore, an early concept including the constraint points is necessary.

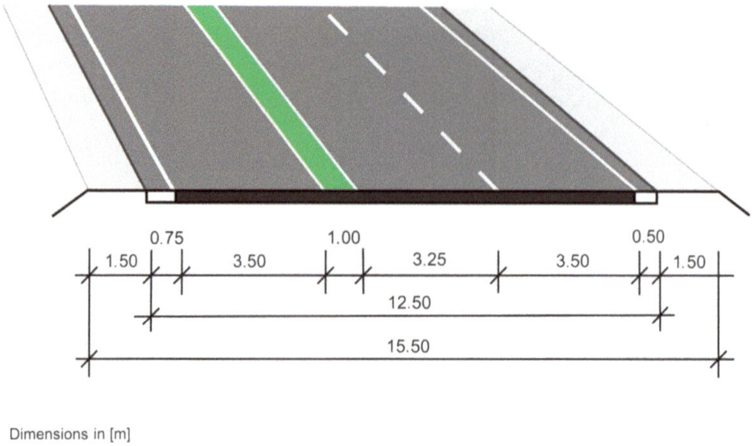

Fig. 9.7 Standard cross-section RQ 15.5, dimensions in [m]. (FGSV 2012a)

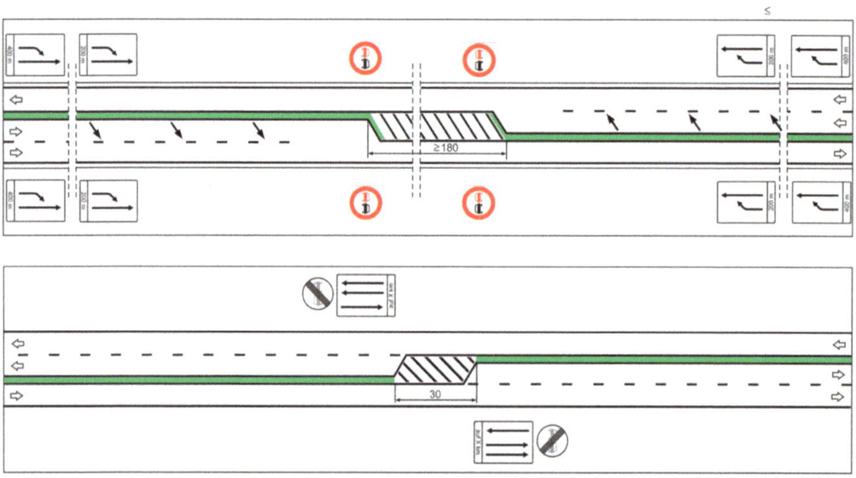

Fig. 9.8 Critical (**a**) and non-critical (**b**) changeover points (According to FGSV 2012a)

A distinction is made between critical changes where the overtaking lanes are retracted from both sides and uncritical changes at the beginning of the widening for the overtaking lanes (Fig. 9.8).

The ghost islands of the critical changeover points are 180 m long. The long length is intended to prevent frontal collisions if the keepout areas are crossed from both sides. On the other hand, there are fears that in awareness of the long restricted areas, critical overtaking maneuvers at the end of the overtaking lanes could also be encouraged. At critical changeover points, the end of the overtaking lanes should be clearly indicated by the signs and markings (arrows "end of a lane").

As very high speeds are driven on the overtaking lanes, the following instructions must be observed for the arrangement:

- Changeover points should be clearly arranged and not be located in areas with poor drainage or where there is a risk of slipperiness (e.g. on bridges).
- Overtaking lanes should not end at the transition to the existing road construction at the end of an improvement section if a significantly lower standard of development is found there.
- In hilly terrain, the overtaking lanes should be laid out in the direction of the gradient or, if it does not fit into the system, in both directions (four-lane section according to RQ 21).
- Emergency stop bays shall be provided in the single-lane sections (Fig. 9.9).

For the RQ 15.5, the emergency stop bays must be regularly positioned in the middle of the single-lane sections or at locations convenient for operation (e.g. at bridges or rain-water retention basins). Emergency stop bays are 3.00 m wide and including the extentions on both sides with a length of 84 m or, in the case of the required safety barrier, 112 m long (Fig. 9.9).

Roads of design class EKL 1 with the RQ 15.5 usually have grade-separated or partially graded separated intersections. Interchanges of overtaking line are often placed in the intersection areas. Some basic principles must be observed here:

- In a lane addition, all vehicles must change to the right-hand lane. In addition, heavy goods vehicles often use such overtaking lanes to overtake other heavy good vehicles. Therefore, the adjoining overtaking lanes should be as long as possible. This is especially important for high proportions of heavy goods vehicles or high longitudinal slopes. At the beginning of the overtaking lane, the two lanes are first separated by a 40 m long lane boundary.
- The right-hand lane of an overtaking section should not change into an exit lane by lane subtraction due to the need to change lanes for all vehicles.
- If the junction is not located in a changeover point, the overtaking lane may have to be extended, since it should be at least 800 m long before the diverging lane and at least 600 m after the merging lane. In the area of the diverging lane, structural elements such as a small guide marker (sign 605 StVO) in the traffic-technical central reservation may then be appropriate.

9.3.2 Standard Cross-sections for Roads of EKL 2

The standard cross-section for the EKL 2 is the RQ 11.5+ (Fig. 9.10). The RQ 11.5+ is a single-carriageway, two-lane cross-section, which is widened in sections to three lanes by an additional overtaking lane. The aim is to concentrate overtaking in these traffic-secure sections and to avoid overtaking where the oncoming lane has to be used as far as possible. This should also be supported as far as possible by two-lane boundaries in the two-lane sections. The problem here is that the arrangement of the lane boundaries only complies

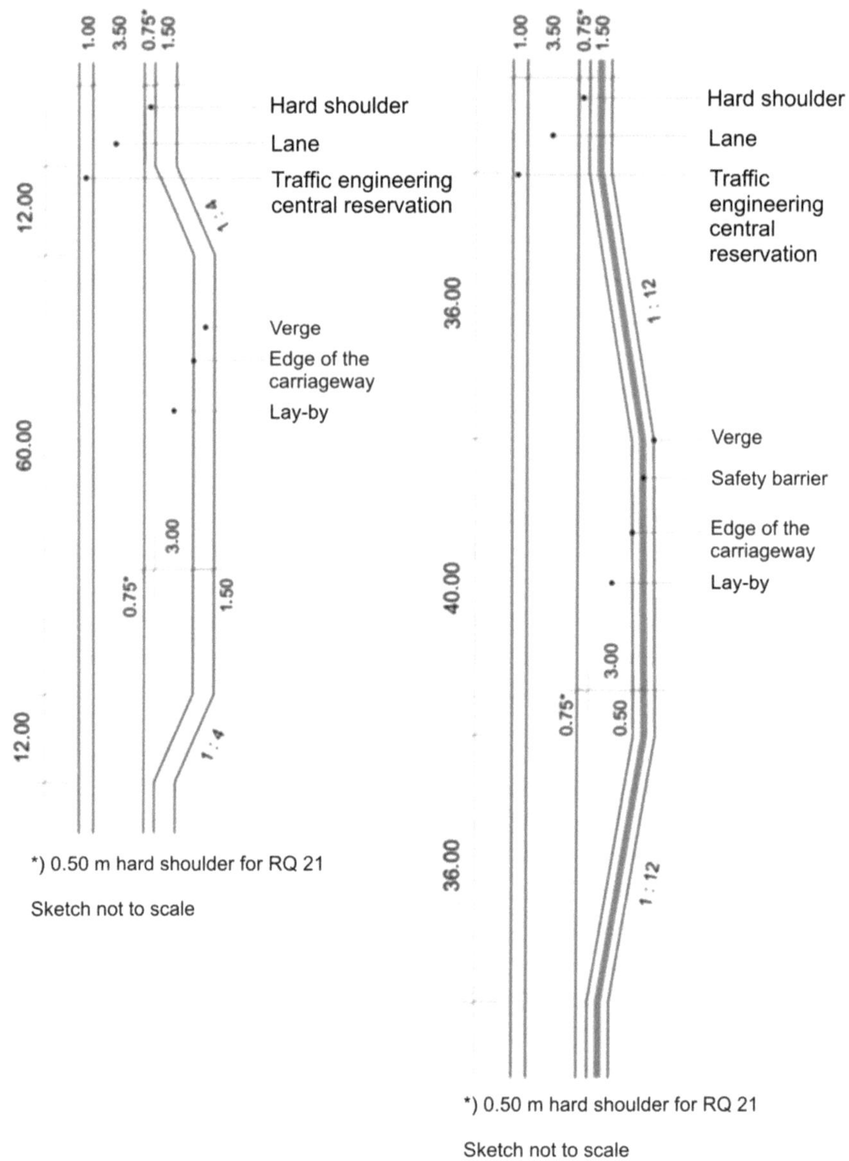

Fig. 9.9 Emergency Stop Bay without (**a**) and with (**b**) vehicle restraint systems (FGSV 2012a)

with the StVO if there is a special danger situation. Due to the large number and the severity of overtaking accidents on two-lane country roads, the author believes that this dangerous situation exists, but individual traffic authorities may see this differently. If a section is to be cleared for overtaking with sufficient overtaking visibility, the two directions of travel are separated from each other by a double broken line if the cross-sectional dimensions are the same. In order to reduce the overtaking pressure and to avoid sections in

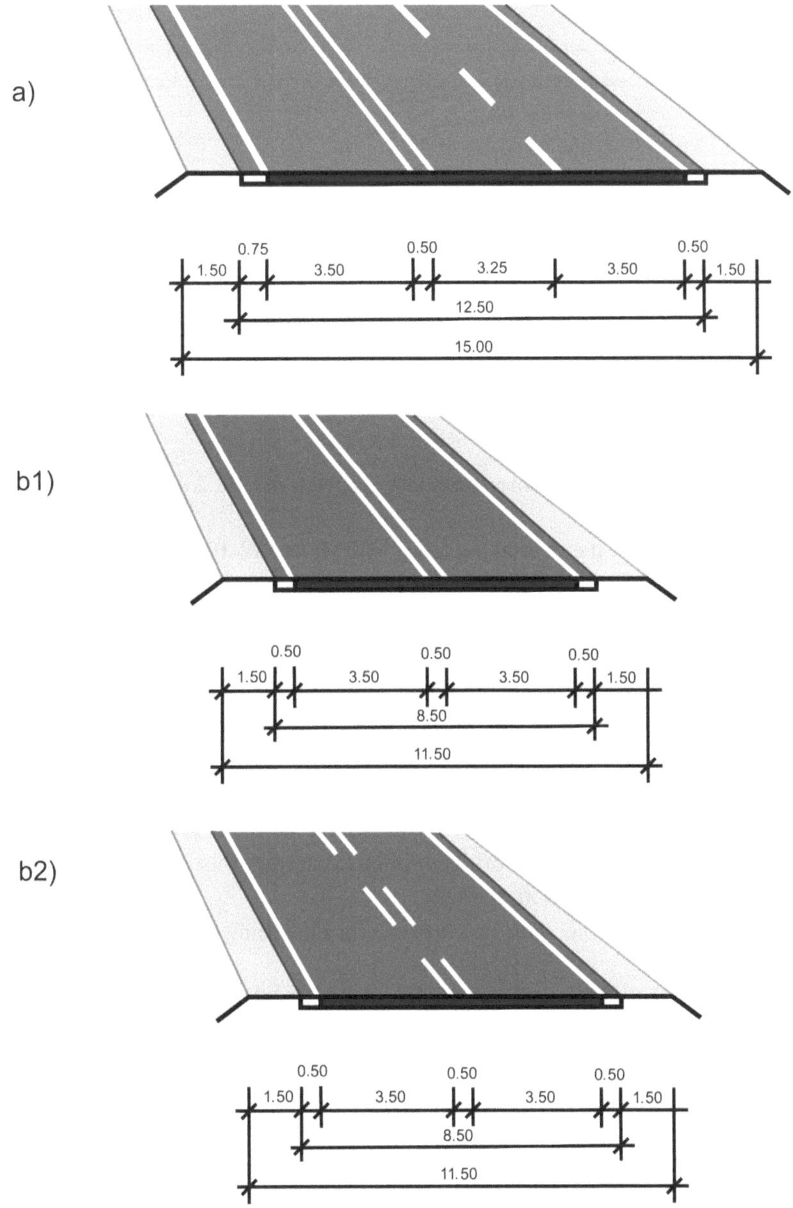

Fig. 9.10 Standard cross section RQ 11.5+, dimensions in [m]. (FGSV 2012a)

two-lane areas that are cleared for overtaking (double guideline), traffic-secure overtaking facilities should be available for each direction of travel on at least 20% of the road length.

Due to the network function and the high speeds to be expected, operation as a motor road also makes sense for the RQ 11.5+. Bike traffic on the roadway must not be permitted. A separate cycle traffic facility must be installed if there is a proven need for it.

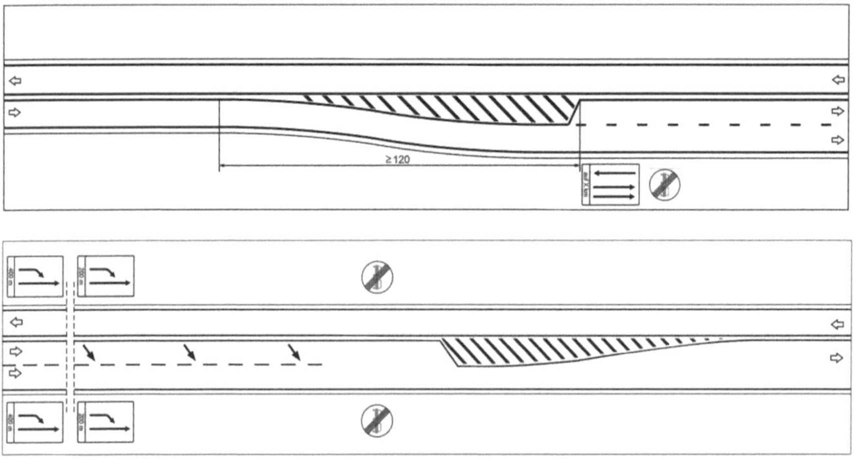

Fig. 9.11 Marking and signage on overtaking lanes on roads of EKL 2 (FGSV 2012a)

The announcement of the overtaking lanes with signs 542-10 StVO (expansion sign) in conjunction with additional signs 1004-35 (2 km off) can reduce the overtaking pressure and promote relaxed driving (see Fig. 9.11).

In principle, the desired quality of the traffic flow must be demonstrated in accordance with the procedures of the HBS and the specifications of the RIN. However, it is not necessary to prove whether the overtaking lanes are necessary for reasons of traffic quality, as they are mainly arranged for reasons of traffic safety. If the desired quality of traffic flow is not achieved, the overtaking lanes can be compacted.

In principle, the overtaking lanes should be distributed as evenly as possible. Junctions and other compulsory points (e.g. bridges) often contradict this. Overtaking lanes of the EKL 2 are between 600 and 1500 m long.

The following instructions must be observed for the arrangement:

- In hilly terrain, overtaking lanes should be located as far as possible in the direction of the ascending gradient.
- In the case of an overlap, the overtaking lane with the lower longitudinal slope should end earlier or, if this is not practical, a four-lane section should be established on the hilltop. The arrangement of small guide markers (sign 605 StVO) between the lane boundaries is recommended to clearly separate the driving directions.
- Overtaking lanes should not end before junctions, cross-town links or at the transition to the existing road if a significantly lower level of development is found there.
- Due to the high speeds on the overtaking lanes, they should not be located in tight right-hand curves if at all possible, in order to prevent driving in oncoming traffic.
- In the single-lane sections, emergency stop bays are to be constructed analogous to EKL 1.
- The length of the widening and narrowing of the carriageway is 120 m long.

- The change of the cross-section in the marking should be abrupt (length of the warp 10 m) for better recognition.
- The narrowing is to be announced by three arrows "end of a traffic lane" Z 297.1 StVO.
- The length of the passing lane shall be shown on the traffic control panel at the beginning of the passing lane.

Roads of design class EKL 2 with the RQ 11.5+ usually have ad-grade or partially ad-grade junctions with traffic signals. As a matter of principle, overtaking lanes should be laid out outside junctions if possible. They must not end abruptly at the junction. They must then be drawn in in enough distance and the left turning lane begins behind the ghost island at the end of the overtaking lane. If it is not possible otherwise, overtaking lanes can start at the intersection. In order to avoid lane addition and to improve the traffic quality, the widening should, if possible, already be carried out at the junction entrance.

9.3.3 Standard Cross-sections for Roads of EKL 3

The standard cross-section for EKL 3 is RQ 11 (Fig. 9.12). The RQ 11 is a single-carriageway, two-lane cross-section with the directions of travel separated by a broken marking. In order to avoid overtaking accidents, overtaking should also be prevented on the RQ 11 in areas where overtaking seems too unsafe (e.g. insufficient overtaking sight distance or in the area of access roads) by means of a simple lane boundary (permanent marking) or by Z 276.

EKL 3 rural roads often have low traffic volumes, especially in heavy traffic. In the case of heavy traffic up to 300 vehicle/24 h, the lane width can be reduced while maintaining the same hard shoulder width. Table 9.3 shows that accident cost rates increase with

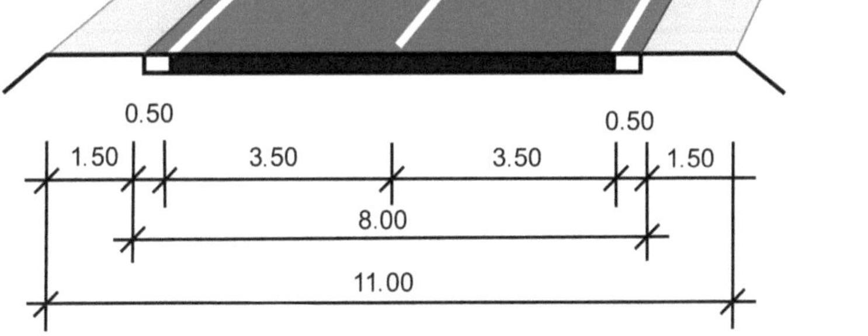

Fig. 9.12 Standard cross-section RQ 11, dimensions in [m] (FGSV 2012a)

smaller lane widths. The reduction in lane width must therefore be carefully considered. Labor law regulations for workplaces on road construction sites that may require the full closure of narrow lanes must also be taken into account.

Overtaking lanes may also be useful on rural roads of EKL 3 to improve traffic quality on ascending gradients or highly congested or unclear sections of the road. These are then to be designed in the same way as for EKL 2.

9.3.4 Standard Cross-sections for Roads of the EKL 4

The standard cross-section for the EKL 4 is the RQ 9 (Fig. 9.13). The RQ 9 is a single-carriageway cross-section without center marking to separate the two directions of travel. Instead of an uninterrupted edge marking, the RQ 9 receives a broken marking at both edges at a distance of 0.50 m from the edge of the paved surface with a line to gap ratio of 1:1. The roadway width and the broken marking at the edge of the roadway require attentive driving if necessary using the full roadway width. For this reason, the range of applications according to RAL ends at traffic volumes of 3000 vehicle/24 h and a heavy traffic volume of 150 vehicle/24 h.

9.3.5 Standard Cross-section for Roads from EKL 1 to EKL 3 with Very High Traffic Demand

The standard cross-section for EKL 1 to EKL 3 with high traffic demand is RQ 21 (Fig. 9.14). The range of application does not only depend on the high traffic demand but

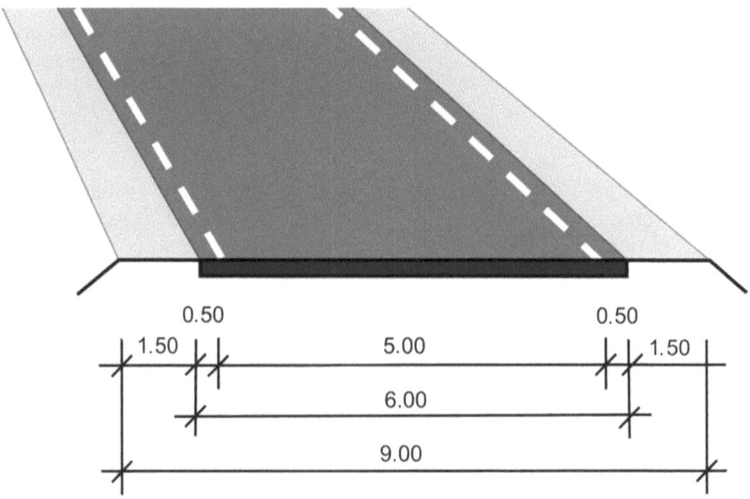

Fig. 9.13 Standard cross section RQ 9, dimensions in [m] (FGSV 2012a)

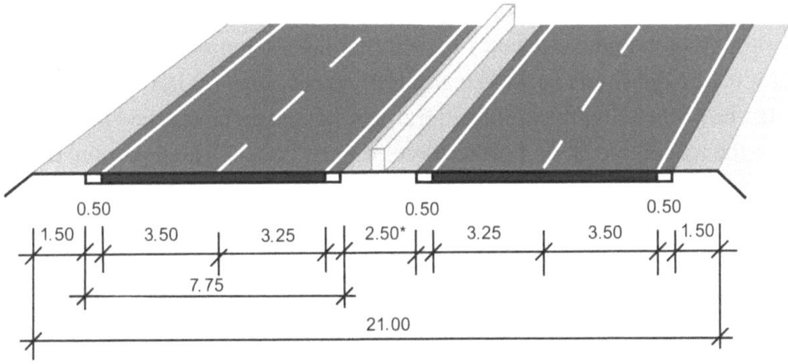

Fig. 9.14 Standard cross-section RQ 21, dimensions in [m] (FGSV 2012a)

starts where the desired quality of traffic flow or the desired average travel speeds can no longer be achieved with the standard cross-sections of EKL 1 to EKL 3. The range of application ends, for instance, at traffic volumes of 30.000 Veh./24 h, because at higher traffic volumes, 2 + 0 lanes in construction sites is no longer possible and a hard shoulder would be indicated. The RQ 21 is also only suitable for short road sections (up to about 15 km length) in otherwise single-carriageway roads.

Due to the high speeds to be expected, it makes sense to operate the RQ 21 as a vehicle-only road. Bike traffic on the roadway must not be permitted. A separate cycle traffic facility must be installed if there is a proven need for it. Agricultural traffic could be permitted if no alternative routes are available. Agricultural traffic is less problematic than with RQ 15.5 and RQ 11.5+ because two lanes per direction are available throughout. A high-risk potential is essentially due to the large speed difference.

Approximately every 500 to 1000 m emergency stop bays shall be provided on both directional carriageways in accordance with Sect. 9.3.1.

The standard width of the structural central reservation is 2.50 m. This width places special demands on the safety barriers. Double-sided, eccentrically arranged steel crash barriers or concrete barriers are often required. The eccentricity generally improves visibility on the inside of the curve.

Since transverse slopes to the outside of the curve can only be arranged at radii greater than 3000 m, drainage facilities often have to be placed in the central reservation. The depth of the pipes and the arrangement of the manholes must be adapted to the safety barriers. If this is not possible or if fixtures such as bridge piers have to be accommodated, the central reservation must be widened.

9.3.6 Design of Standard Cross-sections for Rural Roads

The use of the standard cross-sections in the standard areas of application usually leads to adequate traffic quality. Nevertheless, the desired quality of the traffic flow and the desired average speed of the cars must be confirmed for each design. This also serves, among other things, the proof of economic efficiency by determining the improvements in accessibility.

As a rule, traffic quality is verified using HBS procedures (FGSV 2015a). On country roads, it must be proven that the predicted traffic volumes do not exceed the traffic density specified for the respective QSV. Furthermore, it must be confirmed that the desired average car speeds are achieved, taking into account the time lost at junctions. A summary consideration of network sections is given in Chap. 11, where the sections without junctions are considered separately.

The traffic volume, the design traffic volume q_B [Veh./h], which is decisive for the design, is derived from the expected (forecast) traffic demand for both directions separately for the relevant hour. The proportion of heavy goods vehicles SV [%] must always be shown separately. The basis for the dimensioning of rural roads is given in the RIN (FGSV 2008e) by the road category and the target values for the average car speed assigned to it.

On gradients with a large longitudinal ascending gradient ($s > 3\%$), the driving speeds of all vehicles, especially heavy goods vehicles, are reduced. Therefore, the gradient sections are differentiated into gradient classes and taken into account in the calculation procedure.

9.3.6.1 Quality Criteria

The RIN (FGSV 2008e) provide quality specifications for the target car driving speed (see Chap. 2). Since, for the same traffic volume, the attainable car travel speed is significantly lower on sections of road with steep gradients and/or tight curves, although freedom of movement as a measure of traffic quality can be very high, the attainable average car travel speed is not suitable as the sole quality criterion. The HBS, therefore, uses traffic density as a quality criterion, which describes the degree of freedom of movement under the traffic conditions. The formula is not entirely correct, since the traffic density includes all motor vehicles, but the average driving speed only includes passenger cars. Therefore, traffic density is an artificial value to describe traffic quality. For single-carriageway roads, the traffic density is calculated for each direction of travel:

$$k = \frac{q}{V_F}$$

with

k: Traffic density [Veh./km],

q_b: Traffic volume [Veh./h],

V_F: Average car driving speed [km/h].

For dual-carriageway roads, the directional traffic density is calculated as a function of the number of lanes:

$$k_{FS} = \frac{q}{m \cdot V_F}$$

with
k_{FS}: Traffic density [Veh./km],
q_b: Traffic volume [Veh./h],
V_F: Average car driving speed [km/h],
m: Number of lanes in the direction.

9.3.6.2 Influencing Factors

Cross-section
As with motorways, the number of lanes has the greatest influence on traffic quality. For this reason, the HBS differentiates between the different standard cross-sections. The width of the lanes has no influence up to RQ 9, but there is no procedure for this in the HBS. Due to the load limit at 3000 vehicles/24 h, no procedure is necessary.

9.3.6.3 Length of Double and Single Lane Sections on Three-lane Roads

Since the attainable travel speeds increase with the increasing length of the two-lane sections and decrease with the increasing length of the single-lane sections, their length must be considered and the travel directions must be considered separately.

9.3.6.4 Line Management

With high longitudinal slopes and high curvature, the speeds of heavy goods vehicles, in particular, are reduced. On single-lane roads, this also has a negative effect on the attainable speed and traffic quality of passenger cars. These effects are taken into account by dividing the road sections into gradient classes and curve classes.

9.3.6.5 Longitudinal Slope

To take into account the influence of the longitudinal slope, sections are formed with an approximately constant slope in each case. This results in line sections of a certain length with a constant (possibly averaged) longitudinal slope between the tangent intersections in the contour plan without taking into account the rounding of crests and troughs. Each section is assigned to the gradient classes.

9.3.6.6 Curvature and Overtaking Possibility

On two-lane roads, the curvature of a road has a considerable influence on the driving speed. This influence is expressed by the curvature KU [gon/km]. This is the sum of the absolute values of the angle changes of the route in relation to its length. Furthermore, in case of the low curvature and low traffic intensity, there is the possibility to overtake slower vehicles. This was still taken into account in the HBS 2001/2009. The possibility of overtaking may also be restricted by overtaking bans (e.g. on access roads or in hilly areas). The influence this has on the speed of cars was taken into account at the time by adding a curve surcharge. This surcharge has been dropped in the HBS 2015 because overtaking in oncoming traffic is no longer a basic principle of highway design.

$$KU_i = \frac{\sum_{i=1}^{j} |\gamma_i|}{L_i}$$

with
KU_i: Curvature of the section i [gon/km],
γ_i: Angular change in the site plan within the alignment element i [gon],
L_i: Length of subsection i [km],
j: Number of alignment elements within the considered subsection [–],
L: Length of the country road section.
The HBS 2015 also takes into account the fact that hardly any (legal) overtakings are possible at the high traffic volumes in the peak hour and that, due to the high risk, these overtakings should also be increasingly restricted in the future under traffic law. The influence of the limited overtaking possibilities at high curvature is already taken into account in the curve classes.

9.3.6.7 Control Conditions

The limitation of the permissible maximum speed is only taken into account for dual-carriage roads.

9.3.6.8 Traffic Intensity and Composition

Traffic volumes and the proportion of heavy goods vehicles have a major influence on traffic quality and are taken into account accordingly. The share of slow-moving vehicles (i.e. agricultural traffic) is not taken into account.

9.3.6.9 Environmental Conditions

As with motorways, the traffic qualities are determined to apply to dry roads and daylight.

9.3.6.10 Speed Flow Relation

The average travel speeds of passenger cars on the sections of the route can be taken from the HBS depending on the traffic load, the heavy goods vehicle proportion, the gradient class and other environmental and control conditions.

With the help of these **speed flow relation** diagrams, it is possible to check that the desired speed V_F is reached for a given design traffic volume q_B and to read off the quality level of the traffic flow. Figure 9.15 shows the example of a single carriageway, a two-lane country road with one lane for the considered direction with low longitudinal slopes (gradient class 1).

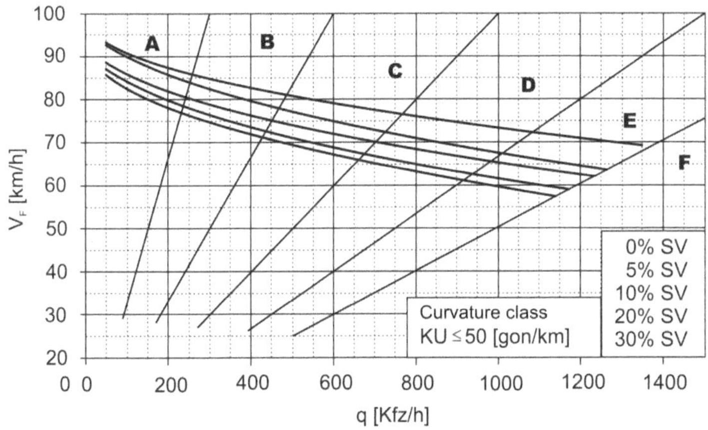

Fig. 9.15 Example of a diagram for determining the average car speed for sections of single carriageway rural roads with one lane of traffic in the direction under consideration with lower longitudinal slopes (Gradient class 1)

9.3.6.11 Average Car Driving Speed

If a section of a country road consists of several subsections that differ from each other in terms of gradient or curvature, different speeds will generally be reached on the subsections. In this case, the average car speed V_F over a section of length L is obtained from the average of the n individual speeds $V_{F,i}$ weighted by the n section lengths L_i according to the following equation:

$$V_F = \frac{L}{\sum_{i=1}^{m} \frac{L_i}{V_{F,i}}}$$

with

V_F: Average car speed on the considered route [km/h],

$V_{F,i}$: Average car speed on the section i [km/h],

L: Length of the considered route [km],

L_i: Length of section i [km].

This average car driving speed indicates the extent to which the requirements of the RIN (FGSV 2008e) for the connecting functions of the rural road required by regional planning are fulfilled. In order to determine the quality of the traffic flow applicable to the entire section, the average traffic density k must be calculated analogously using the equation below.

$$k = \frac{\sum_{i=1}^{n} k_i \cdot L_i}{L}$$

with
k: Average traffic density on the considered route [vehicle/km],
L_i: Length of subsection i [km],
L: Length of the considered route [km],
k_i: Traffic density on section i [vehicle/km].

9.4 Cross-sectional Elements

In addition to the lanes, the hard shoulders and the central reservation, the unpaved cross-sectional elements alongside the carriageways are also very important for the safe operation of the roads. Next to the carriageway, 1.50 m wide verges are arranged (in cuts next to 2.00 m wide drainage troughs also 1.00 m possible). The guide systems, traffic signs and, if necessary, safety barriers (see Chap. 18) are accommodated on the banquettes. Furthermore, damaged vehicles can be parked there and the road maintenance service can use these areas. For safety reasons, the banquettes are constructed for all road categories in a stable manner, for example with gravel lawn. If surface water is drained off from the roadway via the roadway, the verges are inclined outwards at 12%, otherwise at 6%.

In addition to the verges, the drainage facilities and then the embankment follow in incisions. The embankments follow directly on embankments and the drainage facilities are arranged at the foot of the embankment. Embankments receive a standard inclination of 1:1.5 if they are higher than 2.00 m. The edge between the embankment and the terrain is then rounded off with a tangent length of T = 3.00 m.

On flat embankments (h < 2.00 m), this design would result in an unsteady drainage trough or ditch and the boundary of the property would be very close to the carriageway. In these cases, a uniform slope width of 3.00 m is chosen and the slope rounding is executed with a tangent length $T = 1.5 \cdot h$.

Note

1. RiZ-ING: Reference drawings for engineering structures, available for free download from the homepage of the Federal Highway Research Institute (www.bast.de).

Intersections

10

Summary

A building structure that serves to link roads is called an intersection. The roads connecting to the intersection are called intersection arms. Intersections with at least four intersection arms, of which two each form a continuous road, are called crossroads. An intersection with only one continuously passable road to which another road is connected is called a T-junction. Intersection and T-junctions are the standard forms of intersection design.

Crossing processes can be avoided in whole or in part by routing traffic flows at different levels. These are then intersections on several levels, intersections of unequal height or grade-separated intersections. Intersections at which all crossing traffic flows are routed at one level are accordingly called intersections on one level, or at-grade intersections.

10.1 State of the Art

Since intersections on rural roads are among the road infrastructures with a relatively high number of serious accidents, much research has been done on traffic safety and also on the traffic flow at intersections.

10.1.1 Road Safety

Only a few research results are available for grade-separated intersections that deal with the overall system (Brilon and Betz 2008; Brilon et al. 2007). The results did not provide the occasion to fundamentally revise the regulations in the RAA (FGSV 2008a). Individual

© Springer Fachmedien Wiesbaden GmbH, part of Springer Nature 2023

T. Richter, *Road Planning - Freeways and Country Roads*,
https://doi.org/10.1007/978-3-658-35189-2_10

design elements of grade-separated intersections have been the subject of much research, including current research.

Entrances, exits and ramps have been examined in current research projects with regard to their traffic safety and traffic flow (Hoffmann et al. 2006, 2008; Weiser et al. 2006; Zimmermann et al. 2013). The results largely confirm the regulations in the RAA. At ramps, however, it is pointed out that smaller radii should not follow initial radii because this can obviously lead to increased accidents (Zimmermann et al. 2013).

In the 1990s, Richter and Kölle carried out research on at-grade intersections on rural roads, which dealt with the different types of intersections as a system and in interaction with the sections outside of intersections. In the research work "Design standards for intersections on bypasses", the types of intersections crossing with traffic lights and partially at-grade crossing with and without traffic lights (partially at-grade according to today's terminology) were examined (Richter 1993). Signalised intersections had clear safety advantages, especially when left-turning traffic was secured by signalling. It was also possible to prove an interaction between the type of the intersections and the speed behaviour on the sections without intersections. In follow-up work, the intersection types crossing without traffic lights and crossings with right offset were included in the investigations (Richter and Schnüll 1994). Here the crossing with the right offset showed clear advantages. The safety characteristics of grade-separated intersections on two-lane rural roads completed a series of investigations (Kölle and Schnüll 1999). Grade-separated intersections had a very high safety level, with a high-speed level being observed on the sections of the road without intersections. Finally, roundabouts on two-lane rural roads were examined (Richter and Hüsken 1996). Here the results already known for roundabouts within built-up areas confirmed that the roundabouts had a very high safety level.

The then ISK (now UDV) presented a study in 2002 which essentially confirmed the results of Richter and Kölle (Meewes and Eckstein 2002). For the different types of intersections, they come to the following results (Table 10.1).

Meewes and Eckstein also found an influence of the type of intersection on the accident occurrence of the subsequent intersection-free sections of the route and took this into account by reducing the accident cost rate. However, in contrast to Richter/Kölle, they assume fixed lengths of influence between 0 and 700 m.

In 2011 Vieten presented two research reports commissioned by the BASt to examine various elements of rural roads with regard to their effects on road safety (Vieten et al. 2010, 2011). For the different types of intersections, it comes to the following results (Table 10.2).

The comparison of Meewes and Vieten shows partly very different results. However, it should be noted that, for example, the rate of accidents involving personal injury on federal roads outside built-up areas fell from 0.39 to 0.28 accidents/million vehicle-km between 1995 and 2005. Nevertheless, clear safety advantages have been observed in Vieten of at-grade intersections with 2-phase traffic signal systems compared to intersections with right-of-way control. Similar trends have been confirmed for T-intersections. In principle, RAL is based on the more recent studies of Vieten et al. (2010).

Table 10.1 Accident cost rates of the different intersection types, price level 2000, accidents from 1989 to 1995 (Meewes and Eckstein 2002)

Intersection basic form[a]	Operational regulation	Accident cost rate UKR [EUR/1000 Kfz · km]
Intersection, at-grade	Right of way	66
Intersection, at-grade	Traffic lights, 2 phases	66
Intersection, at-grade	Traffic lights, 4 phases	34
Intersection, at-grade	Right of way, with speed monitoring	18
Partly at-grade	Right of way	35
Partly at-grade	Traffic lights, 3 phases/right of way	16
Partly at-grade	Right of way with speed monitoring/right of way	13
Roundabout	Right of way	9
Partly at-grade	Merging/diverging/right of way	17
Partly at-grade	Merging/diverging/traffic lights, 3 phases	11
Partly at-grade	Merging/diverging/right of way with speed monitoring	10
At-grade T-intersection,	Right of way, without speed monitoring	27
At-grade T-intersection	Traffic lights, 2 phases	27
At-grade T-intersection	Traffic lights, 3 phases	12
At-grade T-intersection	Right of way, with speed monitoring	7

[a]Designations according to RAL (FGSV 2012a)

Table 10.2 Adjusted accident cost rates by intersection type, price level 2000, accidents from 2002 to 2006 (Vieten et al. 2010)

Intersection basic form[a]	Operational regulation	Accident cost rate UKR [EUR/1000 Kfz · km]
Intersection, at-grade	Right of way	25.4
Intersection, at-grade	Traffic lights, 2 phases	16.3
Intersection, at-grade	Traffic lights, 4 phases	12.5
Roundabout	Right of way	8.7
T-intersection, at-grade	Right of way, without speed monitoring	13.9
T-intersection at-grade	Traffic lights, 2 phases	10.9
T-intersection at-grade	Traffic lights, 3 phases	5.9

[a]Designations according to RAL (FGSV 2012a)

10.1.2 Traffic Flow

The traffic flow at elements of grade-separated intersections was usually examined parallel to traffic safety. The investigations were mainly used to revise the HBS, which was published in 2015. Brilon and Weiser have studied the overall system of grade-separated intersections (Brilon and Betz 2008; Brilon and Geistefeld 2010; Brilon et al. 2007; Weiser et al. 2006). Hoffmann and Friedrich investigated exits and entrances (Hoffmann et al. 2006, 2008). Zimmermann examined the ramp areas (Zimmermann et al. 2013).

10.1.3 Status of the Guidelines

For motorways, the RAA lists intersection systems for three-way and four-way motorway interchanges and access intersections (FGSV 2008a). The proof of traffic quality in the elements of the grade-separated intersections is provided according to the HBS (FGSV 2015a).

In the RAL, one or a few types of intersection are recommended for each design class (FGSV 2012a). In addition to the standard areas of application, individual intersection types can also be used in other design classes in exceptional cases for which special justification is required. These are listed in the description of the intersection types. The final design is again based on the HBS (FGSV 2015a).

10.2 Intersections for Motorways

10.2.1 Basic Principles

The intersection systems and the associated intersection elements are defined in accordance with the RAA (FGSV 2008a). The aim is to ensure that the motorways are designed as uniformly as possible. In contrast to the standard cross-sections, the intersections are not generally differentiated according to the design classes, as the intersection systems should be as similar as possible. Different design elements for the respective design classes are shown with the elements.

In contrast to at-grade intersections, grade-separated intersections consist of separate sub-areas (continuous carriageways, exit areas, entry areas, interlinking areas, connecting ramps), which follow each other at certain intervals and require special driving procedures. While the characteristics of the continuous roads should remain the same, the connecting ramps are deliberately designed with fewer design elements to restrict freedom of movement and homogenize the traffic flow in order to make clear to road users the safety risk associated with the separation and merging of traffic flows. The principles of recognisability, clarity, comprehensibility and trafficability as well as uniformity of traffic facilities apply primarily to these sub-areas.

In principle, all motorways are grade-separated, which means that entrances and exits are always arranged along the motorways. Connections of motorways with each other (motorway intersections) are generally designed as grade-separated intersections. Connections of motorways with the minor road network (access intersections) can be designed as grade-separated or partially grade-separated intersections.

In the planning process, the most appropriate intersection system for the design class in each individual case shall be determined. Aspects of the

- Land availability,
- Topography (terrain, settlement density),
- Ecology or urban design,
- Traffic safety,
- Capacity and
- Economic efficiency

a role. The planning considerations will initially be based on the standard economic solutions. System reasons are always decisive for these solutions. However, high traffic loads can also make more complex solutions—beyond the system requirements—necessary.

In order to achieve uniformity in the design and operation of motorways, appropriate intersection systems are defined for the motorway, from which there should be no deviation without reason. However, along a section of motorway, the intersection system may vary if this is not expected to affect traffic flow and safety.

On urban motorways, the choice of a suitable intersection system must also take into account the urban characteristics and the resulting requirements of the surrounding area. Intersection systems for urban motorways are often characterised by the fact that the intersections

- have to be distorted, shifted or often incomplete due to the local structural or network situation
- may have more than four intersection arms,
- must be used in combination, and
- follow in close sequence.

This must be taken into account when pre-selecting and determining suitable intersections on urban motorways in individual cases and when designing the individual design elements.

10.2.1.1 Position of the Continuous Roadway

On motorways of the EKA 1 A the main carriageways should not be led across corners. In principle, this also applies to the motorways of EKA 1 B and EKA 2 unless any corner streams require such a solution and the traffic forecast is secured in the long term.

On urban motorways, the location of the continuous carriageway is primarily determined by the urban planning constraints and the high capacities required on these motorways. A load-oriented rather than a network-oriented choice of the continuous carriageway is therefore generally more sensible. If the load-oriented choice of the continuous carriageway is difficult to realise, quality losses can also be accepted for the continuous carriageways. Due to the high density of urban road networks and the small distances between intersections, it is often necessary to dispense with less important, difficult to realize driving relationships in individual intersections.

10.2.1.2 Cross-section of the Continuous Road

The standard cross-sections for motorways are generally used as cross-sections for the continuous carriageways of grade-separated intersections. In individual cases on motorways of EKA 2 and EKA 3 it may be appropriate to dispense with the hard shoulder in the intersection area (e.g. to reduce building costs). However, this must always be weighed up against the loss of safety.

If strong exit and entry streams occur in the intersection area, the subtraction of a lane in the exit or the addition of a lane in the entry must be checked. The arrangement of weaving lanes results in the addition of at least one lane to the respective standard cross-section.

10.2.1.3 Alignment of the Continuous Roadway

Grade-separated intersections should, if possible, be clearly above the limit values in sections with stretched alignment. The radii in the area of the entrances and exits should not be smaller than 800 m in the case of EKA 2 and 400 m in the case of EKA 3, in order to guarantee the threading view and sufficient safety reserves when braking in the curve area. Furthermore, the longitudinal inclination should not exceed 3%, in order to facilitate the threading of heavy goods vehicles and to avoid long lengths of the connecting ramps.

10.2.1.4 The Distance Between Intersections

The desired distance between grade-separated intersections on motorways results from the network function. It is usually determined in network planning. A minimum axis distance of 8.0 km is to be aimed at on EKA 1 A motorways and 5.0 km on EKA 1 B and EKA 2. This results in sufficiently long distances for an uninfluenced traffic flow between the intersections.

The minimum distances from grade-separated intersections resulting from the distance requirements for the signposting and the length of the entry and exit openings are given in Table 10.3.

The axis distance and the effective intersection distance are explained in Fig. 10.1.

If the minimum distances specified in Fig. 10.1 cannot be adhered to due to local peculiarities (mainly within built-up areas) or due to constraining points, the instructions of the

Table 10.3 Effective minimum distances e from grade-separated intersections

Type of intersection following in the direction of travel	Minimum value for the standard direction	Minimum value for individual signposting in special cases according to the guidelines RWBA[a]	Minimum value for isolated intersection planning
Three- or four-way motorway intersection	3000 m	1600 m	600 m
Access intersection	2000 m	1100 m	600 m

[a] If there are more than 2 intersections in close sequence succession

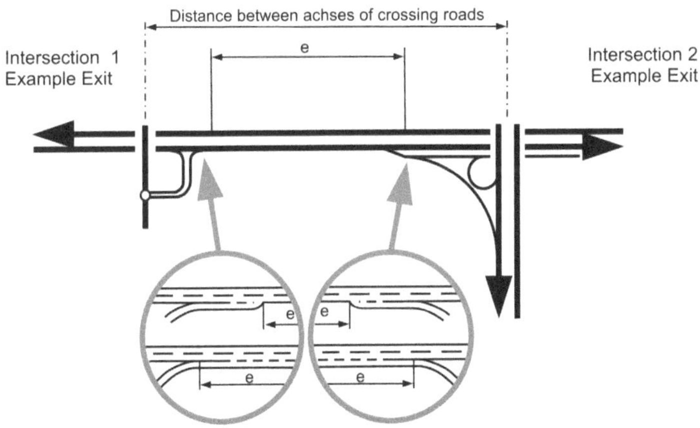

Fig. 10.1 Axis distance and effective intersection distance (FGSV 2008a)

Guidelines for the Signposting of Motorways for the Arrangement of Directional Signs (Beschilderung auf Autobahnen zur Anordnung der wegweisenden Beschilderung) (RWBA) (FGSV 2000a) must be observed. In the case of very short distances, consideration should be given to connecting successive intersections with each other by means of an interlinking road.

On urban motorways, on the other hand, the smallest possible distances between intersections are desirable in terms of planning strategy, so that the desired collection and distribution effect can be achieved. For this reason, special signposting forms with advance signs across several intersections may be required on these motorways in order to be able to keep below the minimum intersection distances resulting from the signposting.

As a rule, the decisive criterion for the permissible minimum distances from intersections on urban motorways is the capacity of the continuous carriageway, which can be considerably disturbed by a dense sequence of uncontrolled entries and exits. Modified ramp layouts, weaving sections between successive intersections and the use of operational measures in grade-separated intersections are therefore often necessary on urban motorways (Fig. 10.2).

10.2.2 Four-way Motorway Intersections

In principle, the choice of a suitable intersection system for four-way motorway intersections depends on the location of the strong corner streams (Fig. 10.3).

The basic cloverleaf system (Fig. 10.4) is the most economical solution for a four-arm motorway intersection, since it requires only one bridge. In the cloverleaf system, swivelled distributor lanes are advantageous, as they make the exit areas into the tangential ramps visually clear, increase the distance between the intersection points in entry ramps and thus facilitate the installation of merging lanes within the distributor carriageway.

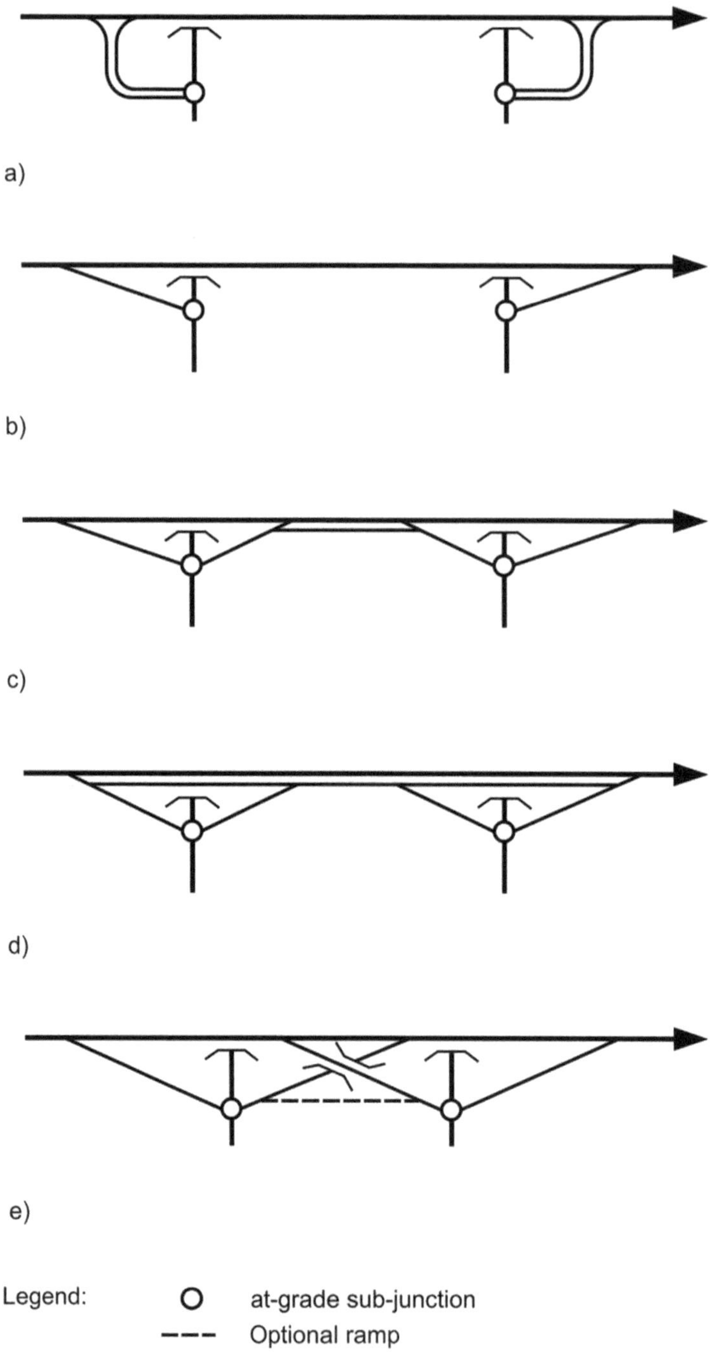

Fig. 10.2 Remedial action if the distance between intersections is too small (FGSV 2008a)

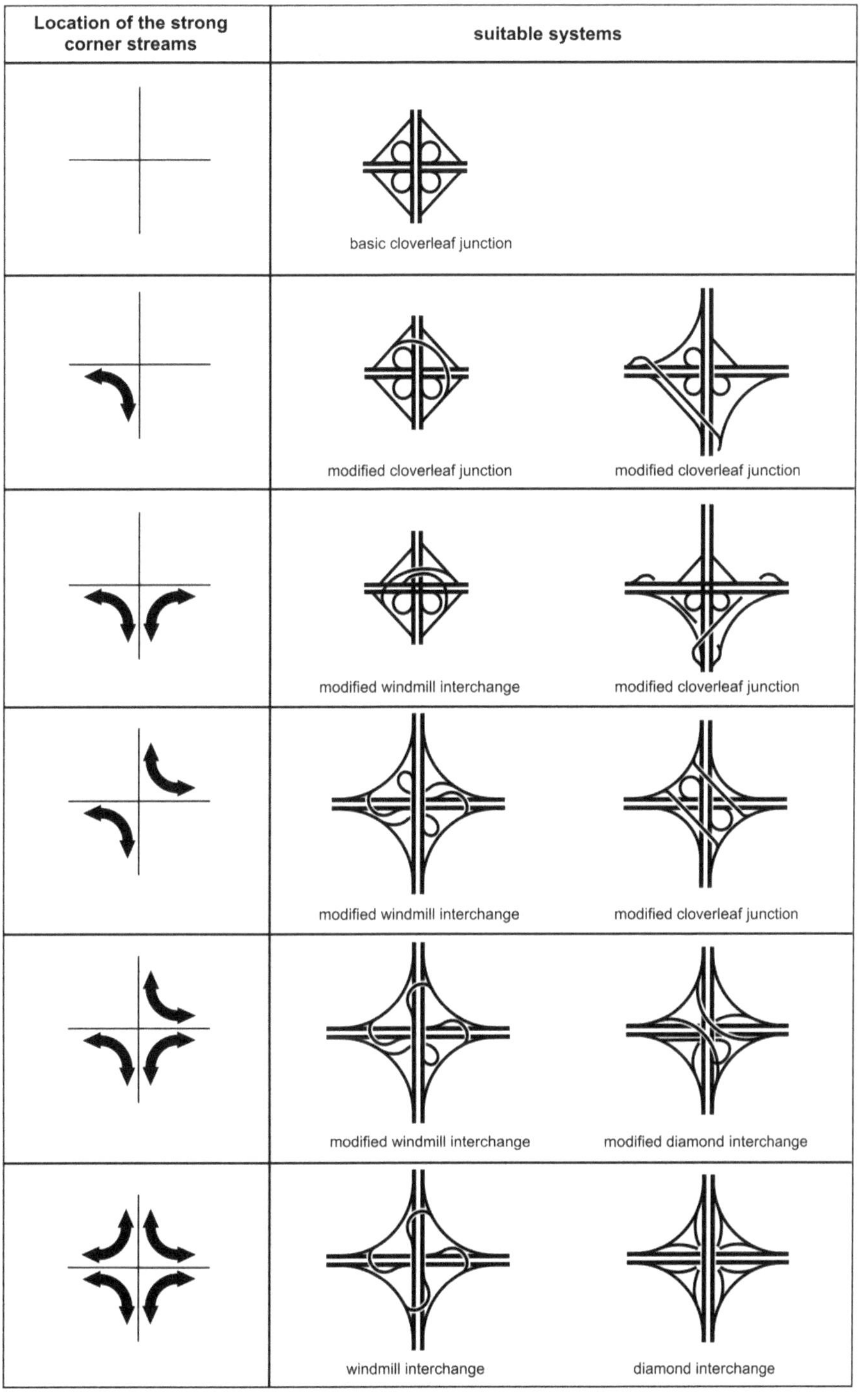

Fig. 10.3 Use of intersection systems of motorway intersections (FGSV 2008a)

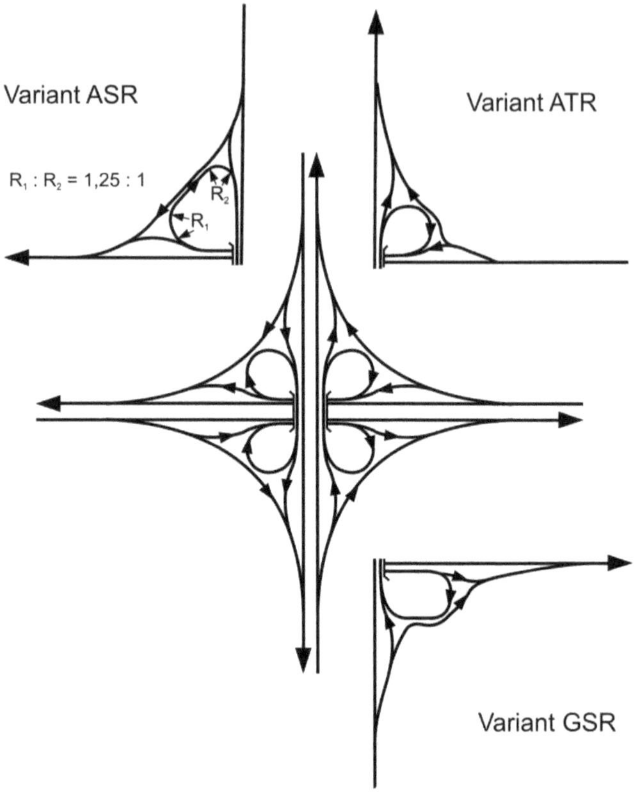

Advantages:
- Only 1 bridge
- Loop ramp with relatively large radii
- Relatively smooth ride
- Turning possibilities, convenient for operational and winter services

Disadvantages :
- 4 weaving sections
- Capacity limitation for corner traffic
- Hugh surface requirement (for sufficiently long weaving sections)
- Relatively short weaving sections (with limited surface availability)
- Bridges very large

Variant GSR: Stretched loop ramp (EKA1)
Goals:
- longer weaving sections (in one axis)
- longer ramp development, necessary with trailing longitudal slope

Variant ASR: Adapted loop ramp (EKA 2 and 3)
Goals:
- longer weaving sections (with smaller radii in the loop ramps)
- shorter intersection areas

Variant ATR: Adapted tangential ramp
Goals:
- less surface area required
- shorter intersection areas

Fig. 10.4 Cloverleaf with variants of the ramp guidance (FGSV 2008a)

The length development in the weaving section between cloverleaf loops and the entire distribution carriageway can be designed more favourably if the double-sided swivels are drawn into the weaving section and the distribution carriageway is only guided parallel to the mainline carriageway in the area of the crossing bridge. In this case, special attention must be paid to the vertical alignment.

The cloverleaf with circular loop ramps and non-adapted tangential ramps allows for a smooth ride and the application of the largest possible radii. A disadvantage is the large area requirement.

The cloverleaf with adapted (depressed) loop ramps and adapted tangential ramps leads to longer weaving sections in the distribution carriageways, usually to a saving of space and to a shortening of the intersection area. In order to reduce the disadvantage of changing curve radii in the loop ramps, a radius ratio of R_1: R_2 = 1.25: 1 should not be exceeded. A small radius must be accepted in the entry area if the space requirement is not to increase compared to the circular loop ramp. In the adapted tangential ramps, visually unsatisfactory road edge alignment and drainage difficulties can occur when using counter-curves due to the necessary changes in cross slope.

Stretched loop ramps can be used when a long weaving section is to be created, when long development lengths are required for the ramps due to steep (rectified) longitudinal inclinations in the motorways to be linked and when local conditions (topography, building development, protection of certain areas) make it necessary.

The cloverleaf without distribution carriageway reduces the construction costs for the central bridge and the land use. It is applicable if weaving lanes are provided on the mainline carriageway, the capacity of the weaving lanes is sufficient, the maximum speed allowed in the continuous carriageway is limited to 100 km/h and the entry radius of the loop ramp exceeds 30.0 m (better 50.0 m). In these cases, special care must be taken with the signposting. In principle, the application must be assessed critically for safety reasons. It should therefore only be used when there is a particular shortage of space and then only for EKA 2 and EKA 3.

The basic system of modified cloverleaf (Figs. 10.5 and 10.6) is used when the capacity of one or more weaving sections in a cloverleaf is no longer sufficient, the quality of the traffic flow is to be improved in comparison with the cloverleaf for several turn-off relationships by faster ramp guidance or topographical reasons speak against a normal cloverleaf. The semi-direct guides require higher construction costs and a larger area consumption due to additional bridges. It should therefore always be checked whether the quality improvements can also be achieved by the use of operational measures in selected intersection elements (e.g. lane signalling or ramp metering). If at modified cloverleaves the distribution lanes directly merge into loop ramps, special precautions must be taken there for safety reasons (lane markings, directional signs, speed limits, the upstream connection of a counter curve).

The basic windmill system (Fig. 10.7) and modified windmills (with individual loop ramps) are practical intersection system when semi-direct connecting ramps cannot be used for space reasons. The small routing parameters to be applied in such cases may

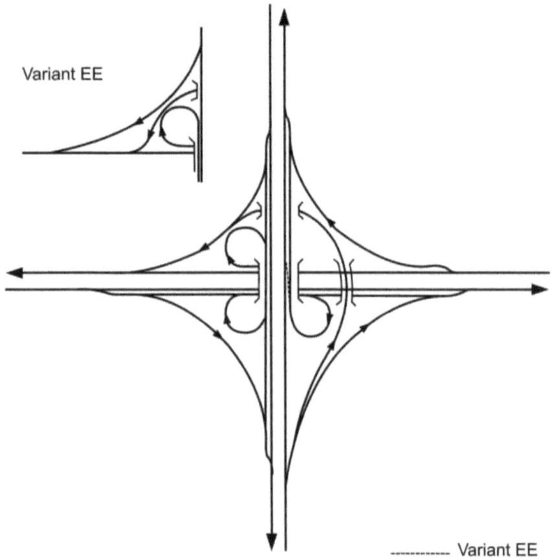

Advantages:
- Elimination of weaving sections
- More speedy ramp guidance than with the basic form
- Better traffic quality in the semi-direct-ramp

Disadvantages:
- 3 Bridges
- more surface area required than of the basic form
- Limited turning possibilities

Variants EE: Entrance type EE
Goal:
- Better traffic flow with heavily loaded connecting ramps

Fig. 10.5 Modified cloverleaf with semi-direct guidance of a left-turning stream (FGSV 2008a)

result in large longitudinal inclinations and unfavourable visibility conditions in the crest areas. For this reason, windmills must be aligned very carefully in the site plan and elevation plan. Due to the fact that there are up to five bridges, the construction effort is very high. In connection with the unfavourable alignment parameters, the application must therefore be checked particularly critically.

The basic system stack interchange (Fig. 10.8) has a lower construction effort than the windmill. Under certain circumstances, however, a larger area is required. The four-level bridge leads to an impairment of the landscape or townscape. Stack interchanges should therefore only be used in exceptional cases for special compulsory points.

10.2.3 Three-way Motorway Intersections

Due to system disadvantages of three-armed intersections (the continuous directional lane of the ending motorway changes abruptly into an entry), motorway networks should be designed in such a way that three-armed intersections are avoided wherever possible and

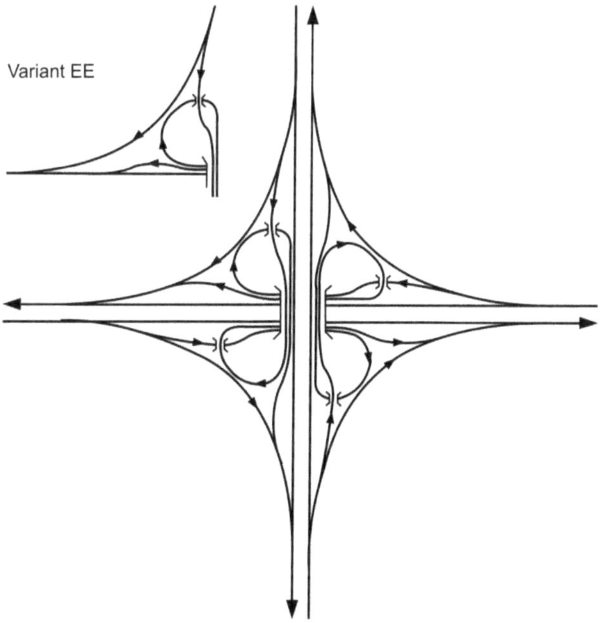

Advantages:
- Elimination of weaving sections
- Conversion under traffic

Disadvantages:
- Unfourable loop ramp guidance (reduced visibility, lost gradients, longer lanes)
- 4 additional bridges
- Large surface area requirement
- Large central bridge

Variant EE: Entrance Type EE
Goal:
- Better traffic flow with heavily loaded connecting ramps

Modified cloverleaf intersection with grade-free guidance of the crossing corner streams (for reconstruction and extension)

Fig. 10.6 Modified cloverleaf with rapid semi-direct guidance of a left-turning stream (FGSV 2008a)

four-armed intersections are preferred instead. Figure 10.9 shows the recommended basic systems as a function of the design classes.

The basic trumpet system is the most economical solution for a three-arm motorway intersection. Due to the system, all vehicles have to decelerate strongly on the non-continuous directional lanes, which requires timely and clear recognition. The main ramp curve should therefore start before the bridge. In addition, an upstream counter curve or an adapted cut slope can improve the recognisability of the ramp.

The more favourable solution is the left-facing trumpet (Fig. 10.10). Compared to the right-facing trumpet, it has the advantage that the left-hand turnouts from the incoming motorway are led into the larger curve with a constant radius. Turners from the continuous motorway are more likely to expect an access intersection with a smaller curve.

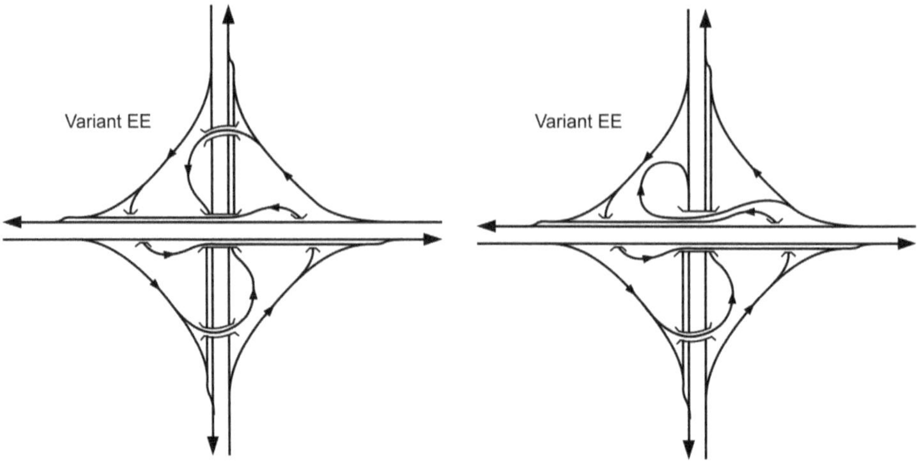

Advantages: - No weaving sections
- Relatively small surface area requirement

Disadvantages: - 5 bridges
- Relatively large longitudal slopes and unfourable visibility on the summt curves due to small alignment elements in the semi-direct ramps
- No turning possibilities, not suitable for operational and winter services

Modified system with 1 loop ramp
Goal: - Savings of 1 bridge

Variant EE: Entrance type EE
Goal: - Better traffic flow with heavily loaded connecting ramps

Fig. 10.7 Windmill (FGSV 2008a)

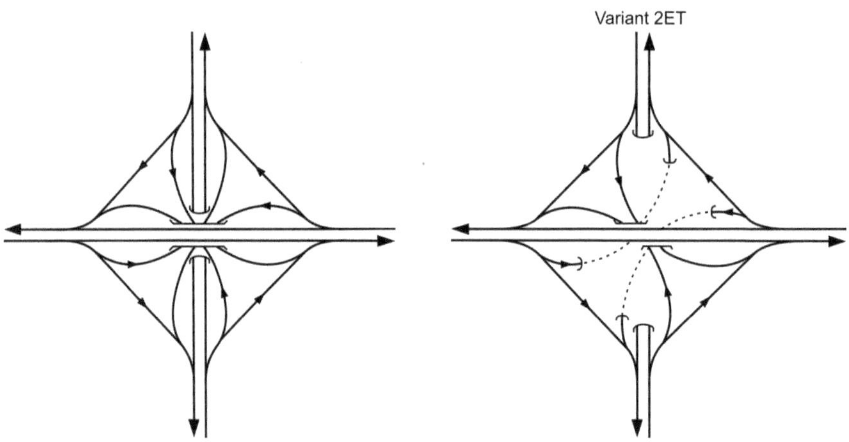

Advantages: - Speedy direct traffic guidance in all directions
- Large corner current capacities

Disadvantages: - Central bridge with 4 levels
- Long junction areas
- No turning possibilities, not suitable for operational and winter services

Variant 2ET: 2 levels in tunnel position

Goal: - Lower high development
- Better urban planning compatibility

Disadvantages: - Higher construction and operating costs
- Slightly larger surface area requirement

Fig. 10.8 Maltese Cross (FGSV 2008a)

Design class of the throughgoing motorway		EKA 2	EKA 2	EKA 3
Design class of the connected motorway ("third intersection leg")		EKA 2	EKA 3	EKA 3
left trumpet intersection	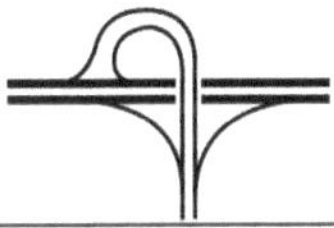	+	+	+
right trumpet intersection	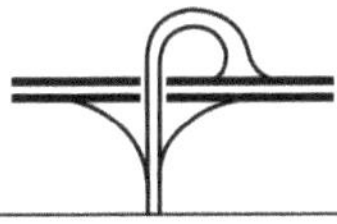	-	-	•
Pear intersection	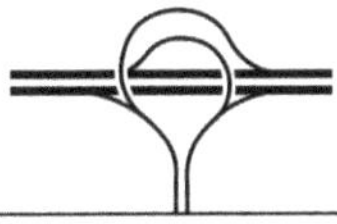	•	•	•
Triangle intersection with one bridge	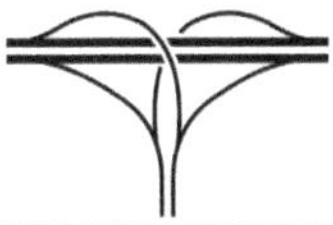	+	+	+
Triangle intersection with three uniform bridges		+	+	+
triangles without uniform definition of the main roadways		-	-	-

Fig. 10.9 Basic systems of motorway intersections with recommendations for use (FGSV 2008a)

Right-facing trumpets have the system-related disadvantage that the left-hand turners from the motorway entering the motorway have to follow a decreasing sequence of arcs with a small loop ramp radius at the end of the sequence of curves. If these cannot be avoided due to constraining points, accompanying measures (direction signs, limitation of the maximum permitted speed, etc.) should be arranged.

The basic pear intersection system (Fig. 10.11) allows the disadvantage of trumpets, the unavoidable loop ramps, to be avoided. A disadvantage for economic reasons is the two bridges required.

The most generous solution for motorway intersections is the basic T-intersection system. Here, all left-turning and left-incoming vehicles are guided semi-directly, whereby excessively generous ramps should be avoided to avoid excessive speeds of incoming and outgoing vehicles. All in all, the construction effort and the space consumption are greater than with the trumpet. Triangles with a central three-level bridge (Fig. 10.12) require less

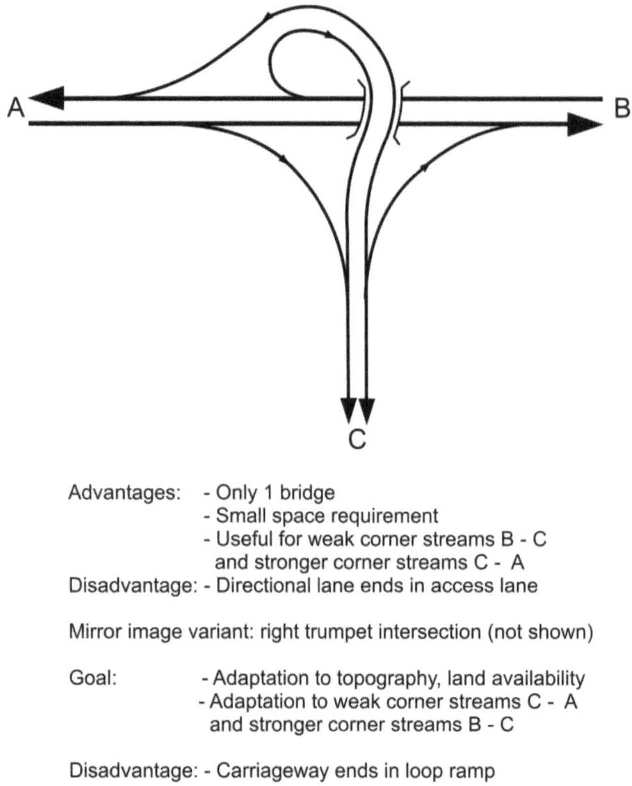

Fig. 10.10 Basic system left-facing trumpet (FGSV 2008a)

construction work than a T-intersection with three two-level bridges (Fig. 10.13). Due to their vertical extension, however, they are less easy to integrate into the landscape.

In the case of motorway triangles with a sharp angle, it is often possible to dispense with a corner relationship because this is guaranteed by intersections located in the inflow. The fork intersection (Fig. 10.14) is a T-intersection with a missing corner relationship.

10.2.4 Access Intersections

Intersections (partially grade-separated intersections) are defined by their grade-separated routing on the major motorway (entrances and exits) and the at-grade connection to the minor rural road/city road. The at-grade sub-intersection can be designed as a level T-intersection (or intersection) with or without traffic lights or as a roundabout. The areas of application of the at-grade sub-intersection are determined in the RAL (FGSV 2012a) according to Sect. 10.3. The intersection systems for access intersections (Fig. 10.15) differ in the ramp routing, the position of the connecting ramps to the minor road, the width of the bridges, the capacity of the ramp connections to the minor road and the extent of the

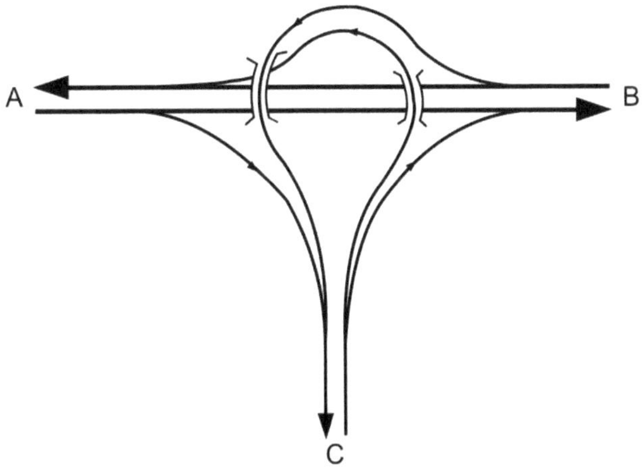

Advantages: - Relatively small space requirement
- 2 bridges
- Useful for strong corner streams B - C and weaker corner streams A - C

Disadvantages: - Parallel guidance of a rising and a falling ramp
- Carriageway ends in access lane

Mirror image variant: pear intersection with preferred guidance of the corner streams A - C (not shown)

Goal: - Adaptation to traffic load pattern of corner streams

Fig. 10.11 Basic pear intersection system (FGSV 2008a)

intersection area on the minor road. In principle, the ramp areas of intersections adjacent to the motorway should be designed with design elements that allow for a continuous change of driving behaviour from the motorway to the minor road and the necessary speed reduction.

Basic systems for four-armed connection points are half cloverleaf, diamond intersection and access intersections point with distribution roundabout. With the half cloverleaf, the two intersecting streets are connected by ramps, which are arranged in two quadrants. The position and shape of the connecting ramps shall be determined by traffic engineering considerations, in particular the direction of the strongest vehicle flows, local conditions and space availability, and the height of the roads to be connected. By arranging the connecting ramps in the corresponding quadrants in a way that is adapted to the traffic load, such intersections can be designed relatively efficiently even without traffic light systems. However, the significantly lower traffic safety of at-grade intersections without traffic lights compared to at-grade intersections with traffic lights or roundabouts must be taken into account.

In the minor street, orientation is made more difficult by the fact that, in comparison with diamond intersections, the necessary entry and exit movements are not always possible in a targeted manner and the space requirement is greater.

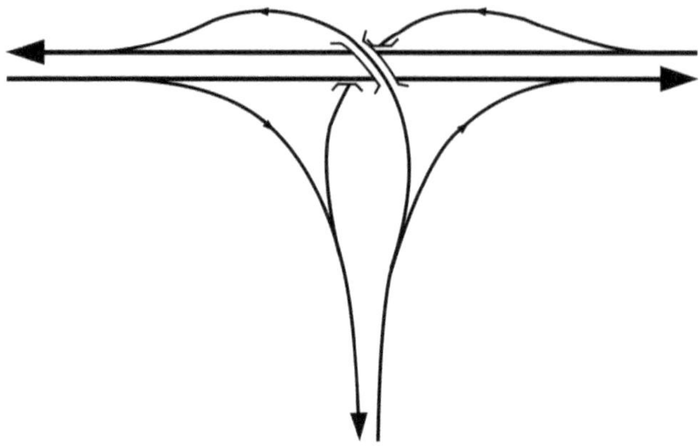

Advantages:
- Useful with approximately equal, strong corner streams
- More space-saving than the triangle intersection with 3 bridges

Disadvantages:
- Bridge with 3 levels
- Altitude development
- More area-intensive than the pear intersection
- Carriageway ends in access lane

Fig. 10.12 T-Intersection with one three-level bridge (FGSV 2008a)

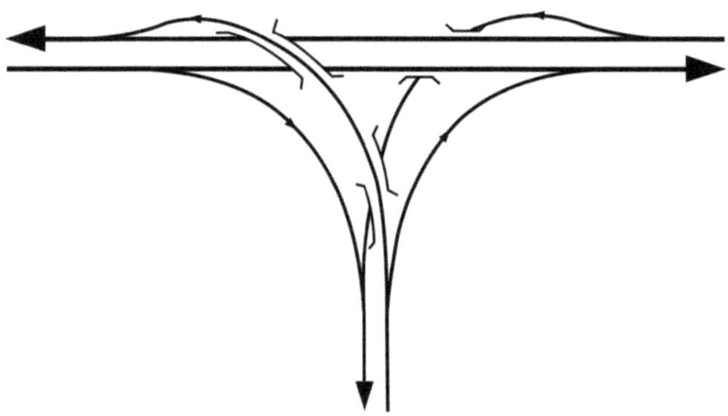

Advantages:
- Useful for about equal very strong corner streams
- Speedy and direct guiding
- High capacity

Disadvantages:
- 3 bridges with acute crossing angle
- Large area requirement
- Carriageway ends in access lane

Fig. 10.13 T-Intersection with three single-level bridges (FGSV 2008a)

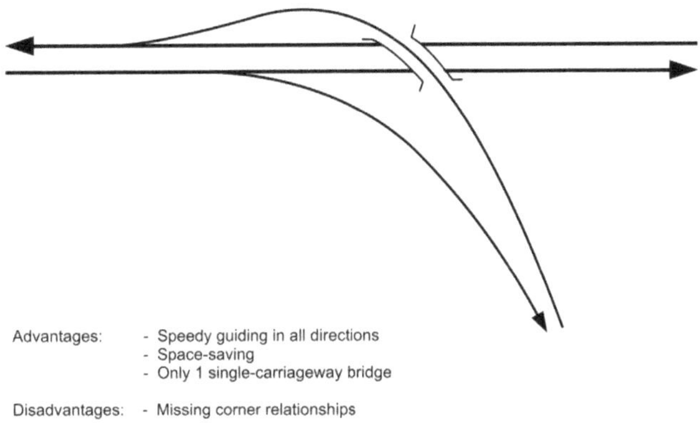

Fig. 10.14 Fork (branch) (FGSV 2008a)

Access point systems			EKA 1	EKA 2	EKA 3
four-arm systems	Diagonal half cloverleaf intersection with exit in front of the bridge	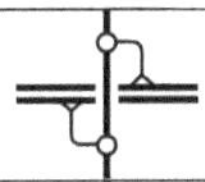	+	+	•
	Diagonal half cloverleaf intersection with exit behind the bridge	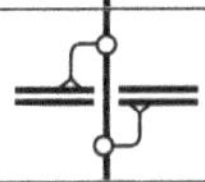	•	+	•
	Symmetrical half cloverleaf intersection	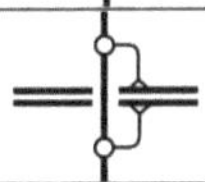	+	+	•
	Diamond intersection with two intersection	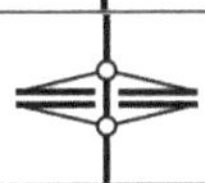	-	•	+
	Diamond intersection with one intersection	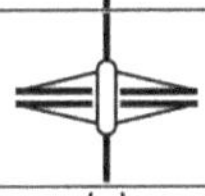	-	-	+
	Diamond intersection with biaxial widened intersection	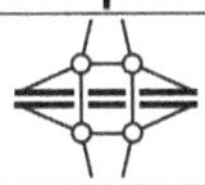	-	•	+
	Diamond intersection with distribution roundabout	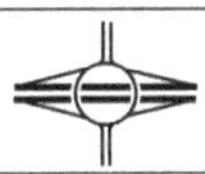	-	•	+
	Special systems (mixed forms)		•	+	+
	Access point in trumpet intersection form	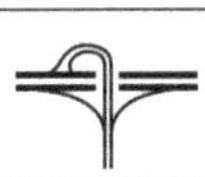	•	+	+
	half cloverleaf (three-way)	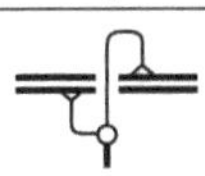	-	-	+ As temporary solution
	Diamond intersection (three-way)	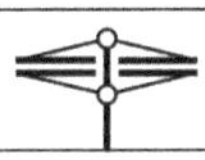	-	-	+ As temporary solution

Legend: O at-grade sub-intersection + suitable • conditionally suitable - not suitable

Fig. 10.15 Intersection systems for intersections and their areas of application (FGSV 2008a)

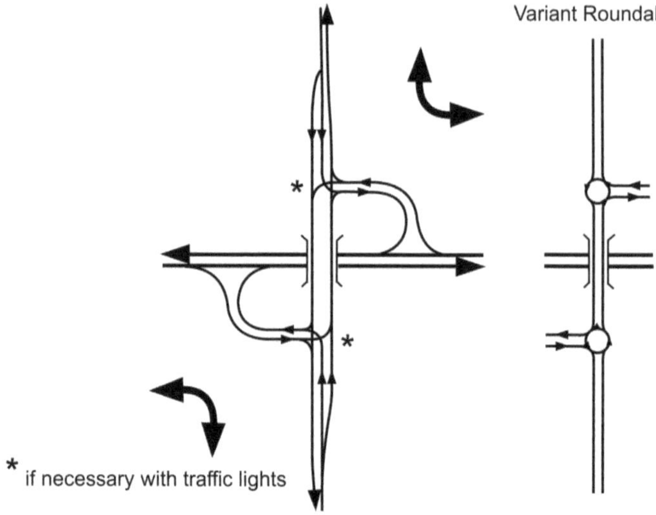

Advantages:
- Favourable driving dynamics (fast, rising) exit ramp corner relations (double arrows)
- Unlimited extension of the left turn lane in the minor road (not applicable for roundabout version)
- Useful for strong corner relations (double arrows)
- Relatively narrow bridge
- Turning possibilities, convenient for operational and winter services

Disadvantages:
- Large junction expansion in the minor road (not applicable for variant roundabout)

Variant roundabout: Small roundabout

Goal:
- Coverage of the capacity range between T-junction without and with traffic lights
- Avoidance of left turning lanes
- Minimizing the extent of the junction expansion in the minor road
- Minimum bridge width

Fig. 10.16 Diagonal half cloverleaf with outside left-hand turn stripes and with an exit in front of crossing structure (FGSV 2008a)

A diagonal half cloverleaf with exits in front of the bridge (Fig. 10.16) is most favourable, as these are favourable in terms of driving dynamics and allow a relatively narrow bridge. The double arrows indicate the corner relations which can be handled most favourably and should therefore be the strongest.

Diagonal half cloverleaves with exits after the bridge (Fig. 10.17) are less favourable in terms of driving dynamics and require left turning lanes on the inside behind or next to each other and thus wider bridges (not when using roundabouts).

Symmetrical half cloverleaves (Fig. 10.18) should only be used if the local conditions do not permit any other solution or if capacitive aspects are not important. Special care must be taken in the arrangement of the signposts and the design of the direction signs because of the close sequence of left-turning strips and the difficulty of orientation.

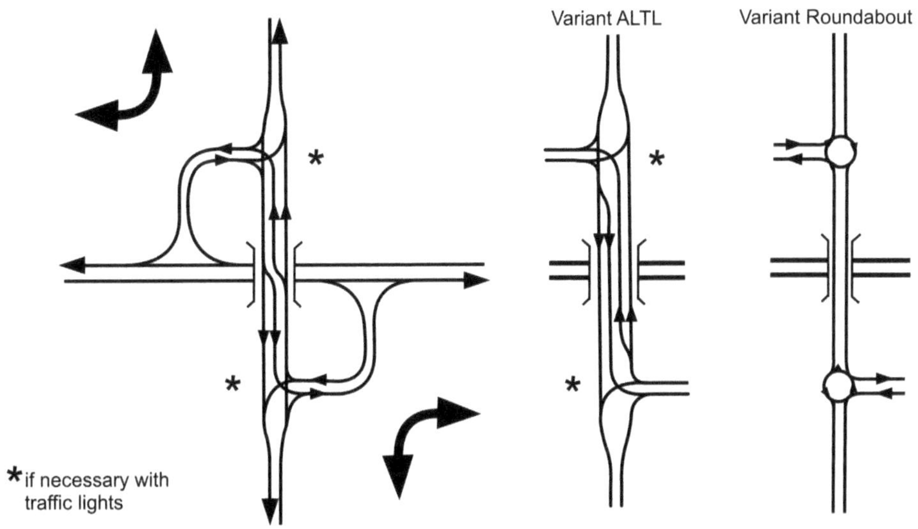

Advantages:
- If right-turning lanes are omitted, low junction expansion in the minor road (not applicable for variant roundabout)
- Useful in case of strong corner relations (double arrows) and/or area restrictions in the free quadrants
- Turning possibilities, convenient for operational and winter services

Disadvantages:
- Unfourable driving dynamics
- No or only limited possibility to extend the left turning lane in the downstream road (not applicable for variant roundabout)
- Relatively wide bridge (not applicable for variant roundabout)

Variant ALTL: Adjacent left turn lanes

Goals:
- Higher capacity
- Shorter distance between the ramp connection points

Disadvantages:
- Very large bridge width

Variant roundabout: Small roundabout

Goals:
- Coverage of the capacity range between T-junction without and with traffic lights
- Avoidance of left turning lanes
- Enabling a shorter distance between the ramp connection points
- Minimum bridge width

Disadvantages:
- Congestion in exit ramp

Fig. 10.17 Diagonal half cloverleaf with inside left-hand turn stripes and with an exit after the crossing structure (FGSV 2008a)

In the basic diamond intersection system (Fig. 10.19), the two intersecting streets in each quadrant are connected by a direct ramp used in one-way traffic. Because of the low space requirement, the small extent of the intersection area in the minor road and the favourable conditions for controlling the traffic flow with traffic light systems, if correctly designed, diamond intersections are particularly suitable for intersections on EKA 3 motorways with high traffic load.

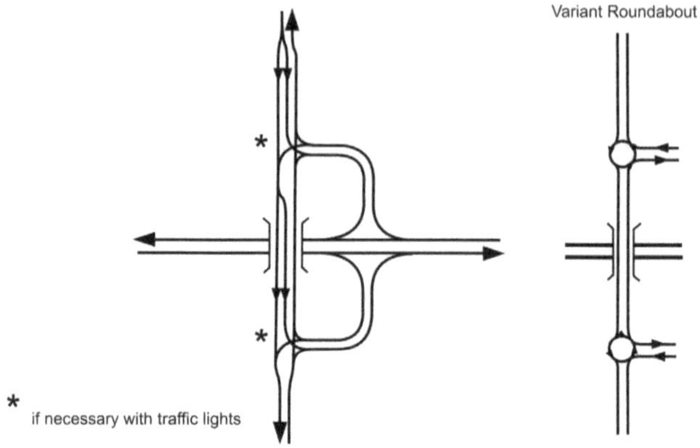

Advantages:
- Useful in case of one-sided area restriction along the minor road
- Turning possibilities, convenient for operational and winter services

Disadvantages:
- No preferred corner relationship in both directions
- Without traffic lights only applicable for low left turn or left turning-in streams
- One exit ramp unfourable in terms of driving dynamics
- Relatively wide bridge, if the length of the second left turn lane is not limited

Variant Roundabout: Small roundabout
Goals:
- Coverage of the capacity range between T-junctions without or with traffic lights
- Avoidance of left-hand turning lanes
- Minimum bridge width

Fig. 10.18 Symmetrical half cloverleaf (FGSV 2008a)

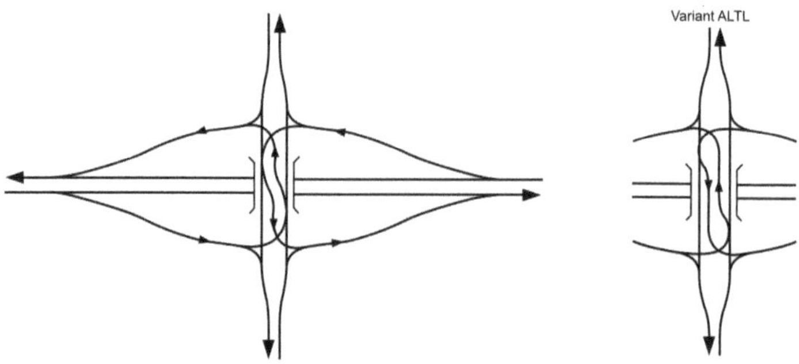

Advantages:
- Driving dynamics favourable for motorway traffic
- Very efficient with traffic lights
- Less space requirement than half cloverleaf intersections
- Small extent of the junction in the minor road
- Turning possibilities, convenient for operational and winter services

Disadvantages:
- Because of the short crossing distance only with traffic lights
- Susceptible to wrong-way driving
- Four quadrants covered with ramps
-

Variant ALTL: adjacent left turn lanes
Goal:
- Higher capacity
- Enabling a shorter distance between the ramp connection points

Fig. 10.19 Basic rhombus system with two crossings and internal left turn strips (FGSV 2008a)

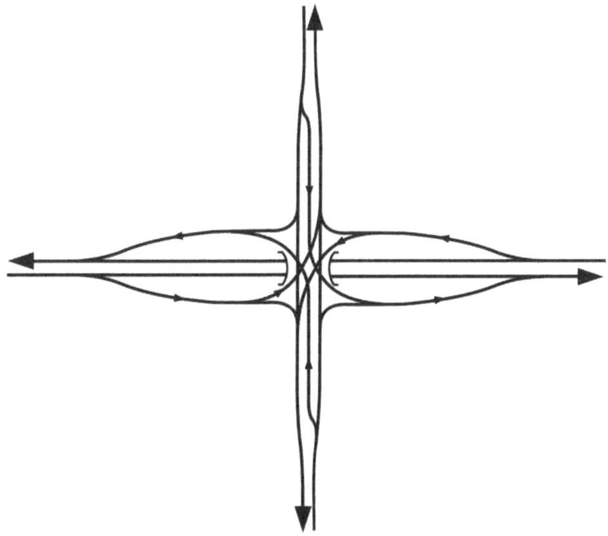

Advantages:
- Favourable driving dynamics
- More efficient than the diamond intersection with 2 junctions
- Lower space requirement
- Smaller junction extent in the following road than the diamond intersection with 2 junctions

Disadvantages:
- Only with traffic lights
- 4 quadrants covered with ramps
- With wide motorway cross sections, this results in very large bridge
- No turning possibilities

Fig. 10.20 Basic diamond intersection system with one crossing and external left turning lanes (FGSV 2008a)

The intersection area on the minor street can be divided into two intersections (Fig. 10.19) or combined into one intersection (Fig. 10.20). The division into two intersections requires internal left-turning lanes. Diamond intersections with inside and adjacent left turning lanes are more efficient than diamond intersections with inside and behind each other left turning lanes.

Diamond intersections with an intersection (Fig. 10.20) are usually more efficient and less susceptible to congestion because the left turn lanes are on the outside. However, a wider bridge structure is required.

Special solutions (mixed systems) of access intersections are created, for example, when half cloverleafs have to be combined, either wholly or in part, with individual direct ramps in limited conditions or when several strong turning flows are involved. Due to the larger number of ramp connections on the minor road in comparison to half cloverleafs, it can be difficult to accommodate a comprehensible signposting system and the number of possible wrong turns can increase. Such special systems should therefore only be used on

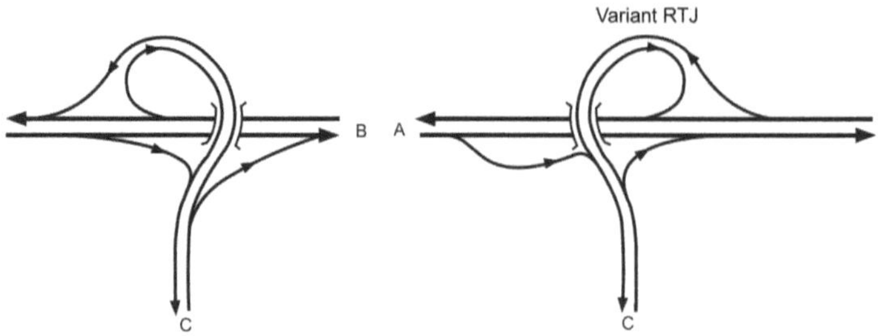

Advantage:	- Useful for stronger corner relationship A - C
Disadvantages:	- Minor road ends in ramp
	- Relatively large space requirement (also applies to variant RLT)
	- Wrong-way driving not excluded (also applies to variant RLT)
	- Limited turning possibilities (also applies to variant RLT)
	- 1 exit ramp unfavorable in terms of driving dynamics
Variant RTJ:	Right trumpet junction
Goal:	- adaption to topography
	- adaption to stronger corner relationship B - C
Disadvantages:	- Carriageway ends in loop ramp

Fig. 10.21 Three-armed intersection on two-lane rural roads as trumpets (FGSV 2012d)

urban motorways (EKA 3) if no satisfactory problem solution can be achieved with a basic system.

The basic system for a three-arm access intersection is the trumpet (see Sect. 10.2.3). Here, the connecting ramp for the right-turning road should become a stretched alignment, in the case of a two-lane cross-section of the minor road, connected to it at right angles as a parallel ramp (Fig. 10.21). An offset of the ramp connections (entry ramp before exit ramp) improves the orientation and reduces the probability of driving the wrong way if there is no separation of directions in the minor street. The use of a diamond intersection is recommended when only narrow strips are available along the motorway and requires additional measures at the end of the minor road to improve orientation.

10.2.5 Design Elements for Motorway Intersections

10.2.5.1 Ramps

Ramps can be distribution lanes next to the motorway or connecting ramps. Distribution lanes are used to keep the interlinking processes away from the main roadway or to combine closely following intersections. Connecting ramps are used to connect corner traffic with the main roadways.

The characteristics of the intersection-free sections of the road shall not be continued in connecting ramps. The principles of three-dimension alignment, therefore, apply only insofar as they concern the creation of sufficient visibility and the avoidance of optical illusions. Optical discontinuities must often be consciously accepted in order to make unavoidable minimum elements sufficiently visible. Depending on the routing of connecting ramps, a distinction can be made between ramp groups and ramp types.

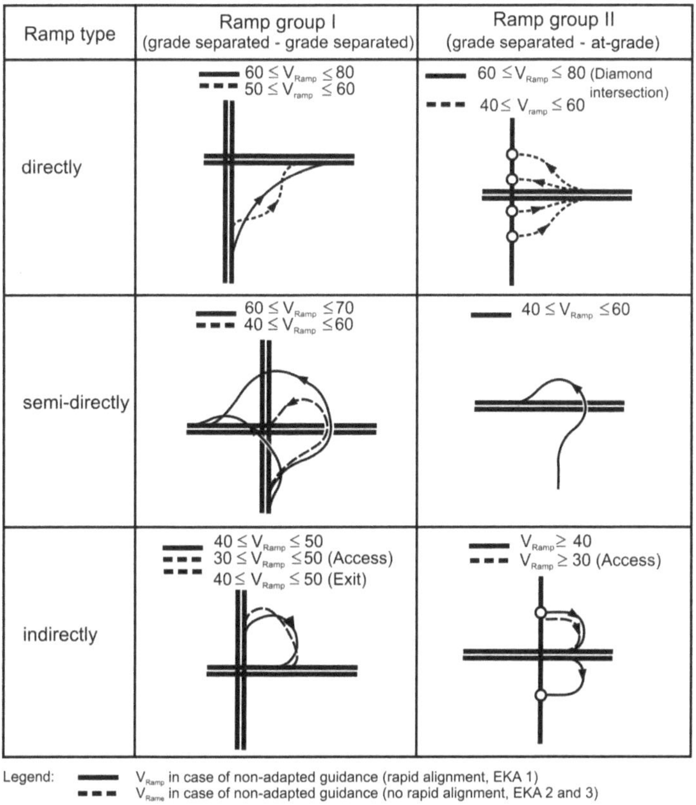

Fig. 10.22 Ramp groups and ramp types (FGSV 2008a)

Connecting ramps, which start with an exit and end with an entrance, form ramp group I (grade-separated—grade-separated). Their guidance can be direct, semi-direct or indirect. The alignment of the connecting ramps can be non-adapted or adapted. Even indirectly guided connecting ramps (loop ramps) do not have to have a continuous even curvature (Fig. 10.22).

Ramp group II (grade-separated—at-grade) comprises connecting ramps that start with an exit and end in an at-grade sub-intersection (exit ramps), as well as connecting ramps that start in an at-grade intersection and end in an entrance (entrance ramps). Their guidance can be direct, semi-direct or indirect. The alignment of the connecting ramps can be non-adapted or adapted. Even indirectly guided connecting ramps do not have to have a continuous even curvature (Fig. 10.22).

With the alignment of the ramps, the requirements of the road signs (FGSV 2000a, b) must be taken into account. The ramp cross-sections to be used for connecting ramps are given in Fig. 10.23. On urban motorways, the lane widths for ramp cross-sections Q2 and Q3 can be reduced to 3.25 m.

		Areas of application
Q 1	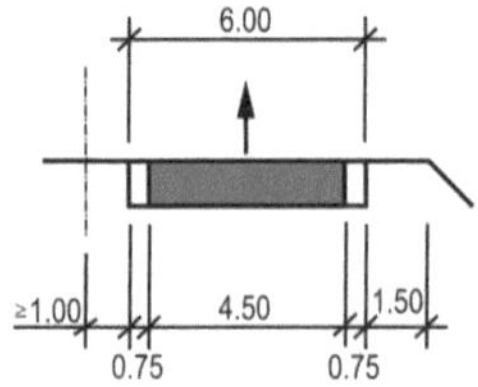	in ramp group I: q_{Ramp} ≤ 1350 vehicle/h and l_{Ramp} ≤ 500 m in ramp group II: separately guided access and exit ramps with $l_{Parallel\ guidance}$ ≤ 125 m
Q 2	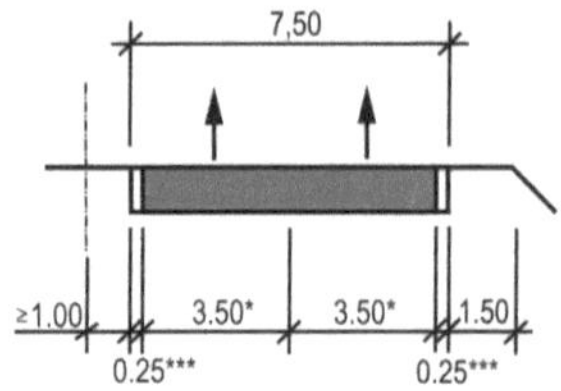	in ramp group I: q_{Rame} ≤ 1350 vehicle/h and l_{Ramp} > 500 m furthermore: two-lane weaving sections without hard shoulder in ramp group II: q_{Ramp} > 1350 vehicle/h
Q 3	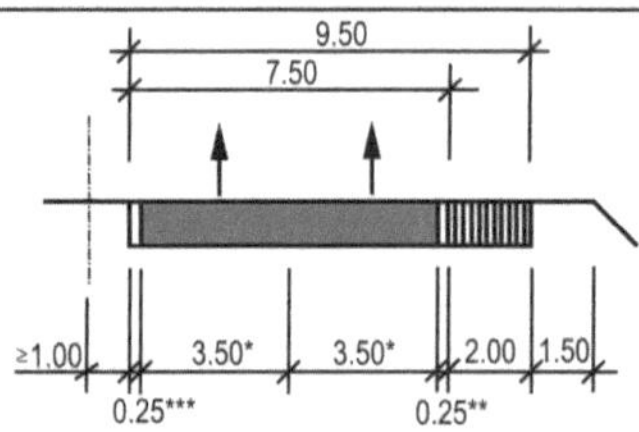	only in ramp group I: q_{Ramp} > 1350 vehicle/h and l_{Ramp} > 500 m furthermore: two-lane weaving sections without hard shoulder
Q 4	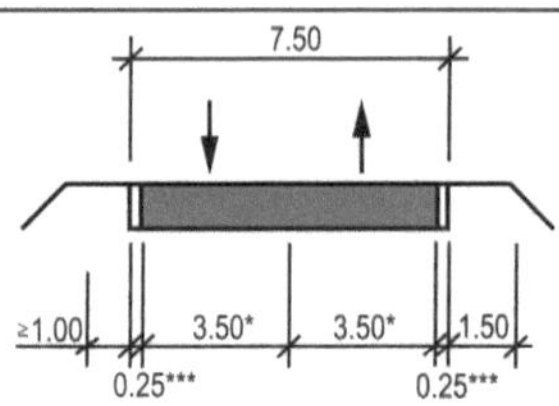	only in ramp group II: jointly guided exit and access ramps with $l_{Parallel\ guidance}$ > 125 m

Fig. 10.23 Cross sections and application limits of ramps. *) For EKA 3 and stretched lines, reduction of the lane width to 3.25 m permitted. **) The marking (broad line) is at the expense of the hard shoulder. ***) In the course of bridge construction, the hard strip width is 0.50 m (FGSV 2008a)

The standard cross-section Q4 of the RAA will probably be adapted to the wider standard cross-section RRQ2 of the RAL. In the 2008 RAA version, a single lane marking is used instead of the double lane marking with a width of 0.50 m and the lanes/hard strips are 3.50 m/0.25 m instead of 3.25 m/0.50 m wide.

Ramp Group I

The areas of application of the connecting ramps are based on the length and traffic volume. In ramp group, I, the Q1, Q2 and Q3 cross-sections are mainly used for all motorways.

The Q1 cross-section is to be used if a connection ramp is shorter than 500 m and the traffic volume does not exceed 1350 Kfz/h. The distance between the island tip of the exit (or lane subtraction) and the island tip of the following entry (or lane addition) for the vehicle stream under consideration shall be taken as the length of the connection ramp.

The cross-section Q2 is to be used if the traffic volume does not exceed 1350 Kfz/h but the length of the connecting ramp is more than 500 m. This enables overtaking even in case of traffic volumes below the capacity of a single-lane connection ramp ("comfort cross-section"). However, a recent study by KIT (Zimmermann et al. 2013) has shown that the Q2 cross-section has few advantages if it is combined with single-lane entrances or exits and could therefore be dispensed with.

The cross-section Q3 is to be used if the traffic volume exceeds 1350 Kfz/h. In addition, connecting ramps are given the cross-section Q2 or Q3 if they combine several intersections and if a weaving lane is also available in weaving sections. In the case of three-armed intersections, those connecting ramps which arise directly from or merge into the intersection-free section receive the standard cross-section of a carriageway in one direction instead of the ramp cross-section Q2 or Q3. The 2.50 m wide hard shoulder is also used in the exit and entrance areas.

In the exit and entry areas of ramp group I at the height of the tip of the island, the same cross-section must be provided as in the connecting ramp. The cross-sectional constrictions required in the entry area for cross-sections Q1 and Q2 are only made by road markings.

Ramp Group II

In ramp group II, the Q1, Q2 and Q4 cross-sections are applicable for motorways. The cross-section Q2 is to be chosen if the traffic volume at the intersection to the minor road requires more than one lane per direction over a longer length and the two lanes can avoid otherwise possible congestion into the motorway. In this case, a 2.00 m wide central reservation is to be provided in the area where the entrance and exit ramps meet. Alternatively, in these cases, in order to avoid congestions into the mainline carriageway, cross-section Q1 can be used with a congestion monitoring system in the connecting ramp. If cross-section Q2 is not required, cross-sections Q1 (with directional separation strip) or Q4 (oncoming traffic lane) are used in the area where entry and exit ramps are parallel on motorways. For economic reasons, cross-section Q4 should only be used if the resulting oncoming traffic area exceeds a length of 125 m.

Ramp Design Elements

The following limit values of the design elements refer only to the connecting ramps and to the entry and exit areas. The at-grade ramp connections to the minor road are designed according to RAL (FGSV 2012a).

For the transition curves, clothoids with parameters $R/3 \leq A \leq R$ are used analogous to the intersection-free sections. For the curve lying in the area of the island tip, the smaller value should be aimed at for reasons of recognizability. On loop-shaped exit ramps with main curve radii from $R = 40.0$ to 60.0 m, on the other hand, clothoid parameters $A \approx R$ are often necessary to be able to twist the roadway in the transition curve. In order to improve visibility, departure angles of at least 12 gon should be aimed for. In parallel ramps and distribution lanes, compliance with this condition will often require pivotings in the layout plan. If no pivoting is possible, the exit angle must be at least 6 gon. Entry ramps are to be connected to the mainline lanes with the smallest possible entry angle (3 gon to 5 gon) to improve acceleration and early parallelism.

In a recent KIT study on ramp types on motorways (Zimmermann et al. 2013), it is recommended to use hardened verges for narrow ramp radii (R about 30 m) and safety barriers set back by about 2.0 m (if necessary).

The maximum longitudinal inclination is 6% (ascending gradient) or 7% (descending gradient) for all ramp groups and ramp types. If the roads to be linked have a steep longitudinal inclination and the connecting ramps are subsequently inclined, the use of larger values is permissible (in the case of a descending gradient, only with stretched alignment). In any case, the inclination to be observed is 10%.

Crest and sag curves are rounded off with curves, which can be replaced with sufficient precision by square parabolas. The crest and sag curve radii specified in Table 10.4 may be undercut provided that the minimum stop sight distances are observed when superimposing the curvatures of the site and longitudinal planes and that the fields of vision are kept clear. The cross fall in connecting ramps is always one-sided and in curves, it is directed towards the inside.

The superelevation areas in the entry and exit ramps at the beginning of the entrances and the end of the exits are often problematic, especially when conversion and extension or subsequently construction of an intersection, because the longitudinal inclination of the main carriageway and thus the forced gradient of the entries and exits is often less than 0.7%. If, due to a curve in the main roadway, the crossfall is directed towards the central

Table 10.4 Ramp design elements for motorways

Ramp speed	V	[km/h]	30	40	50	60	70	80
Crown radius of the ramp	min R	[m]	30	50	80	125	180	250
Minimum crest diameter	min H_K	[m]	1000	1500	2000	2800	3000	3500
Minimum sag diameter	min H_W	[m]	500	750	1000	1400	2000	2600
Required stop sight distance	min s_h	[m]	30	40	55	75	100	115
Limits of the longitudinal inclination	max s	[%]	+6.00 (ascending gradient)					
	min s	[%]	−7.00 (descending gradient)					
Minimum cross slope outside of superelevation areas	min q	[%]	2.50					
Maximum crossfall	max q	[%]	6.00					
Minimum radius at q = −2.5%	min R	[m]	1000					
Minimum relative gradient	min Δs	[%]	0.1 · a a [m]: Distance axis of rotation from the edge of the road					
Maximum diagonal gradient	max p	[%]	9.00					

reservation, a standard superelevation in the clothoid between the last ramp element (right-hand curve) and the entry/exit is not possible because a zone with weak drainage is created. In this case, it should be checked whether a crossfall to the outside of the curve can be permitted by increasing the radii to min $R \geq 1000$ m. Otherwise, a diagonal superelevation must be used (cost-effective) or the area can be equipped with open-pored asphalt (cost-intensive, maintenance-intensive).

10.2.5.2 Exits

Exit zones are deceleration areas and should therefore always be executed with a parallel diverging lane. They should be designed as uniformly as possible by using standardized exit types. Exits can be arranged on mainline carriageways or connecting ramps. The connecting ramps can lead to the Weaving section or the mainline carriageway of another motorway or an at-grade intersection with a minor road.

In addition to the design, the function and capacity of an exit is largely determined by signposting. It also influences the dimensioning of the required length of the exit opening (l_A), which is differentiated for the different design classes.

Design of the Exit Areas

The hard shoulder should only be carried along next to the diverging lane if the cross-section in the connecting ramp is executed with a hard shoulder. In all other cases, it is sufficient to set the safety barriers back far enough (≥ 2.00 m) next to the diverging lane that emergency stops are possible on the then hardened verge.

The diverging lanes are given the same width as the adjacent mainline lanes. The hard shoulder is 0.50 m wide. This results in a roadway widening of 0.50 m for the RQ 31 without the hard shoulders carried along.

The length of the diverging lanes is given in Table 10.5.

The island head between the main roadway and the ramp is given a width of 1.50 m. This results in a triangular ghost island with a maximum width of 4.50 m between the tip of the island and the island head on the RQ 31 (Fig. 10.24).

Particularly on left-hand curves, there is a ridge in the cross-section of the diverging lane between this lane and the mainline carriageway, since the ramp cannot usually accommodate the ramping and the superelevation in the transition curve completely. Therefore, a ridge of maximum $\Delta q = 5.0\%$ is permitted in the area regularly crossed by motor vehicles. In the area of the roadway boundary marking (sign 295 StVO) in front of the tip of

Table 10.5 Lengths l_A and l_Z at exits (FGSV 2008a)

Size	Extension type	EKA 1/EKA 2	EKA 3
l_A	All A-types	250	150
	AR 1, AR 3 (Q 2), AR 4 (Q 2)	150	100
	AR 1- additional type for EKA 3	–	100
	AR 3 (Q 3), AR 4 (Q 3)	200	125
l_Z	All types	60	30

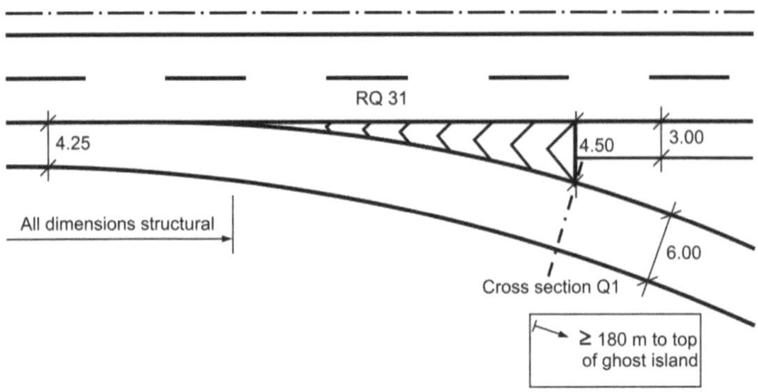

Fig. 10.24 Formation of the island head at exits (FGSV 2008a)

the island, the ridge may also be larger. If this principle cannot be adhered to, taking into account the maximum relative grade of 2.0%, the roadway boundary marking must be extended accordingly.

The standard length of the carriageway boundary marking is $0.1 \cdot l_A$, i.e. 25.0 m for all A types in EKA 1 and EKA 2. The same principles apply to the avoidance of poor drainage zones in the superelevation areas as for the sections of the sections without intersections. For better integration of the superelevation, the superelevation section may be pulled forward into the diverging/merging lanes with sufficient longitudinal inclination so far that the crossfall $q = 0.0\%$ is already reached at the beginning/end of the transition curve.

The exit types described below on mainline carriageways and on connecting ramps can basically be used on all motorways. The regulations of the RWBA (FGSV 2000a) must be taken into account when determining the minimum distance between consecutive exits on mainline carriageways and on connecting ramps.

On mainline carriageways, the ramp cross-section is important for the design of an exit or the selection of the appropriate exit type. The exit types applicable to mainline carriageways are shown in Fig. 10.25. All exit types can also be used on three- or four-lane carriageways.

Table 10.6 shows the limits of use of the extension types and the required distances to the following intersection.

On connecting ramps, the ramp cross-sections provided in front of and behind the separation point are decisive for the formation of exits. The exit types within connecting ramps are shown in Fig. 10.26.

10.2.5.3 Interweaving

Interweaving areas are created by superimposing incoming and outgoing vehicle streams in grade-separated intersections or a dense sequence of intersections. In grade-separated intersections, weaving areas can be arranged both on mainline carriageways and on connecting ramps. They can be used on all motorways.

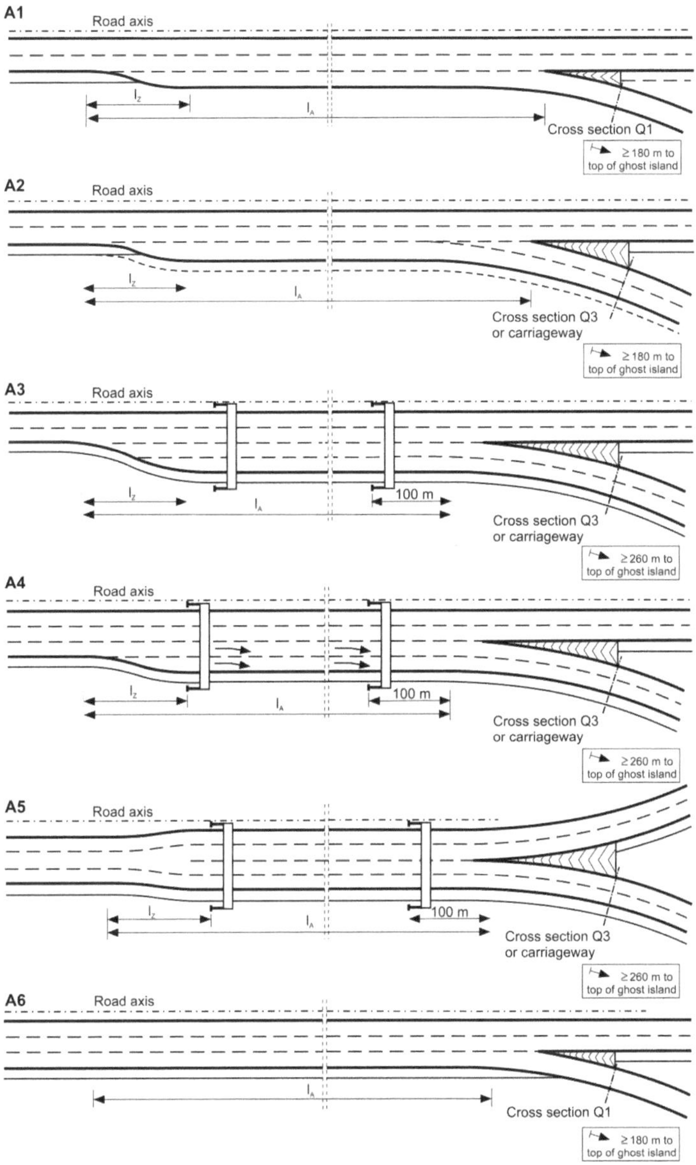

Fig. 10.25 Types of exits on mainline main carriageways (FGSV 2008a)

Weaving areas must be of sufficient length (I_v) to carry out the weaving operations and to accommodate the signposting. For the quality of traffic flow and the capacity of weaving areas, their length is less important than the number of weaving lanes and mainline lanes and the type of marking.

Table 10.6 Application limits for extension types on main carriageways (FGSV 2008a)

Distance to the next intersection	Number of lanes of the main carriageway before/after the exit	Traffic volume of the exit [car/]		
		≤ 1350	≤ 2300	> 2300
≥ 250 m Or no further separation existing	2/2, 3/3, 4/4	A 1, A 2	A 2	A 3
	3/2, 4/3	A 5, A 6	A 5	
	4/2	–	A 7	
< 250 m	2/2, 3/3, 4/4	A 1	A 3	
	3/2, 4/3	A 6	A 4	
	4/2	–	A 7	

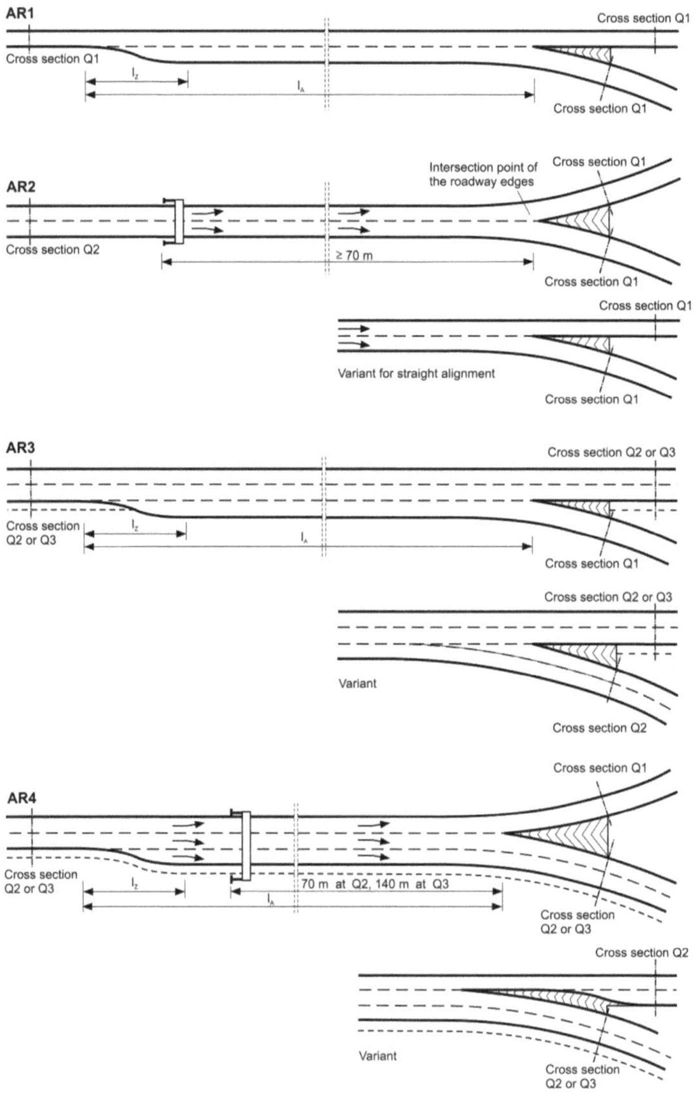

Fig. 10.26 Types of exits on-ramp systems (FGSV 2008a)

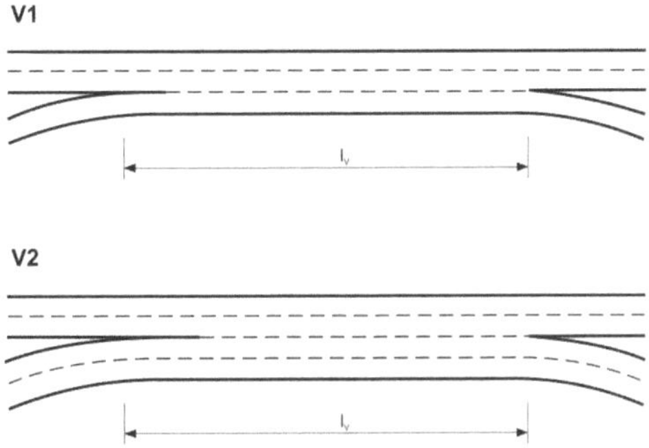

Fig. 10.27 Types of interweaving areas on mainline carriageways and in ramp systems. (FGSV 2008a)

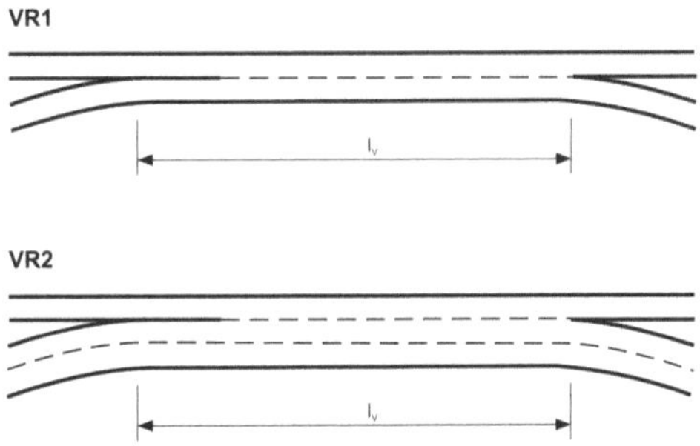

Fig. 10.28 Types of interweaving areas within connection ramps (FGSV 2008a)

Type V weaving areas are universally applicable on mainline carriageways and in ramp systems (Fig. 10.27). They allow an adaptation to the volume of the weaving vehicle streams and to the cross-sectional traffic volume that occur. In the case of significantly different traffic volumes in the incoming and outgoing connecting ramps, asymmetrical interweaving areas can also be used.

Interweaving areas on connecting ramps with cross-section Q 1 (type VR) are shown in Fig. 10.28. In principle, they correspond to the types suitable for mainline carriageways, but have different dimensions and are usually adjacent to a single-lane distribution carriageway. Asymmetrical interweaving areas can also be used on connecting ramps in case of significantly different traffic volumes in the incoming and outgoing connecting ramps.

According to HBS (FGSV 2015a), the capacity of a weaving area as part of a grade-separated intersection is determined by the maximum weaving traffic volume, by the capacity of the exit ramp and by the capacity of the connecting ramp or the mainline carriageway below the weaving area (Fig. 10.29).

10.2.5.4 Entries

Entries should be designed as uniformly as possible by using standardized entry types. Entries can be arranged on mainline carriageways or on connecting ramps. Connecting ramps from other motorways or minor roads end at them. Entries at connecting ramps bring connecting ramps together. The types of entries on mainline carriageways and on connecting ramps described below can be used on all motorways.

	Motorways	
Weaving sections on continuous carriageway	Type V 1	Type V 2
Length of the weaving section l_v m	250	250
Weaving sections on connecting ramps	Type VR 1	Type VR 2
Length of the weaving section l_v m	200	200
	Urban highways	
Weaving sections on continuous carriageway	Type V 1	Type V 2
Length of the weaving section l_v m	200	200
Weaving sections on connecting ramps	Type VR 1	Type VR 2
Length of the weaving section l_v m	170	170

Fig. 10.29 Dimensions of interweaving areas (FGSV 2008a)

When designing entries on mainline carriageways and on connecting ramps, it should be noted that the speed differences between entering and passing vehicles should be as small as possible in order to achieve a high level of safety and capacity for merging. Entries must therefore always be designed with parallel merging lanes. The entry types applicable to mainline carriageways are shown in Fig. 10.30. The entry types can also be used on three- or four-lane carriageways. For entry types, E 2, E 4 and E 5, the cross-section of a carriageway in one direction must be used instead of the cross-sections Q2 and Q3 if the carriageway of another motorway passes into the entry ramp without an exit.

In principle, the range of application of an entrance type is based on the cross-sectional traffic volume which ensures an appropriate quality of traffic flow in the connecting ramp and in the carriageway below the entrance. This must be verified in accordance with HBS (FGSV 2015a).

Where ramp streams are merging individually into the mainline carriageway, the types of successive entries shown in Fig. 10.31 shall be used. The individual entries are to be designed as standard types in compliance with the application limits. The specified distances between the two entrances should not be undercut. In order to minimise problems in the merging of traffic flows from the outset, the use of lane signals for successive entrances should be checked.

On urban motorways (EKA 3), the entrance types shown in Fig. 10.32 can also be used on mainline carriageways.

On connecting ramps, the ramp cross-sections provided in front of and behind the point of entry are decisive for the design of entries. The possible entry types are shown in Fig. 10.33. In addition to the design with a merging lane, the design as a lane addition is also possible.

The standard dimensions for the lengths of the entries and the lengths of the merging lanes are given in Table 10.7.

The hard shoulder of the mainline carriageway shall be guided next to the merging lane only if the ramp cross-section has a cross-section with hard shoulders. In this case, the hard shoulder must be 2.50 m wide. If no hard shoulder is included, the verge must be designed to be stable and widened to 2.00 m or, in the case of safety barriers, widened so that a 2.00 m wide stable verge is available in front of the safety barrier.

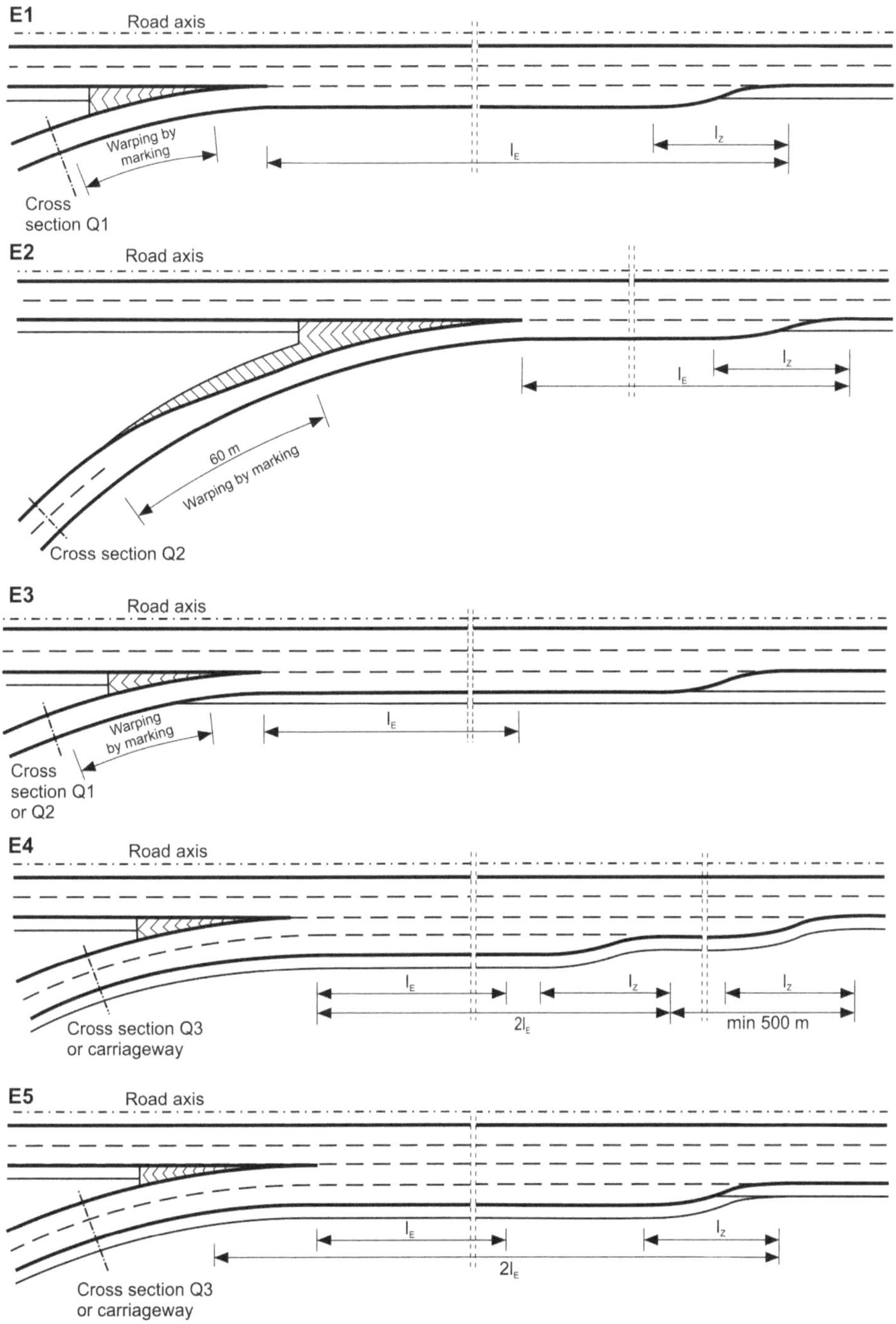

Fig. 10.30 Types of entries on mainline carriageways (FGSV 2008a)

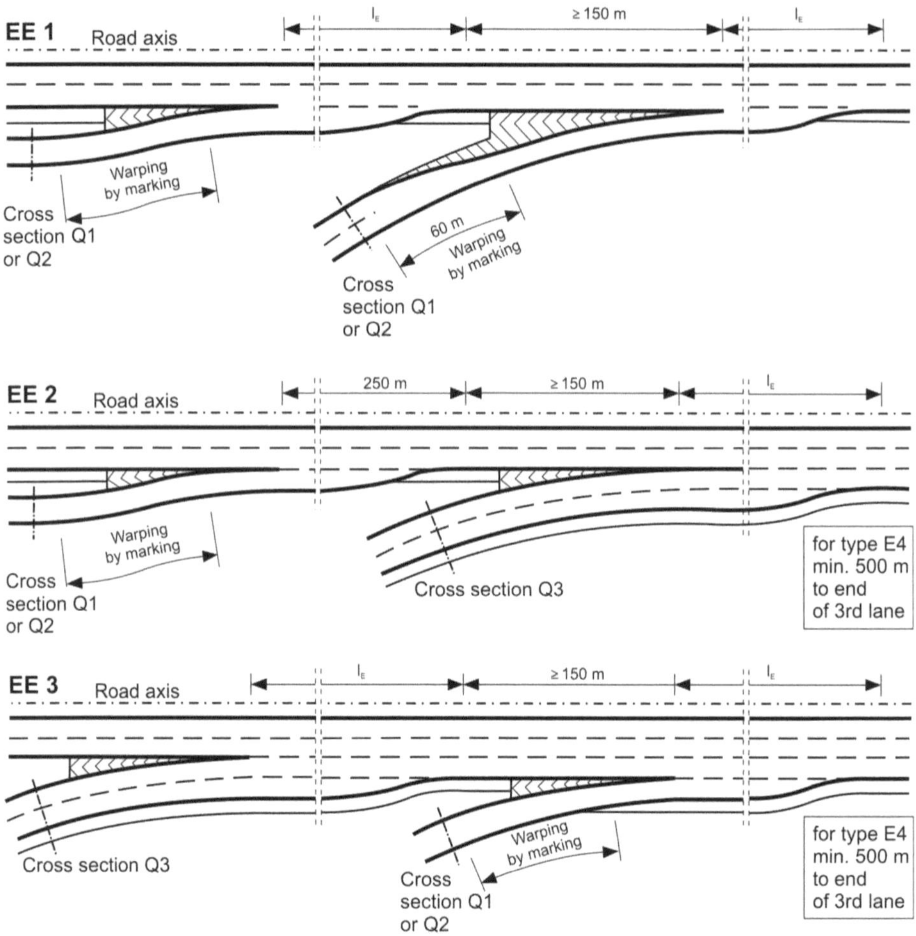

Fig. 10.31 Types of successive entries on mainline carriageways (FGSV 2008a)

The merging lanes are as wide as the adjacent mainline lane (for RQ 31: b = 3.75 m). The hard strip is 0.50 m wide.

The transition of the width of the lane and the right-hand hard strip starting from the dimensions of the ramp is done by marking according to the RMS (FGSV 1993). The island head is 1.50 m wide between the hard shoulder and the edge of the ramp. Thus the entire island is 4.50 m wide at RQ 31.

Particularly with left-hand curves, there is a ridge in the cross-section at the merging lane between these and the mainline carriageway, since the ramp cannot usually accommodate the ramping and the superelevation in the transition curve completely in the ramp. Therefore, a ridge of maximum $\Delta q = 5.0\%$ is permitted in the area regularly crossed by motor vehicles. In the area of the roadway boundary marking (sign 295 StVO) in front of the tip of the island, the ridge may also be larger. If this principle cannot be adhered to,

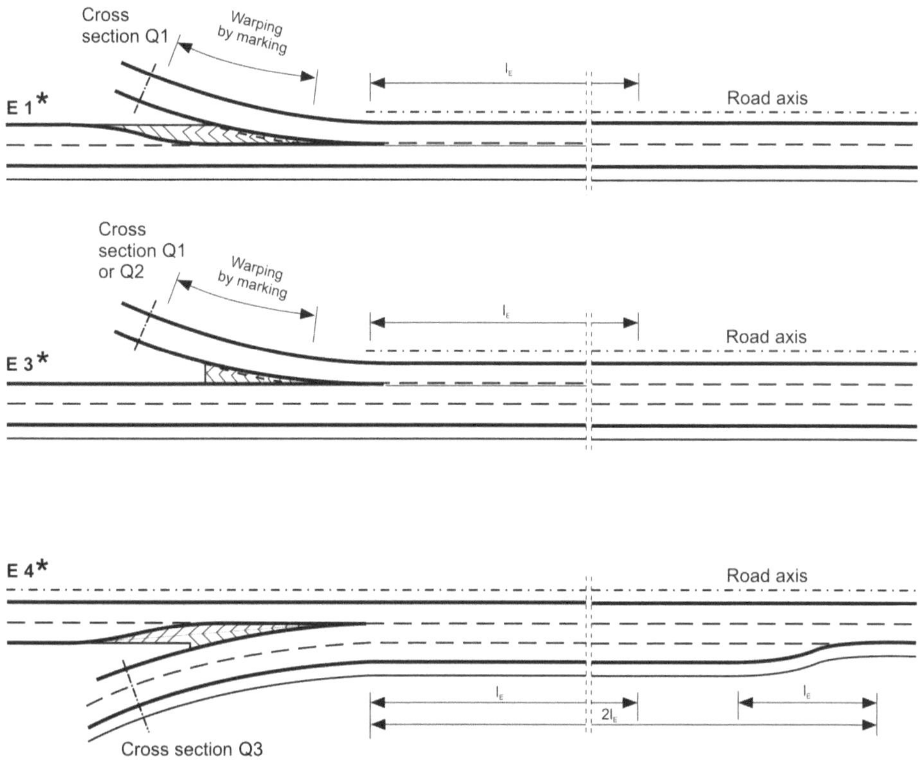

Fig. 10.32 Additional types of entries on mainline urban motorways (FGSV 2008a)

taking into account the maximum ramp gradient of 2.0%, the roadway boundary marking must be extended accordingly.

The standard length of the roadway boundary marking is $0.2 \cdot l_A$, i.e. for all A types in EKA 1 and EKA 2 50.0 m. The same principles apply to the avoidance of poor drainage zones in the superelevation areas as for the sections of the road sections without intersections. For better integration of the superelevation, the superelevation section may be pulled forward into the merging and diverging lanes with sufficient longitudinal inclination so that the crossfall q = 0.0% is already reached at the beginning/end of the transition curve.

10.2.6 Dimensioning of Intersection Elements for Motorways

Grade-separated intersections consist of the subintersections exit, entrance and weaving sections. The sufficient quality of the traffic flow must be proven for each sub-intersection. The worst quality of the sub-intersections is decisive for the overall quality. In addition, the quality of the traffic flow of the sub-intersections and the adjoining intersection-free

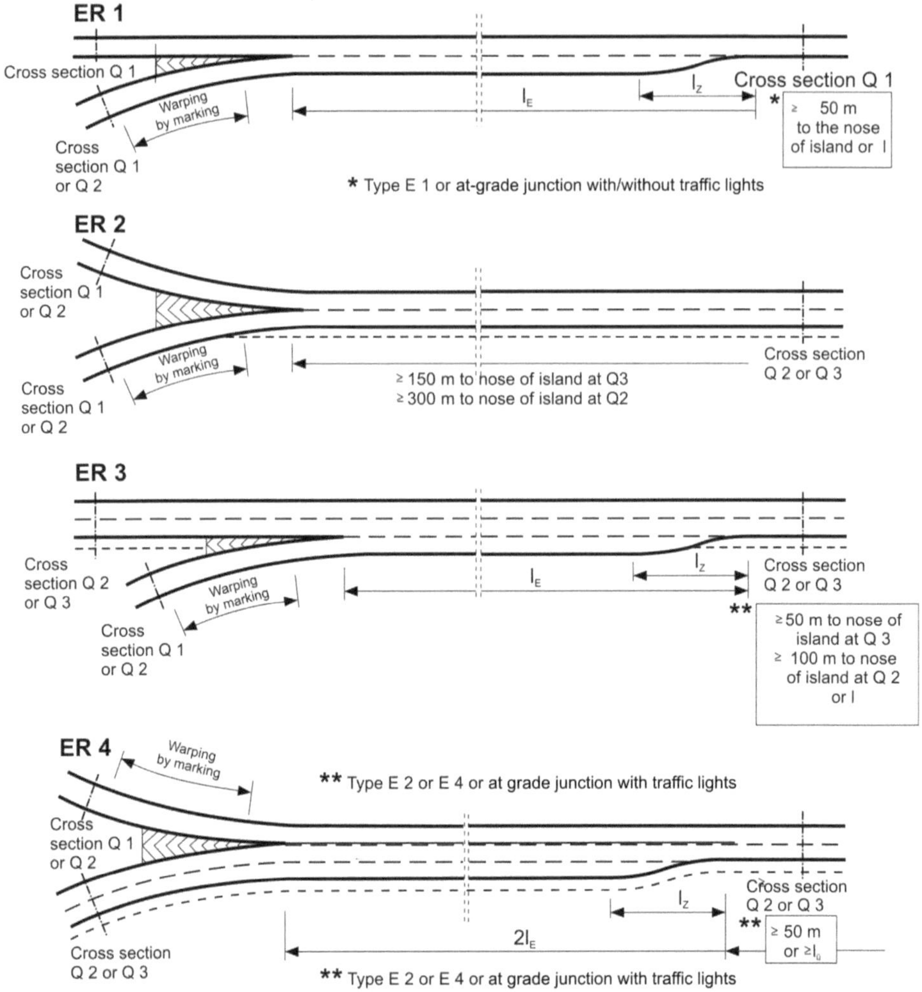

Fig. 10.33 Types of entries on connecting ramps (FGSV 2008a)

Table 10.7 Dimensions of entrances. (FGSV 2008a)

Size	Entry type	EKA 1 and EKA 2	EKA 3
l_E [m]	All E and E[a] types All EE types	250[a]	150
	HE 1, HE 3, HE 4	150	100
l_Z [m]	All types	60	30

[a]Extend the gradient sections with s > 4% if necessary

Table 10.8 Limit values of the utilization levels for the quality grades (FGSV 2015a)

Quality level of the traffic flow	A	B	C	D	E	F
Capacity utilisation rate a [−]	≤ 0.30	≤ 0.55	≤ 0.75	≤ 0.90[a]	≤ 1.00	> 1.00

[a] 0.92 for inflow control (for entry types E 1 and E 2)

road sections should be coordinated. As a rule, the procedures of the HBS (FGSV 2015a) are used (Table 10.8).

The quality criterion for grade-separated intersections is the degree of capacity utilization, analogous to the intersection-free road sections.

$$a = \frac{q_B}{C}$$

with
a: Degree of utilization [−],
q_b: Rated traffic volume [vehicle/h],
C: Capacity [car/h].
For the dimensioning of the sub-intersections, the traffic volumes of the occurring vehicle flows are required in the unit [vehicle/h] and the share of heavy goods vehicles b_{sv} in [%]. The alignment in the site plan and the elevation plan is usually not taken into account. Also not considered is the location of the intersection inside or outside of conurbations, although the high number of drivers with local knowledge of the area means that faster driving manoeuvres can be expected. However, there is currently no reliable knowledge about this influence.

10.2.6.1 Entries

The RAA (FGSV 2008a) provide for five standard types of two and three-lane motorways for entrances to mainline carriageways (Fig. 10.34).

The capacity of an entrance as part of a grade-separated intersection is determined by the capacity of the merging lanes and the capacity of the main roadway below the entrance, whereby proof must be provided for both sections (q_m and q_{HU}). The verification for the main carriageway is carried out according to Sect. 9.2.4. The traffic quality of the different types of merging lanes can be taken from the diagrams of the HBS (FGSV 2015a).

10.2.6.2 Weaving Sections

According to the RAA (FGSV 2008a), weaving sections occur on distribution lanes (VR types) and directly on the mainline lanes (V types) (Fig. 10.35).

The individual capacities of the three critical areas are decisive for the capacities of a weaving section:

- Maximum interweaving traffic volume (q_M),
- Capacity of the distribution lane or the main lane below the weaving section (q_{HU}/q_{VU}),
- Capacity of the exit ramp.

The traffic quality of the different types of weaving sections can be seen in the diagrams of the HBS (FGSV 2015a).

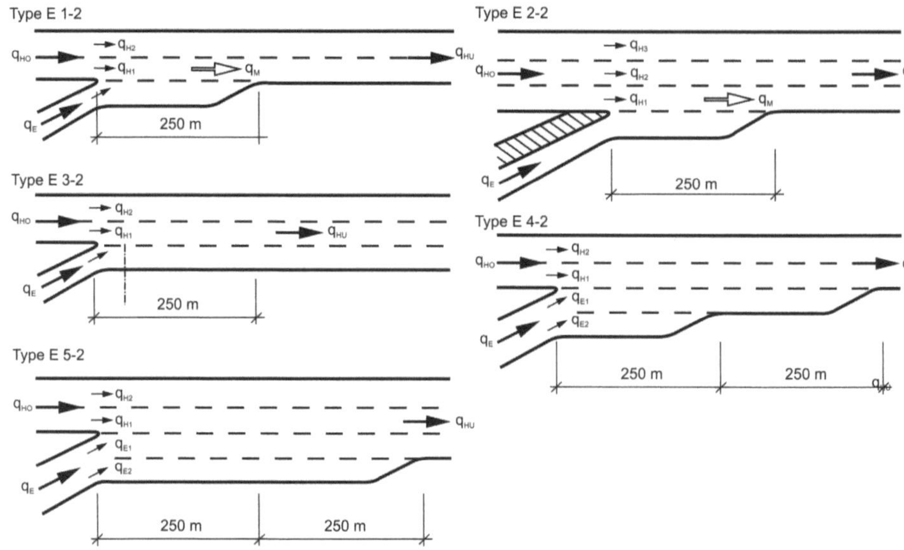

Fig. 10.34 Types of entry (schematic) and designation of the vehicle streams, with q_E traffic volume of the vehicle stream entering [vehicle/h]/[car E/h], q_{HO} traffic volume on the main roadway above (island tip) [vehicle/h]/[car E//h], q_{Hi} traffic volume on a mainline lane of an n-lane main road above (island tip) [car/h]/[car-units/hour], q_{HU} traffic volume on the mainline main road below the subintersection [car/h]/[car-units/hour], q_M threading traffic volume [car-units/hour], car-units: Passenger car units with 1 SV = 2 Pkw-units (2.5 Pkw-units with rising indirect ramps) and 1 Pkw = 1 Pkw-Unit (Extract from FGSV 2015a)

10.2.6.3 Exits

The RAA (FGSV 2008a) provide for the four standard types A 1, A 2, A 3 and A 4 for exits (Fig. 10.36).

The capacities of an exit are determined by the individual capacities of the three critical areas:

- Maximum weaving traffic volume in the diverging area (q_M),
- Capacity of the exit ramp (q_A),
- Capacity of the main carriageway below the exit with lane subtraction (q_{HU}).

The traffic quality of the different types of exits can be taken from the diagrams of the HBS (FGSV 2015a).

10.3 Intersection Types for Rural Roads

10.3.1 Types of Intersections

The intersection systems and the associated intersection elements are defined according to RAL (FGSV 2012a). The aim is to achieve the most uniform possible design of the design classes to support recognisability. In contrast to the standard cross-sections, the

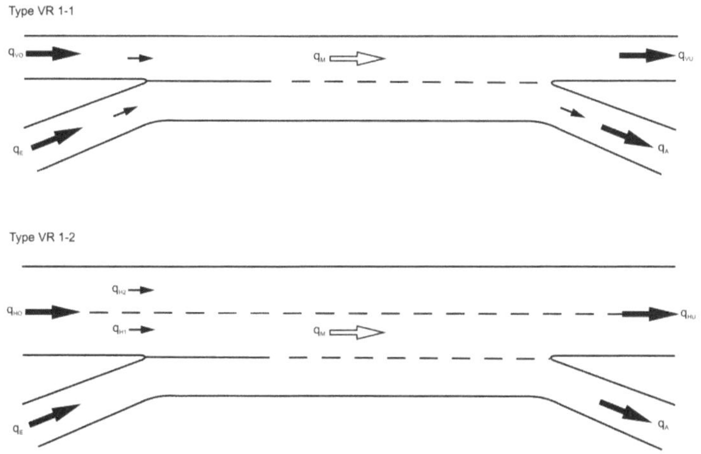

Fig. 10.35 Types of weaving sections (schematic) and designation of vehicle streams (FGSV 2015a)

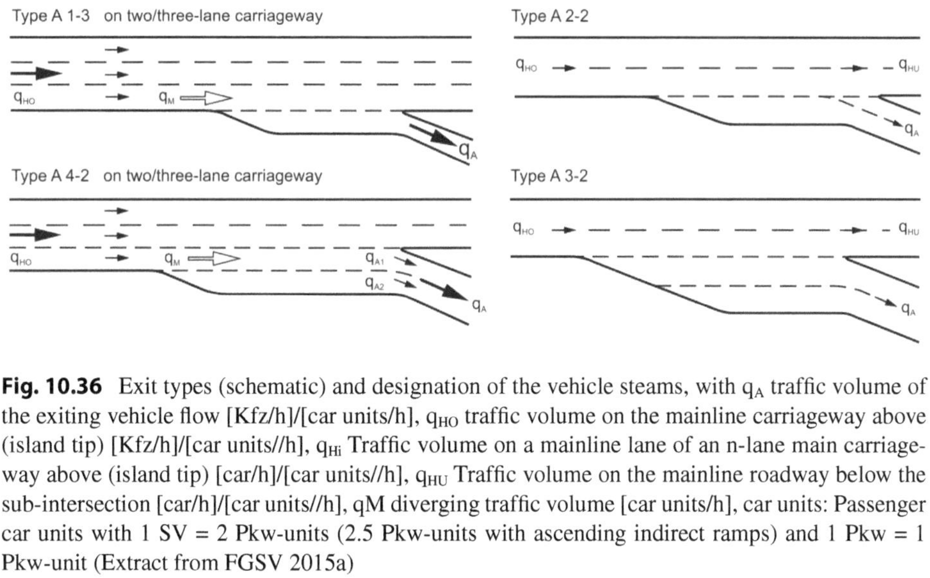

Fig. 10.36 Exit types (schematic) and designation of the vehicle steams, with q_A traffic volume of the exiting vehicle flow [Kfz/h]/[car units/h], q_{HO} traffic volume on the mainline carriageway above (island tip) [Kfz/h]/[car units//h], q_{Hi} Traffic volume on a mainline lane of an n-lane main carriageway above (island tip) [car/h]/[car units//h], q_{HU} Traffic volume on the mainline roadway below the sub-intersection [car/h]/[car units//h], qM diverging traffic volume [car units/h], car units: Passenger car units with 1 SV = 2 Pkw-units (2.5 Pkw-units with ascending indirect ramps) and 1 Pkw = 1 Pkw-unit (Extract from FGSV 2015a)

intersection types are not generally limited to one-design class but are also designed depending on the design classes and the traffic volumes of the intersecting roads. However, it is important to ensure that the intersection types are as similar as possible on a route.

Grade-separated, partially grade-separated and partially at-grade intersections, each of which requires one or more bridges, and at-grade intersections or crossings, as well as roundabouts, are possible. The at-grade T-junctions or crossings, either separately or as partial intersections, can be operated with right-of-way control with traffic signs or with traffic lights.

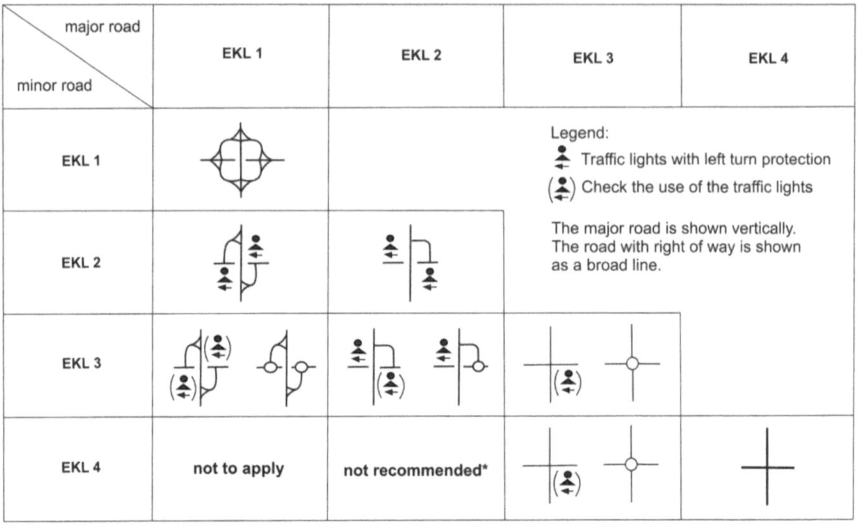

Fig. 10.37 Intersection types for four-armed intersections on rural roads (FGSV 2012a)

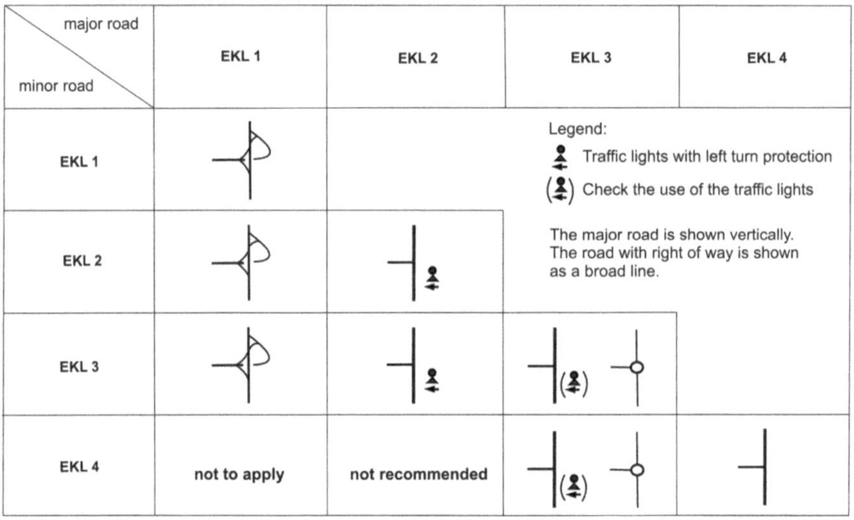

Fig. 10.38 Intersection types for three-armed intersections on rural roads (FGSV 2012a, b, c, d)

For roads of one-design class, only certain types of intersections are normally provided (Figs. 10.37 and 10.38). However, in exceptional cases, traffic requirements and local conditions (area availability, topography, etc.) may make it advisable to select a different type of intersection. This shall be weighed against the desired uniformity of road characteristics in a design class, taking into account the objectives of traffic safety, traffic quality, environmental compatibility and construction costs, in particular. In addition to the single intersection consideration, the neighbouring intersections must also be taken into account. Connections of heavily loaded property access roads are treated like connections of roads of EKL 4.

The basic requirements of safe road design

- Recognizability,
- Clarity,
- Comprehensibility and
- Rideability (passability)

must be given special consideration in the intersections, as there are many points of conflict and high-speed differences occur.

Intersections interrupt or influence the flow of traffic. Therefore, as few intersections as possible should be arranged. On the other hand, many intersections improve the bundling function of rural roads with a high connection function. The number and design of the intersections should be selected so that the network function can be achieved with the desired average driving speeds. This means that the connection of local roads (EKL 4) to roads of EKL 1 and, if possible, also to roads of EKL 2 should be avoided, although this will be difficult to implement, especially in the case of conversion and extension measures in existing buildings. A connection of the agricultural road network is to be avoided for roads of EKL 1 and EKL 2 and limited to the necessary minimum for roads of EKL 3.

The traffic quality of intersections has to be proven according to the procedures of the HBS (FGSV 2015a). In addition, it shall be checked whether, in the case of a sequence of intersections with traffic lights or roundabouts on the network sections of the major road, the average car speed aimed for in accordance with the RIN for the road category can be maintained.

10.3.2 Requirements for the Intersection Design

10.3.2.1 Intersection Distances

Intersections restrict traffic safety and the average speed aimed for in network planning. The distance between intersections should therefore be as large as possible. On EKL 1 roads, intersection distances of more than three kilometres and on EKL 2 roads of more than two kilometres should be aimed for. If smaller distances are unavoidable due to the network (which is usual in the existing network and cannot be avoided, especially during reconstruction and expansion), it should be checked whether two adjacent intersections can be combined into one.

10.3.2.2 Major Road

Roads with the higher design class are major. If the design class is the same, the traffic significance and the design of the adjacent intersections must be taken into account. The traffic load can be used to weigh up the traffic importance. With the same traffic importance, roundabouts are particularly suitable as an intersection type because they do not prefer any direction. In the case of at-grade T-junctions s, the continuous road should be given priority in terms of the right of way.

10.3.2.3 Alignment in Intersection Areas

On rural roads, especially on the higher-level road, the recommended value ranges for the alignment (see Chaps. 12, 13, 14, 15, and 16) should not be undercut wherever possible because of the recognisability and clarity in the intersection area for the roads that meet there. Intersections should be perceptible from a distance of 500 to 600 m and should be clearly recognizable on roads of EKL 1 and 2 from distances of ≥ 300 m and on roads of EKL 3 and 4 from distances of ≥ 200 m, especially with regard to right-of-way control. This applies in particular to intersections and T-junctions without traffic lights.

The alignment in the intersection area can be supported and clarified by the boundaries of the driving area (planting, ramparts) as well as by guidance systems and signs.

When forming the narrower intersection area, the alignment in the site and elevation plan, as well as the crossfall, generally remain unchanged on the major road. The road surfaces of the lower-level intersection arms must be adapted to this.

At intersections, the choice of the design vehicle and the case of encounter to be used as a basis for the intersection design depends on the location of the intersection, the decisive function of the roads, the traffic volume and the traffic composition. The choice of a particular encounter case or design vehicle does not exclude the possibility that larger vehicles may use the intersection when entering or exiting with the use of opposite lanes. At intersections of rural roads, this will generally be appropriate or necessary at least for the largest vehicle approved according to the StVZO (truck and trailer, articulated lorry, articulated bus). It is possible to accept slight overdriving of the adjacent left-hand turning lane of the main road by right-hand turning trucks in order to avoid the visual impairment for turning passenger cars and the associated accident risks associated with a larger corner rounding.

10.3.2.4 Intersection Site Plan

The axes of the roads meeting at the intersection should intersect at level crossings at an angle $\alpha = 80$ to 120 gon, as otherwise, unfavourable visibility conditions may arise for the vehicles turning in, especially in inside curves. If the axles do not intersect at this angle, it must be checked whether the axle of the minor road can be bent or offset. This can also clarify the waiting obligation in the minor intersection arms. The design as a right offset also increases the capacity and traffic safety of the intersection if there is no traffic light system.

If intersections are located at turning points on the site plan, visibility conditions are generally good. However, it is problematic to ensure sufficient drainage in the superelevation areas, so that intersections at generous radii are more favourable.

In general, intersections in outside curves are more favourable than in inside curves, because especially in narrow inside curves the visibility is unfavourable for the car driver waiting. When entering outer curves on rural roads, however, it should be noted that the road surface of the higher-level road with a one-sided crossfall is difficult to recognise by the car driver waiting when approaching the intersection and the speed of the priority vehicles may be difficult for the car driver waiting to estimate. If the visibility conditions and the expected speeds cannot be reconciled, a light signal system may be required.

At roundabouts, the axes of the roads to be linked should be directed as radially as possible towards the centre of the roundabout.

10.3.2.5 Intersection Longitudinal Section

The visibility of an intersection and the visibility conditions are best when the roads meeting at the intersection are in a sag vertical curve. Intersections should not be designed so that both roads in the intersection area form a crest curve.

If it cannot be avoided that one of the two roads runs on a crest curve, the visibility often has to be improved by flanking measures in the intersection (e.g. early right turn lanes or directional islands) or in the surroundings (e.g. planting). Furthermore, a traffic light system should always be checked.

On fast rural roads, the longitudinal inclination of the higher-level road in the intersection area should not exceed approximately 4% (maximum 6%), as otherwise large "wrong" crossfalls can occur for those turning off, making it more difficult for heavy goods vehicles to start.

The longitudinal inclination of the minor intersection arms should not have maximum values because of the recognisability of the intersection and because of the starting and braking processes on a distance of about 25.0 m from the edge of the major road. For rural roads, a longitudinal inclination of the minor road of a maximum of 2.5% should be aimed for. However, these values cannot always be achieved during reconstruction and upgrading.

For the connection of the gradients of the minor intersection arms (Fig. 10.39), a tangential connection should be selected if possible, especially in design classes EKL 2 and EKL 3. A strong kink (e.g. 5%) should always be rounded out if—as for example at intersection arms with traffic lights—a fast crossing must be expected.

For the crest and sag curves, radii $H_K/H_W \geq 500$ m should be selected at the inclination transitions. The tangent length shall be $T \geq 12.5$ m.

If, for example, a tangential connection is not possible due to the gradients of the minor road and the opposite transverse inclination of the major road, a kink can be provided at the edge of the major roadway (Fig. 10.39, cases b and c). Following the kink of > 2.5%, rounding off with $H_K/H_W \geq 500$ m must be carried out (case c). Inclination differences of ≤ 2.5% do not have to be rounded off (case b).

10.3.2.6 Road Drainage and Gradient Transitions

Crossfalls, diagonal inclines and slope transitions are to be designed in such a way that the surface water flows off over the shortest possible distances, even in intersection areas. The following principles apply:

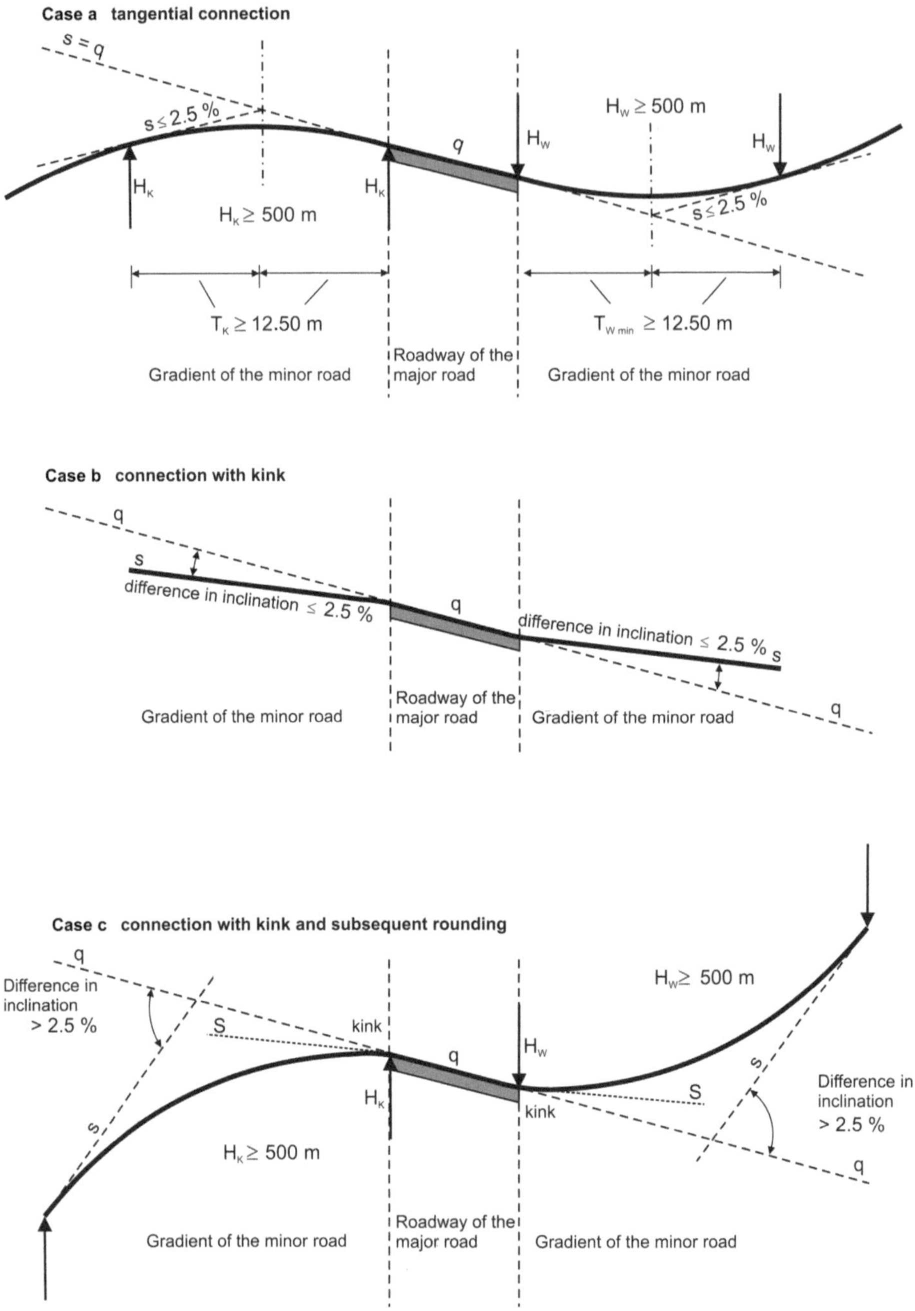

Fig. 10.39 Connecting the gradients of the minor intersection arms

- The inclinations of the major road remain unchanged, the inclinations of the road surfaces of the minor intersection arms are adjusted.
- The surface water on the road surfaces of the intersection arms should not enter other intersection arms or the inner surface of the intersection.
- Drainage aspects are to be given priority over vehicle dynamics because of the low entry and turn-off speeds at at-grade intersections, especially in the minor intersections.
- Low points of sags and high points of crests should, if they cannot be avoided, only be located in areas where a crossfall sufficient for drainage has been reached ($q \geq 2.5\%$).
- A diagonal inclination of 2.0% (in superelevation sections 0.5% at the most) must be maintained.
- Directional islands can facilitate drainage, as they divide the intersection areas into individual areas in terms of drainage technology, promote the formation of "correct" crossfalls in terms of driving dynamics and enable the arrangement of low points with gullies at the edges of the islands.

Due to the complex inclination conditions, a surface longitudinal section should be prepared for the intersections, if necessary on a larger scale. The surface longitudinal section should contain contour lines with fall lines. These serve not only to check the surface longitudinal section but also to determine the low points and to determine any necessary road gullies. The formation of the gradient transitions and low points is influenced by the connection of the gradients of the minor road (with or without bend) and by the presence of directional islands. The guidance of the surface water and the correct position of the road gullies should be checked by drawing the edge of the carriageway as an independent elevation plan.

10.3.2.7 Carriageways

Carriageways consist of continuous lanes at intersections and, in individual cases, additional lanes such as left-turning lanes, right-turning lanes and, if necessary, merging lanes for right turners. The number of continuous lanes in the intersection area, especially at intersections without traffic lights, should generally remain unchanged from the sections of road that are free of intersections. A continuous lane must not suddenly change into a turning lane in an intersection entrance, as otherwise unexpected lane changes can be expected.

At intersections with traffic lights, it may be necessary for reasons of efficiency or to improve traffic quality to increase the number of continuous lanes in the intersection access roads and to move back in after the intersection. In this way, the capacity of the intersections can be brought closer to that of the adjoining intersection-free sections (see Sect. 10.3.4).

Continuous lanes should be just as wide at intersections as the lanes on adjacent intersection-free sections.

10.3.3 Intersections for Roads of EKL 1

Due to the desired high average travel speeds, the EKL 1 is provided with grade-separated intersections. At an intersection or T-junction of two roads of EKL 1, there are grade-separated intersections. An intersection or T-junction of a road of EKL 1 with a road of EKL 2 or EKL 3 (in exceptional cases also with EKL 4) results in partially grade-separated intersections.

10.3.3.1 Grade-Separated Intersections

Grade-separated intersections should always be considered in the same way as motorway intersections (Sect. 10.2.2). The standard solution for a four-armed intersection is the cloverleaf (Fig. 10.4) and for a three-armed intersection the left-faced trumpet (Fig. 10.10). The exits and entrances and the ramps are single-lane, whereby the ramps should be designed adapted (with counter-curve) to save space. At grade-separated intersections in single-lane RQ 15.5, it should be checked whether the traffic-related central reservation should be replaced by a structural separation of directions. This could be useful for safety reasons or to improve drainage. However, the effect on visibility must be examined. The central reservation must then be widened in accordance with the requirements of the safety barriers.

In exceptional cases, grade-separated intersections can also be considered if an EKL 1 road is crossed by a heavily loaded EKL 2 road with long driving distances. In this case, however, it must first be examined whether it would be appropriate to upgrade EKL 2 to EKL 1 (see Sect. 6.2). Furthermore, it is to be checked whether in the course of the EKL 2 road individual grade-separated intersections are compatible with the otherwise intersections with traffic lights.

10.3.3.2 Partly Grade-separated Intersections

Partly grade-separated intersections consist of the entrances and exits on the major road and at-grade sub- intersections (intersection with or without traffic lights or roundabout) on the minor road and connecting ramps in between. The exits and entrances and the ramps are laid out analogously to grade-separated intersections. The minor sub-intersections are designed analogous to the corresponding solutions for the EKL 3.

The standard solution for a four-armed partly grade-separated intersection is the half of the cloverleaf, analogous to motorway intersections (Fig. 10.15). For reasons of efficiency and traffic safety, the connecting ramps (asymmetrical or symmetrical) should be arranged so that the strongest corner streams do not have to turn left.

In case of limited space availability, for example in the run-up to built-up areas, diamond crossing can also be used. The at-grade sub-intersections of the ramps can be designed as at-grade intersections with traffic lights or as roundabouts.

Table 10.9 contains the possible designs of partly grade-separated intersections with specific advantages and disadvantages.

Table 10.9 Management of traffic steams in the at-grade sub-intersections of a partly grade-separated intersection. (FGSV 2012a)

Diagonal, half cloverleaf with an exit in front of the bridge (asymmetrical)	
Advantages: Possibility to extend the left turning lane in the lower street Suitable for strong corner relations Relatively narrow bridge *Disadvantages:* Large intersection extension in the minor street	*Advantages:* Avoidance of left-turning lanes a small extent of the intersection in the minor street Minimum bridge width *Disadvantages:* Cumbersome traffic routing for passing secondary stream
Diagonal, half cloverleaf with an exit after the bridge (asymmetrical)	
Advantages: If there are no right-turning lanes, a small number of intersections in the minor road Suitable for strong corner relations *Disadvantages:* Unfavourable to driving dynamics Only limited possibility to extend the left turning lane in the minor street Very wide bridge	*Advantages:* Avoidance of left-turning lanes a small extent of the intersection in the minor street Minimum bridge width *Disadvantages:* Unfavourable to driving dynamics Congestion in extension ramp possible Bigger space requirement Cumbersome traffic routing for passing secondary steam
Symmetrical, half cloverleaf	
Advantages: Useful for one-sided area restriction along the minor road *Disadvantages:* No preferred corner relationship in either direction An exit ramp unfavourable in terms of driving dynamics Relatively wide bridge	*Advantages:* Avoidance of left-turning lanes Minimum bridge width *Disadvantages:* An exit ramp unfavourable in terms of driving dynamics Congestion in extension ramp possible Bigger space requirement Cumbersome traffic routing for passing secondary stream
Diamond	
Advantages: Favourable driving dynamics (fast, rising exit ramps) when the minor road is elevated Less space requirement than a half cloverleaf Small extent of the intersection in the minor road *Disadvantages:* Relatively wide bridge High speeds in the ramps Poor visibility at the sub-intersections prone to wrong turns	*Advantages:* Favourable driving dynamics (fast, rising exit ramps) when the minor road is elevated avoidance of left-turning lanes Minimum bridge width *Disadvantages:* Larger intersection extension in the minor street Spread parallel ramps required Bigger space requirement Cumbersome traffic routing for passing secondary stream prone to wrong turns

The major road is shown vertically.

In exceptional cases, partly grade-separated intersections can also be used if a heavily loaded EKL 2 road with long distances is connected to a road of EKL 2 or EKL 3. For safety reasons, at-grade EKL 2 road intersections must be signalled.

At the grade-separated sub-intersections in an EKL 1 road (RQ 15.5), it should be examined, analogous to the grade-separated intersections, whether the central reservation with marking should be replaced by a structural separation of directions.

10.3.4 Intersections for Roads of EKL 2

The basic principle for intersections on roads of the EKL 2 is the at-grade guidance with traffic lights. In order to disentangle the traffic streams and to avoid intersections, partly at-grade intersections can also be used on roads of level 2.

10.3.4.1 Partly at-grade Intersections

Partially at-grade intersections consist of a bridge structure for the grade-separated guidance of the crossing streams and a connecting ramp (loop ramp) for all turning relationships (Fig. 10.40). The intersections of the connecting ramp are usually equipped with a traffic light on streets of EKL 2. At intersections in EKL 3 roads, roundabouts can also be used or, in the case of very small turning relationships, intersections without traffic lights.

The position of the connection ramp should be selected so that no large left-turning streams occur. This applies in particular to intersections without traffic lights. Furthermore, it is advantageous if the left turning lanes are arranged on the outside to avoid a three-lane cross-section in the area of the bridge structure.

Partially at-grade intersections can also be used on roads of EKL 3 if the gradients of the intersecting roads show large differences in height due to the topography. Furthermore, intersections with a partially at-grade layout are favourable if a lot of agricultural traffic or pedestrian and bicycle traffic crosses.

10.3.4.2 At-grade Intersections with Traffic Lights

At T-junctions to the EKL 2, level T-junctions with traffic lights are the rule. The traffic light system should only be omitted in exceptional cases where there are unavoidable connections to roads of EKL 4 or to commercial access roads with very few turns.

At intersections of roads of EKL 2 with roads of EKL 2 or EKL 3, EKL 2 or EKL 3 roads can also be intersections with traffic lights (see Sect. 10.3.5.1). This avoids the costly bridge construction and the intersection usually fits better into the landscape. These are particularly advantageous when the traffic volumes in the crossing roads are low so that a good traffic quality can be achieved for the continuous main streams even without partly at-grade intersections.

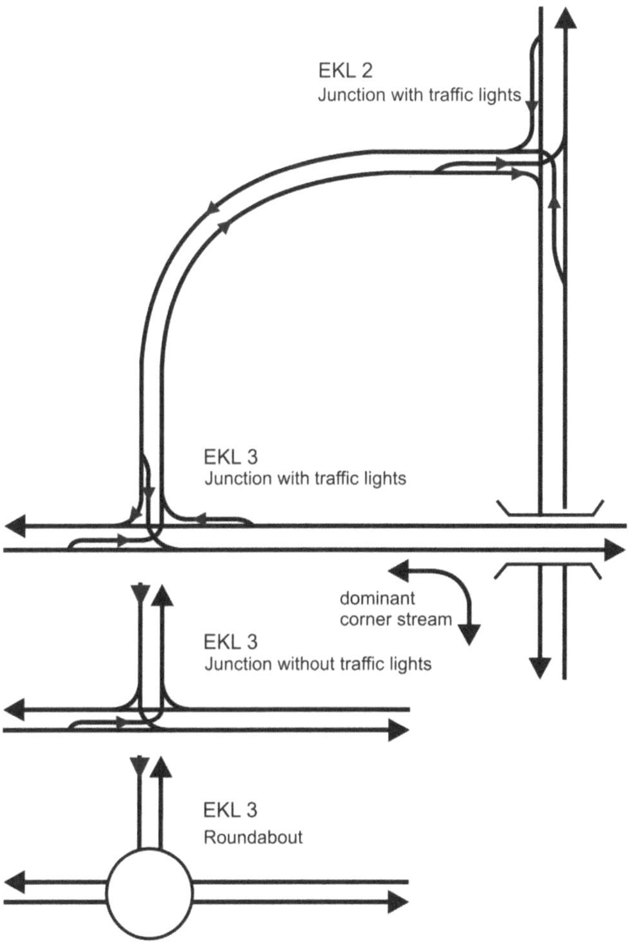

Fig. 10.40 Partly at-grade intersections (sketch)

10.3.5 Intersections for Roads of EKL 3

On roads of EKL 3, at-grade intersections are the rule. At-grade intersections with traffic lights and roundabouts are considered equivalent in the preselection. In this case, it must be examined in detail, taking into account the target fields of traffic safety, traffic flow, environment and costs and in coordination with the most continuous route characteristics possible, which solution is the more advantageous.

10.3.5.1 At-grade Intersections with Traffic Signals

At-grade T-junctions and crossings with traffic lights are used for crossings of streets of EKL 3 or when a street of EKL 3 crosses a street of EKL 4. For safety reasons, traffic lights should only be dispensed with in exceptional cases in the case of low intersection traffic volume of less than approximately 5000 Kfz/24 h (see Sect. 10.1).

Traffic lights are usually traffic dependent controlled. The small additional expenditure for traffic-dependent control is clearly compensated by the saved waiting times. For safety reasons, left turners from the main direction should be protected by a separate phase. This also applies in principle to the secondary direction, although in practice this is often dispensed with in the secondary direction in the event of lower traffic loads, as there is usually no separate turning lane available. This results in at least a 3-phase and better a 4-phase control. When designing the signalling system, the RiLSA (FGSV 2010a) and the VwV-StVO (BMVBS 2012) must be observed. This requires, for example, that the maximum permissible speed in the intersection area is limited to 70 km/h. For safety reasons, the traffic lights should also be operated at night.

To increase capacity and improve traffic quality, the number of continuous lanes can be increased. This also allows heavy goods vehicles to overtake. The additional lane is to be retracted 250 m behind the intersection (intersection of the road axes) with a distortion length of 150 m (see Fig. 10.41).

In principle, T-junctions/intersections with traffic signals must be designed in such a way that they can also be operated safely without traffic signals. Crossing angles of 80 to 120 gon are favourable and avoidance of the situation on a crest or in tight curves. Due to the sometimes high turning speeds, connections of the minor intersection access roads with a kink (see Fig. 10.39) should be avoided.

10.3.5.2 Roundabouts

Roundabouts are, in addition to the at-grade intersection with traffic lights, the standard solution when a street of the EKL 3 crosses a street of the EKL 3 or EKL 4. Especially in the case of evenly distributed traffic volume and low to medium waiting times for all traffic flows, roundabouts can also be used when a street of EKL 2 crosses a street of EKL 2 or EKL 3. However, attention must be paid to the adjacent intersections. A dense sequence of intersections with traffic lights and roundabouts should be avoided.

Roundabouts can also be used as an at-grade sub-intersection at partly grade-separated or partly at-grade intersections. Here again, the main area of application is at T-junctions with a road of EKL 3.

On roads of EKL 3, the use of roundabouts must be considered especially in relation to the at-grade intersection with traffic lights. As there are often shorter waiting times at roundabouts, a lower accident cost rate is to be expected and the construction costs are not

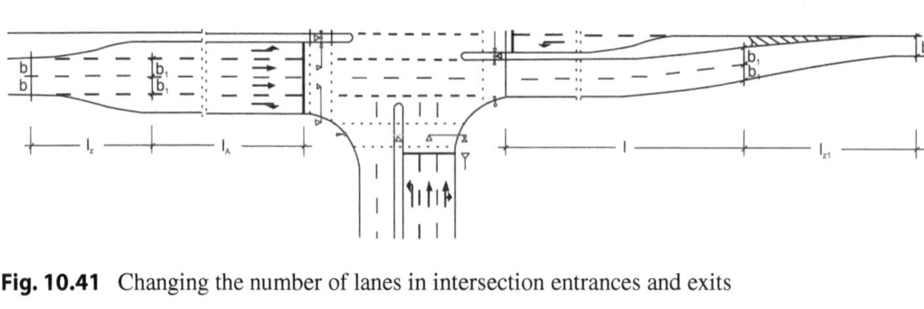

Fig. 10.41 Changing the number of lanes in intersection entrances and exits

significantly higher, the roundabout will often be the more economical solution. Roundabouts are particularly advantageous when the traffic volumes of the intersecting roads are similarly high. The RAL (FGSV 2012a) recommends using roundabouts only if the traffic volume in the less heavily loaded intersection access roads at three-armed roundabouts is at least 15% and at four-armed roundabouts at least 20% (sum of the traffic volume of both intersection access roads of the less heavily loaded road) of the total load of the incoming traffic.

The expected waiting times and thus the quality of the traffic flow must be verified in accordance with the HBS (FGSV 2015a).

If the traffic quality of a single-lane roundabout is not sufficient, an extension can be made in the following order according to the Leaflet for the Construction of Roundabouts (Merkblatt für die Anlage von Kreisverkehren) (FGSV 2008b):

- Structurally separate right turning lane (bypass),
- Circular road with two traffic lanes,
- Two-lane entrances.

A structurally separate right turning lane (bypass) is particularly advantageous in the case of heavy right-turning traffic. This stream is then conducted outside the roundabout and can be neglected in the design. If the installation of one or more bypasses does not bring the desired improvement in the quality of the traffic flow, the roundabout can be designed with two traffic lanes. This brings another capacity increase of about 15%. The last possibility to improve the traffic quality at roundabouts is the installation of two-lane entrances.

The RAL limits the possibilities for rural roads and recommends the installation of only one bypass if possible. Two-lane entrances are only recommended for the conversion and extension of existing roundabouts. This also makes sense because roundabouts are less flexible at high traffic volumes than intersections with traffic-dependent traffic lights, and such intersections with traffic signals can achieve a very large capacity and traffic quality through additional straight or turning lanes.

Another type of design for roundabouts are the turbo roundabouts. This form of roundabout developed in the Netherlands is characterised by partially two-lane entrances and exits, whereby overlapping of lanes is avoided by the inner extension of the lanes in the circular roadway and by appropriate signposting and marking. With a turbo roundabout, very high intersection loads of up to about 40.000 Veh./24 h can be handled. The guidance of pedestrians and cyclists is problematic, especially via the two-lane exits. For this reason, turbo roundabouts should only be used when pedestrian and bicycle traffic is not important or when they can be guided grade-separated or at a distance. Further information is contained in the FGSV regulations (FGSV 2015a, b).

10.3.6 Intersections for Roads of the EKL 4

At intersections of roads of EKL 4 with roads of EKL 4, at-grade T-junction / intersections without traffic lights are the standard solution. They can also be used in case of very low turning streams, taking into account the safety disadvantages, if a road of EKL 3 is connected with a road of EKL 3 or EKL 4. Exceptionally, an intersection without traffic lights is also possible for a connection of a road of EKL 3 or EKL 4 to a road of EKL 2, if this cannot be avoided and if the load is only very low.

At the at-grade sub-intersections of partly grade-separated or partly at-grade intersections, it is possible to use T-junctions without traffic lights, taking into account the safety disadvantages of intersections.

As T-junctions have clear advantages over intersections in terms of capacity and safety, it should be examined whether a right-hand offset can be formed instead of an intersection, particularly in the case of oblique intersections.

For a safe traffic flow at T-junction and intersections without traffic lights, visibility and comprehensibility are of particular importance. Intersections in tight curves are unfavourable. If the minor road is connected to the inner side of the curve, unfavourable visibility conditions arise for drivers who have to wait due to the large visibility circle. If the minor road is connected to the outside of the curve, the intersection is not easily visible to drivers in the secondary access road due to its crossfall to the inside of the curve, especially if it has an incline. It is also more difficult to estimate the speed on the major road.

In addition to the position in narrow curve areas, the position in crests and/or in areas of visual shadow must be avoided due to the unfavourable visibility conditions. Since the location of intersections is often fixed, the location of the intersection in areas with unfavourable visibility conditions is sometimes unavoidable. In such cases, special measures to ensure recognisability, for example through planting, signage or structural elements such as extended directional islands, are indicated. As a matter of principle, entrances of access roads are to be designed as single-lane carriageways.

Due to the high rate of basic accident costs, the large number of points of conflict and the speeds which are often difficult to estimate, the maximum permitted speed should in principle be limited to 70 km/h or possibly lower. This applies in particular to unfavourable visibility conditions and many crossings by cyclists and pedestrians. This recommendation is not in line with the StVO, which provides for restrictions of the permissible maximum speed only in the case of a special danger situation. Due to the significantly higher basic accident rate proven in the literature compared to intersections with traffic lights and roundabouts, the author considers this special danger situation to be given.

10.3.7 Design Elements for Grade-separated and Partly Grade-separated Intersections on Rural Roads

Analogous to motorway intersections, the grade-separated and partly grade-separated intersections on rural roads consist of the elements continuous lanes, entrances, exits and ramps.

10.3.7.1 Continuous Lanes

The continuous lanes of the major road are not to be changed in the intersection area. The carriageway must be warped to create turning lanes. This is done according to Sect. 14.5. Narrow curves in intersection areas must be avoided as a matter of principle. If this is not possible, widen the carriageway as described in Sect. 14.4.

At grade-separated and partially grade-separated intersections with entrances and exits, left-hand curves result in a ridge in the cross-section between these and the continuous carriageway, since usually the ramp and the superelevation in the transition curve cannot be completely accommodated in the ramp. Therefore, a ridge of maximum $\Delta q = 5.0\%$ is allowed in the area regularly crossed by motor vehicles. In the area of the roadway boundary marking (sign 295 StVO) in front of the tip of the ghost island, the ridge may also be larger. If this principle cannot be adhered to, taking into account the maximum ramp slope of 2.0%, the roadway boundary marking must be extended accordingly. The same principles apply to the avoidance of insufficient drainage zones in the torsional areas as for the road sections without intersections. For better integration of the superelevation, the superelevation section may be pulled forward into the merging/diverging lanes if the longitudinal inclination is sufficient so that at the beginning/end of the transition curve the crossfall $q = 0.0\%$ is already reached.

10.3.7.2 Ramps

Connecting ramps are designed analogous to motorways. They can be designed with smaller elements due to the lower driving speeds and the expectations of drivers on one-way rural roads (Table 10.10). A distinction is made between direct, semi-direct and indirect ramps according to the alignment in the site plan. In contrast to the motorways, all ramps will be designed usually with an adapted alignment. The table shows the ramp types with the smallest radii to be used. The reference line for the radius is the inner edge of the carriageway.

Connecting ramps should not contain any further T-junction or intersections. If this cannot be avoided (in practice, some examples are available in advance of built-up areas), the intersections must be announced early enough and conspicuously integrated into the signposting. In addition, a significant reduction in the maximum permitted speed is appropriate. The intersection entries to the at-grade sub-intersections are treated like roads of EKL 3.

The ramp cross-sections (see Table 10.11) can be single-lane or double-lane in oncoming traffic.

The design elements of the ramps can be reduced compared to the intersection-free sections because they should be driven at lower speeds and this is expected by the driver (Table 10.12). Nevertheless, the following aspects must be taken into account when designing the ramps:

Table 10.10 Ramp types with assigned smallest radii. (FGSV 2012a)

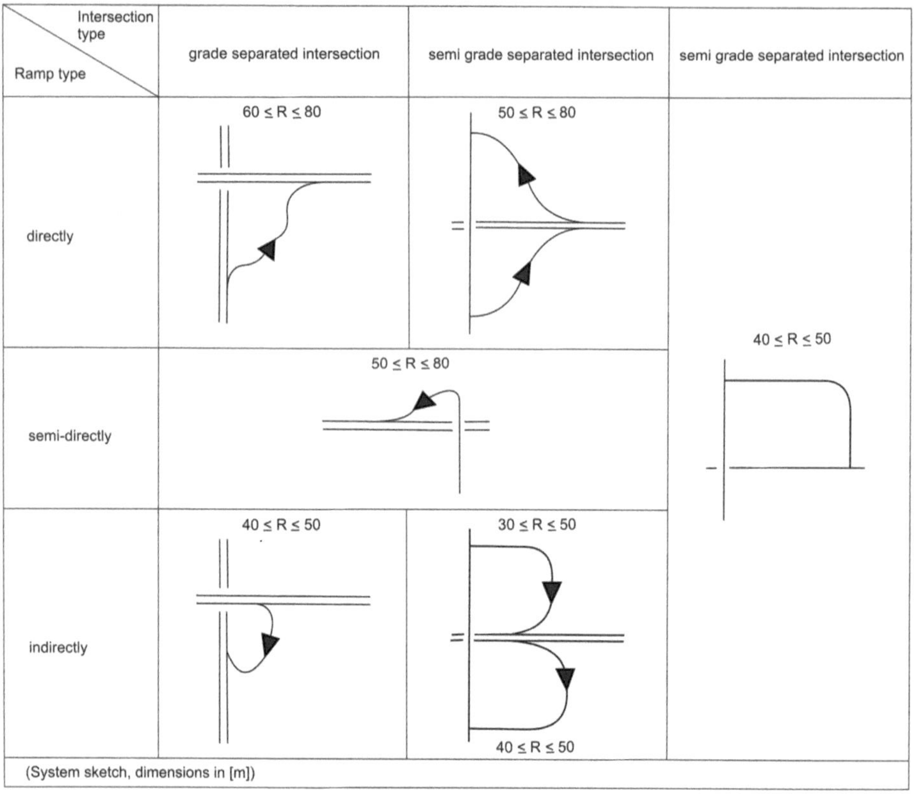

Table 10.11 Application areas of the ramp cross-sections (FGSV 2012a, b, c, d)

Ramp cross-section	Fields of application
	Grade-separated intersections and partly grade-separated intersections with short sections of parallel entry and exit ramps
Dimensions in [m]	Partly grade-separated intersections with jointly designed entry and exit ramps and partly at-grade intersections

- Adequate visibility (stop sight distance according to Table 10.12, reading distances to the signposting) is mandatory. Measures are often required to avoid obstructions to visibility caused by planting.
- The size of the radii must be detected in time and reliably.
- At-grade sub-intersections must be identified in time and timely entering of turning lanes must be ensured. For this reason, the ramp in front of the sub-intersection should be aligned as stretched as possible. A reading path of at least 50 m in front of the signposts is required.

Table 10.12 Limit values of ramp design elements on rural roads. (FGSV 2012a, b, c, d)

Single radius in each case	R	[m]	30	40	50	60	≥ 80
Minimum crest radius	min H_K	[m]	1000	1250	1500	1750	2000
Minimum Sag radius	min H_W	[m]	500	625	750	850	1000
Limits of the longitudinal inclination	max s [%] (gradient)		+6.0				
	min s [%] (slope)		−7.0				
Stop sight distance	S_H	[m]	30	35	40	45	55
Minimum crossfall outside of torsion areas	min q	[%]	2.5				
Maximum gradient	max q	[%]	6.0				
Minimum ramp gradient	min Δs	[%]	0.1 · a a [m]: Distance from the axis of rotation to edge of the road				
Maximum ramp gradient	max Δs	[%]	2.0				
Maximum slope	max p	[%]	10.0				
Lane widening in curves	i in RRQ 2	[m]	2.0	1.0	0.5	–	–

- At at-grade sub-intersections with traffic signals, ramps need waiting lanes. It must be possible to identify these in good time and they must be sufficiently long so that there is no congestion on the exit.
- Small clothoid parameters at the end of the diverging lane improve the recognizability of the following ramp radius because they are correspondingly short and the exit is optically more bent. However, they must be selected in such a way that the maximum ramp slope of 2.0% is not exceeded and that at least 0% cross fall is reached at the end of the transition curve.
- Successive radii should be based on the principle of relational alignment of curves. If possible, the following radius should not be smaller than the previous radius. As this is often unavoidable, it should be only slightly smaller. In the case of consecutive radii with intermediate straights of more than 100 m length, the following radius should be as much larger as possible than the preceding radius (a ratio of R_2:R_1 ≥ 1.5:1should be aimed for).
- If one or both intersecting roads have a steep longitudinal inclination, the ramps can become very long. Therefore, larger longitudinal inclinations can be accepted for the trailing longitudinal inclination of the connecting ramps, provided that the maximum inclination of 10.0% is not exceeded. The minimum inclination of 0.5% must also be observed in ramps.

Lane widening in curves is only necessary in narrow curves in the RRQ 2, as the RRQ 1 has sufficient width for all radii. However, it is then no longer possible for a truck to drive past a damaged truck or service vehicle. In this case, the verge can be used as well. For radii of R = 50 m, the banquet must be free of obstacles and stable over a width of at least 1.00 m and for radii of R = 30 m of 2.00 m. If safety barriers have to be positioned, they shall be moved away accordingly. This is also recommended in a recent study on-ramp types on motorways (Zimmermann et al. 2013). In the case of narrow ramp radii (R about 30 m) hardened verges and safety barriers set back by about 2.0 m (if necessary) should be used.

Exits on rural roads are always planned in the same way as exits on motorways. Exits are always to be designed with a parallel diverging lane. They are 3.50 m wide (including the broken lane marking) plus a 0.50 m wide hard shoulder. The length of the diverging lane l_A is 150 m on single-lane roads and 200 m on double-carriageway roads, measured from the tip of the ghost island in each case. The tip of the ghost island is at the point where the paved width of 5.25 m (lane width of the RRQ 1 with 4.50 m and the width of the right-hand hard strip with 0.75 m) is achieved by aligning the edge of the roadway. The length of the carriageway boundary marking in front of the tip of the ghost island is usually $0.1 \cdot l_A$ and thus 15 m for single-lane roads.

The island head has a dimension between the roadway edges of 2.00 m which is perpendicular to the major road. The offset distance l_Z at the beginning of the exit is 30 m long. Exits should not be arranged as lane subtraction at the end of an overtaking lane, because this provokes dangerous lane changes, especially of heavy traffic (Fig. 10.42).

10.3.7.3 Entrances

Analogous to the exits, entrances to rural roads are always formed with parallel merging lanes. They are (including the broken lane marking) 3.50 m wide plus a 0.50 m wide hard shoulder. The length of the merging lane l_A is 150 m for single-lane roads and 200 m for dual-carriageway roads, in each case starting at the point where the clothoid of the ramp merges into the following routing element of the continuous section. Thus, the beginning and end of entrances and exits are defined differently.

The entrance can be aligned via an axle on the right-hand edge of the ramp. The clothoid between the ramp radius and the horizontal alignment element of the continuous line is buffered between the edge of the ramp and an auxiliary element at a distance of 3.50 m to the right of the right edge of the mainline lane. The distortion of the width of the lane and the right hard strip starting from the dimensions of the RRQ 1 is then only done by marking from the island head. Analogous to the exits, the island head has a dimension between the carriageway edges of 2.00 m, which is perpendicular to the major road.

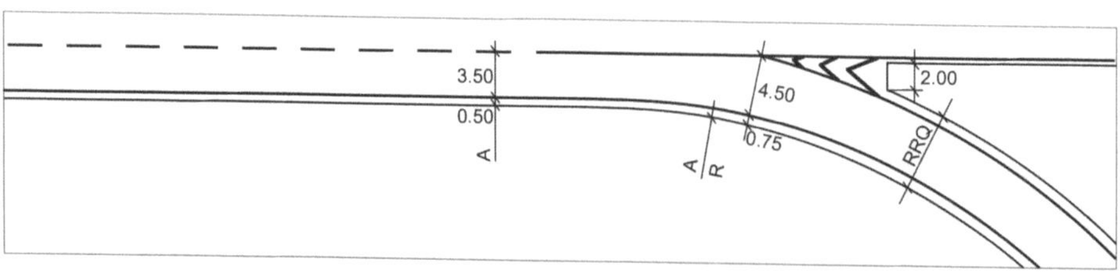

Fig. 10.42 Formation of the island head at an exit

An entrance can also be combined as a lane addition with the start of an overtaking lane. The overtaking lane must then be longer than the standard dimensions specified in Sect. 9.3.1 due to the necessary lane changes and the increased overtaking of heavy goods vehicles.

10.3.8 Design Elements for At-grade Intersections and T-junctions on Rural Roads

For safety reasons, T-junctions and intersections on rural roads must always be equipped with left-turning lanes and directional islands in the minor intersection entrance. In order to improve the recognizability, the design elements in a design class should always be the same.

10.3.8.1 Design Elements for Left-hand Turnings

Left turning lanes separate turning and passing traffic at an early stage. They are an important element for traffic safety and the quality of the traffic flow in the intersection and should therefore be used as far as possible. Only at lightly frequented intersections of farm roads or access roads to properties on roads of the EKL 4 a left turning lane is dispensable. On roads of EKL 2 and EKL 3, left-turning lanes with appropriate marking and signage should not be omitted, because serious accidents of overtaking vehicles with left-turning vehicles are often recorded, which could be avoided by a better recognition of the situation. Otherwise, a ban on overtaking in the area of such intersections should be considered. Basically, four types of left turns (Fig. 10.43) are distinguished according to the EKL from which the turn is made and the type of operation (with or without traffic lights) (Table 10.13).

The left turn type LA1 is the standard type for all intersections with traffic lights. The signal layout plan with the definition of the stopping lines depending on any crossings for cyclists and pedestrians and the locations of signal heads must be drawn up in accordance with RiLSA (FGSV 2010b). The location of the signal heads set up on the left side of the road (i. d. R. Repeat signal for those turning left) can be determined on the directional island in the minor intersection entrance or on the opposite triangular island. Then the stop line can be shifted further into the narrower intersection area and the extent of the intersection can be reduced. Depending on the number of phases and the phase sequence, a simultaneous entry or exit of vehicles must be considered.

The left turn type LA2 is the standard type for intersections without traffic lights on roads of EKL 3 and EKL 2 if the traffic lights are exceptionally not installed there.

The left turn type LA3 is the standard type for intersections without traffic lights on roads of EKL 4, it can also be used if roads of EKL 3 are connected to lightly trafficked roads of EKL 4, roads of LS V, main service roads or factory access roads and the maximum 99% congestion according to HBS (FGSV 2015a) is less than one vehicle. The lane boundaries marking next to the left turning lane or the opposite ghost island begin and end

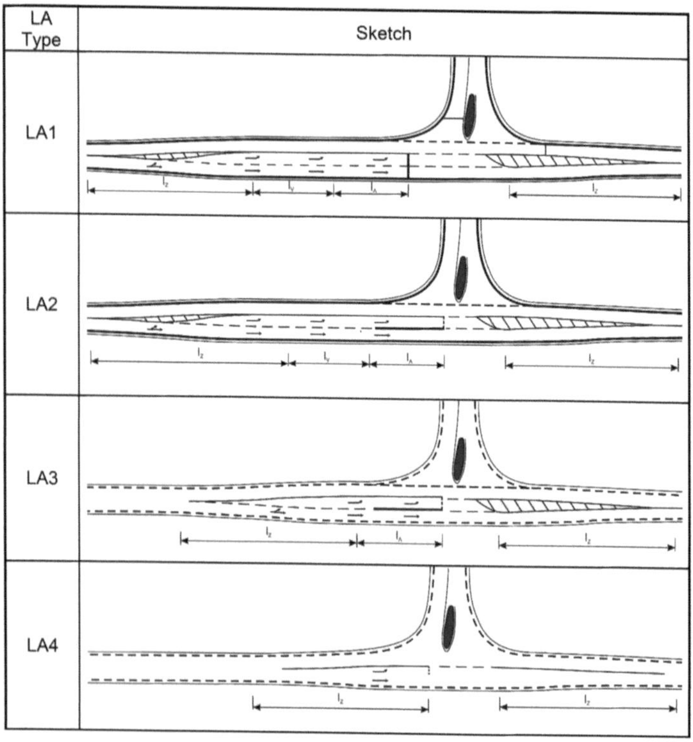

Fig. 10.43 Left turn types

Table 10.13 Design elements of left-turn types (FGSV 2012a)

Design element		LA 1	LA 2	LA 3	LA 4
Queuing space [m] l_A [m]		≥ 20 m[a]	≥ 20 m[a]	≥ 10 m[a]	10
Deceleration distance l_V [m]	EKL 2	40	(20)[c]	–	–
	EKL 3	20	20d	–	–
Distortion section l_Z [m]	Single-side	70	70	70	70
	On both sides	50	50	50	50
Distortion back section [m][b]	Single-side	40	40 (30)[e]	No ghost island	No ghost island
	On both sides	30	30 (20)[e]	No ghost island	No ghost island
Width [m]		3.25	3.25	2.75	4.75[f]

[a]Or longer, if required by the HBS (FGSV 2015a)
[b]Beginning of the retraction of the ghost island
[c]Usually no element of EKL 2
[d]Dispensable with low congestion, the 99% congestion according to the HBS is decisive
[e]Clamp values in exceptional cases to increase the storage space
[f]Incl. adjacent lane

where the lane width (including the 0.50 m wide hard shoulder) is 6.50 m due to the structural widening.

The left turn type LA4 can be used if roads of EKL 4 are connected to roads of EKL 4 with low traffic load, roads of LS V, main service roads or factory access roads and maximum 99% congestion according to HBS (FGSV 2015a) is less than one vehicle. This congestion occurs when the saturation level is 0.1 or less. With a DTV of about 2000 Kfz 24 h in cross-section, the maximum number of vehicles turning left is about 100 at peak hours, so almost all oncoming traffic would have to turn left. Accordingly, this solution is always possible on roads of the EKL 4, but for safety reasons should only be used in exceptional cases, for example during reconstruction and upgrading.

The carriageway is widened in the area of the waiting area to 8.50 m and consists of a lane with a width of 4.75 m (including the waiting area), a lane for the opposite direction with a width of 2.75 m and hard shoulders separated by a broken line marking, each with a width of 0.50 m. The lane boundaries are next to the installation area or the opposite broken line marking begin and end analogous to the left turn type LA3 at the point where the lane width (including the hard shoulder each 0.50 m wide) is 6.50 m due to the structural widening.

On roads of EKL 4, lightly loaded service roads or access roads to properties are regularly connected without structural widening. In exceptional cases, which must be justified, this can also be done for roads of EKL 3 if such accesses cannot be avoided. However, it should be noted that overtaking accidents with turning vehicles are frequently observed on roads of EKL 3. Overtaking should therefore be prohibited in far enough before such intersections.

For all types of left-turning lanes, the vehicle path of the turning vehicles (usually road trains as the decisive design vehicle) must be taken into account when determining the stop lines or waiting lines. The vehicle path of vehicles turning left at the same time must not overlap. If there is a low number of heavy goods vehicles turning (< 2 vehicles/h), the adjacent lanes may also be used to prove rideability with the aid of vehicle path and driving mode 2 may be used as a basis. This is likely to occur more frequently at intersections of roads of EKL 4.

10.3.8.2 Design Elements for Right-hand Turns

On rural roads with fast traffic, those turning right should leave the main lane as quickly as possible in order to avoid hindering and endangering the following straight ahead drivers. Depending on the design class and intersection type, six types of right turns are distinguished (Fig. 10.44).

The right turn types RA1 and RA2 are used at intersections with traffic lights. They are to be combined with the access types KE1 or KE2. The right turn type RA1 is used at roads of EKL 2 when roads of EKL 2 or EKL 3 are crossed or connected. The right turn type RA2 usually occurs on roads of EKL 3, if roads of EKL 3 or EKL 4 are connected with a signalized intersection. RA2 can also be used at low-loaded intersections of EKL 2 if a triangle island is not necessary or not reasonable.

RA Type	Sketch	KE Type
RA1	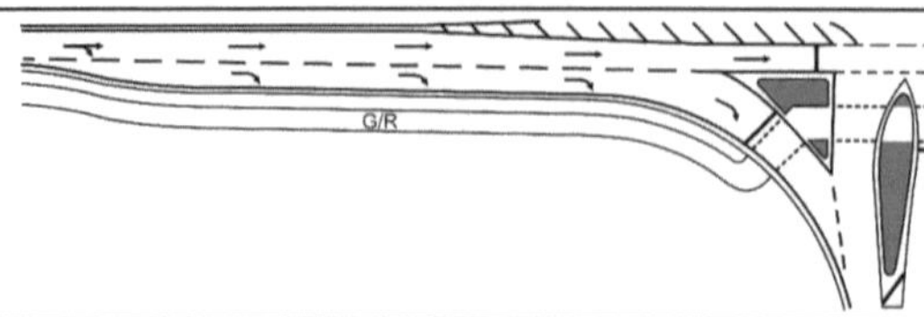	KE1/2
RA2	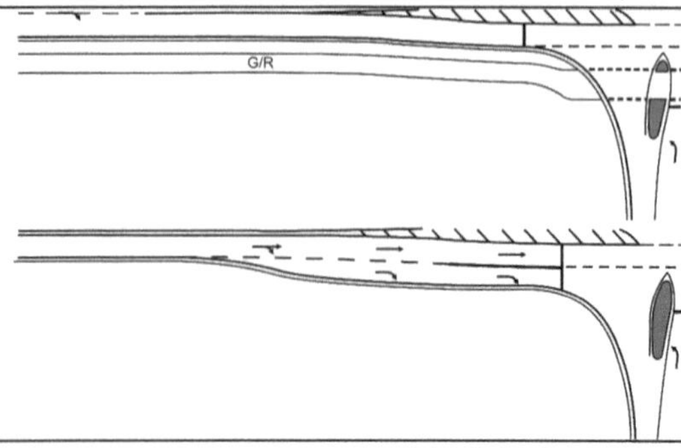	KE2/2
RA3	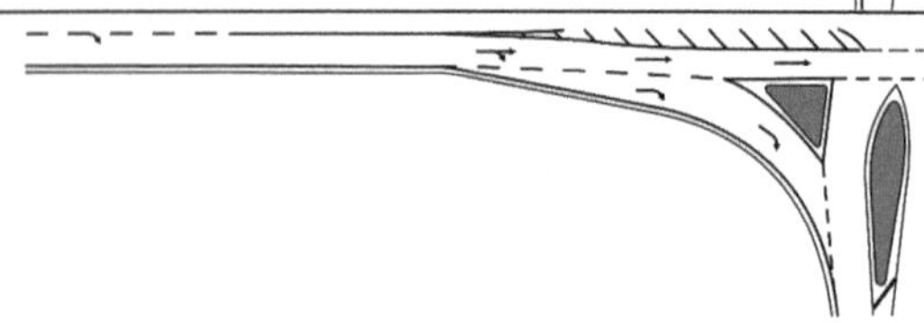	KE3
RA4	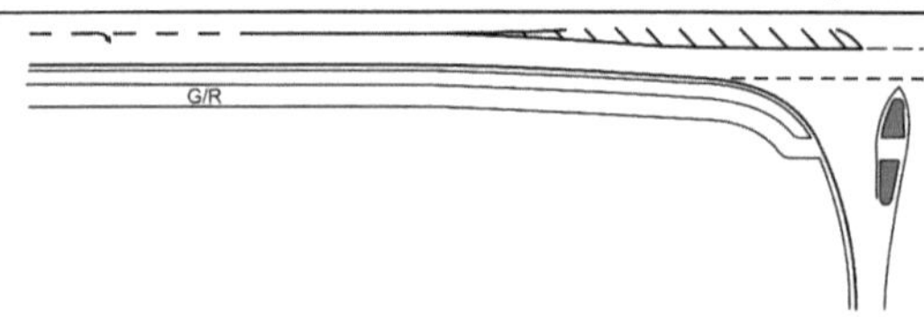	KE4
RA5[1]	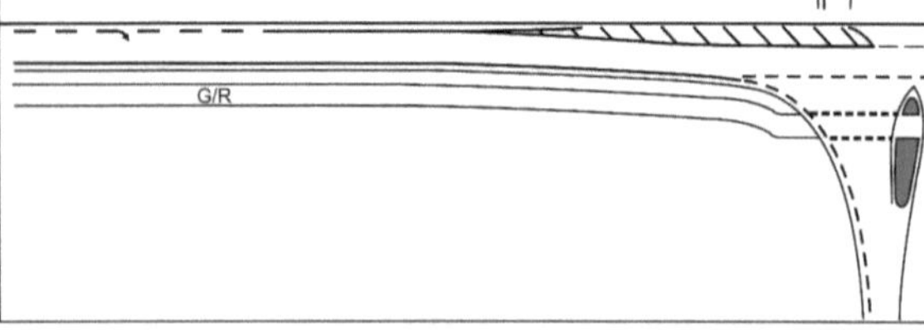	KE5[1]
RA6[1]	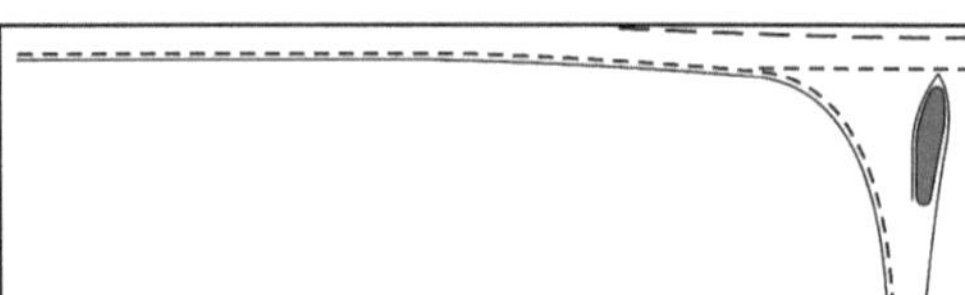	KE6[1]

[1] If a left turn lane is not required according to Section 10.3.8.1, the small directional island can be omitted at junctions, if the recognizability is ensured by signage and, if useful, planting.

Fig. 10.44 Right turn types with associated access types for crossing and turning (KE according to FGSV 2012a)

The right turn type RA1 is characterized by a right-turning strip, a triangular island and a large directional island. The right turning lane consists of a queuing section l_A, which is to be dimensioned according to the HBS (99% congestion, at least 20 m), a deceleration section l_V which is 40 m (EKL 3: 20 m) long and a distortion section l_Z which is 30 m long.

The width of the right turning lane, including the broken lane marking, is 3.25 m plus a 0.50 m wide hard shoulder next to the right turning lane. The width of the carriageway between the triangular island and the corner rounding is at least 5.50 m. This does not yet guarantee the rideability of heavy goods vehicles, especially at scissors intersections. This must be verified with vehicle paths. The rounding of the corner is done with a simple curve, the radius of which is derived from the construction of the triangular island according to Sect. 10.3.8.5.

Right turners should be included in the signal control. This also applies if, according to the procedures of the HBS (FGSV 2015a), the installation of a right-turning lane would not be necessary or there are no footpaths or cycle paths. If there is a footpath and cycle path parallel to the major road, this should be guided at a distance of up to 4.00 m from the edge of the major road via the triangle island and the large directional island.

The right turn type RA2 consists of a corner rounding with a three-part curve sequence (see Sect. 10.3.8.6) and a small directional island (see Sect. 10.3.8.4). The main arc radii are formed according to Sect. 10.3.8.6. If there are footpaths and cycle paths parallel to the major road, they are guided over the small directional island at a distance of up to 4.00 m from the edge of the major road, even if a right turning is arranged. If a right turning lane is required for reasons of capacity and traffic quality according to the HBS (FGSV 2015a), its dimensions should be selected analogous to the right turn type RA1.

The right turn types RA3 to RA6 are used at intersections without traffic lights. They are to be combined with the access types KE3 to KE6, whereby the direct counterpart is to be selected in each case.

The right-hand turn type RA3 is used on roads of EKL 3, when roads of EKL 3 cross or merge and the right-turners should be guided quickly due to the high traffic load of the major road. It is combined with access type KE3. When using RA3, the right-turning stream of crossing bicycle and pedestrian traffic must be excluded.

The right turn type RA3 is marked by a triangle island and a large directional island and a corner rounding with a R = 25 m. The width of the roadway between the triangle island and the corner rounding is at least 5.50 m. The rideability with heavy goods vehicles has to be proved with vehicle path.

There are currently different regulations in Germany regarding the obligation to wait for a right turn behind the triangular island (right-to-left or obligation to wait by waiting line and sign 205 StVO). Only one regulation should be used for standardisation. The RAL (FGSV 2012a) recommends the regulation that right-turners should grant the right of way and this is indicated by sign 205 StVO (give way!) and sign 340 StVO (broken lane marking). The results of a KIT research project (Zimmermann et al. 2014) suggest that the right-turn type RA3 should be dispensed with altogether, as it does not offer any advantages over RA4 at lower traffic loads, but is much less safe. For higher traffic loads, RA3 is out of the question, since these intersections are usually signalled.

The RA4 right turn type is used on roads of EKL 3 when roads of EKL 3 cross or merge. It is combined with access type KE4. Right turn type RA4 is characterized by a small directional island and corner rounding with a three-part curve sequence according to paragraph 10.3.8.6.

At intersections, footpaths and cycle paths are only possible parallel to the major road. These are led at a distance of at least 6.00 m from the edge of the major road, without right of way (without crossing markings) over the small directional island.

The right turn type RA5 is used on roads of EKL 3 when roads of EKL 4, roads of LS V, main service roads, factory access roads or in exceptional cases lightly loaded roads of EKL 3 cross or merge. It is combined with access type KE5. The right turn type RA5 is characterized by a small directional island and a corner rounding with a three-part circular curve sequence according to paragraph 10.3.8.6.

The main difference to the RA4 right turn type is the guidance of cyclists and pedestrians. Cyclists and pedestrians are guided in a distance of up to 4.00 m from the edge of the roadway of the major road with priority on a crossing over the small directional island that is coloured as red as possible.

The problem with the preferential guidance of cyclists and pedestrians is that cyclists travelling against the direction of travel may not be recognised in good time by right turners and cyclists turning right or their right of way may not be taken into account. For this reason, priority should only be given to cyclists if the intersection is very clear (compact and possibly lying in a sag curve) and, due to very low traffic volume, visual obstructions by other vehicles are rare.

The RA6 right turn type is used on EKL 4 roads when EKL 4 roads, LS V roads, main service roads or factory access roads cross or join. It is combined with access type KE6. Right turn type RA6 is characterized by a small directional island and corner rounding with a three-part circular curve sequence according to paragraph 10.3.8.6.

10.3.8.3 Design Elements for Crossing and Turning Vehicles

Analogous to the right-hand turn types, six access types for crossing and turning (KE) are distinguished depending on the design class and intersection type (Fig. 10.45). For all KE types, directional islands are to be constructed according to Sect. 10.3.8.4 to clarify the waiting obligation. There are two opposing requirements for the size of the carriageway area next to the directional islands (queueing area for crossing and turning vehicles). On the one hand, they must be large enough to allow the design vehicles to turn in and out without significant joint use of adjacent lanes. On the other hand, the roadway area should be as small as possible in order to make the waiting obligation clear, to prevent parallel positioning of several vehicles as far as possible and to improve the view to the left by positioning them as radially as possible. Corner rounding is carried out with a three-part circular curve sequence in accordance with Sect. 10.3.8.6. The carriageway width between the directional island and the corner rounding is at least 4.50 m. It consists of a hard strip

Fig. 10.45 Access types for crossing and turning KE with associated right turn types (According to FGSV 2012a)

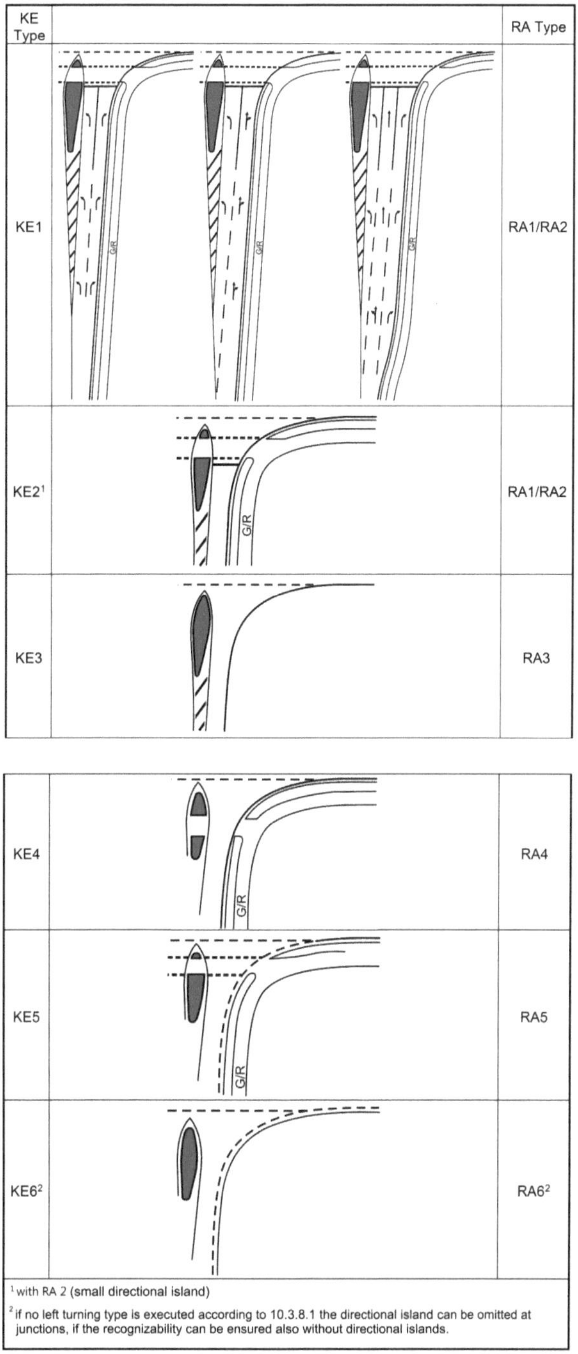

at least 0.25 m wide at the directional island, a lane at least 3.75 m wide and a hard strip 0.50 m wide at the edge of the carriageway. These regulations apply to all KE types since safe trafficability must be ensured even at intersections with traffic lights in the event of failure of the traffic lights. In principle, the lanes for intersecting traffic should be as straight as possible so that no major changes of direction are necessary for the vehicles crossing the intersection.

The crossing and turning in types KE1 and KE2 are used at intersections with traffic lights. They are to be combined with the right turn types RA1 or RA2. At access roads with traffic lights, separate lanes for entering and crossing traffic flows can increase the capacity and traffic quality of the intersection. At the same time, these serve as storage space for vehicles requiring maintenance and reduce the absolute congestion length.

The KE1 access type is used on EKL 2 or EKL 3 roads when EKL 2 or EKL 3 roads are crossed or connected and a high capacity is required. The type KE1 is marked at intersections by separate lanes for the left and right turn. At intersections, a lane for the crossing traffic is added. With only one additional lane, usually, the crossing and the right turning traffic flow are combined.

Lanes for left or right turners are 3.25 m wide. Added to this are the hard strips at the directional island and at the edge of the road. The lengths of the associated installation lanes are determined according to the procedures of the HBS (FGSV 2015a). The required roadway widening for the additional lanes usually develops from the axis of the minor intersection access road to the right. If additional lanes are arranged in both minor intersection accesses, the trafficability for the left turners and for the crossing streams must be checked particularly carefully. To improve the trafficability, the axis of the minor road could be slightly offset. As a lane divider, a large directional island is used on right turn types with triangular island (RA1) and a small directional island on those without triangular island (RA2).

The KE2 access type is used on roads of EKL 2 or EKL 3 if roads of EKL 2 or EKL 3 are crossed or connected and a single-lane queueing area is sufficient to ensure traffic quality. As with access type KE1, a large directional island is used as a lane divider on right turn types with triangular island (RA1) and a small directional island on right turn types without a triangular island (RA2). The arrangement of the crossing for cyclists and pedestrians above the directional island is analogous to the right turn types RA1 or RA2.

The access type KE3 is only possible if no bicycle and pedestrian traffic has to cross the minor road. The lane divider is designed as a large directional island. The associated right turn type is RA3. Only with access type KE3, due to the lack of parallel cycle and pedestrian traffic, it could be considered to install a merging lane with a length l_E of at least 150 m for a right turn with a high proportion of heavy traffic turning into a gradient. For this purpose, the right turners must be guided behind a right-hand triangular island. The advantage of the better turning possibility, especially for heavy traffic, as well as the increase in capacity and traffic quality, must be weighed against the constructional expenditure, the enlargement of the intersection area restricting the clarity and recognizability and the additional conflict potential at the triangle island.

The access type KE4 with the right turn type RA4 is the rule at roads of EKL 3 when a road of EKL 3 crosses or enters. The lane divider is designed as a small directional island. The arrangement of the crossing point for cyclists and pedestrians above the directional island is analogous to the right turn type RA4.

The access type KE5 with the right turn type RA5 is the rule on roads of EKL 3 when a road of EKL 4 crosses or joins. The lane divider is designed as a small directional island. The arrangement of the crossing point for cyclists and pedestrians above the directional island is analogous to the right turn type RA5. Since an EKL 4 road is constructed without a centre marking, a lane boundary line (sign 295 StVO) must start before the directional island at the point in the widening area where the road is 6.50 m wide and be guided around the directional island in accordance with Sect. 10.3.8.4.

The access type KE6 is, together with the right-hand turn type RA6, the standard case at roads of EKL 4, when a road of EKL 4 crosses or enters. The lane divider is designed as a small directional island. Similar to the access type KE5, a lane boundary line (sign 295 StVO) is arranged in front of the directional island at the point in the widening area where the road is 6.50 m wide and is guided around the directional island according to Sect. 10.3.8.4.

Since heavy goods vehicles are relatively rare on roads of the EKL 4, it is acceptable that the traffic areas of oncoming traffic are used for a short time when turning right. Therefore, the corner roundings according to Sect. 10.3.8.6 can be chosen somewhat smaller.

Intersections with LS V roads, main service roads and factory access roads are treated as EKL 4 roads.

10.3.8.4 Lane Dividers: Directional Islands

Lane dividers in the minor intersection accesses improve the recognizability and comprehensibility of the intersections and especially their right-of-way rules. For this reason, lane dividers should be provided at all intersections. Exceptions are only lightly loaded intersections of LS V roads, main service roads and works access roads where, in accordance with Sect. 10.3.8.1, left turns are guided without structural alterations and where recognisability and comprehensibility is ensured by signs and/or planting. Lane dividers are designed with sloping kerbs. Lane dividers at intersections and intersections are designed as large and small directional islands depending on the presence of a triangular island (Fig. 10.46).

The large directional island is used for the right-hand turn types RA1 and RA3 (with triangular island). Otherwise the small directional island is used. When constructing the directional islands, please note that it must be possible to turn left at the same time. The movement ranges of the design vehicles must not overlap. The driving lines are to be checked with trailing curves and, if necessary, the road dividers must be moved further away from the edge of the carriageway of the major road.

If at intersections with traffic signals, it should be possible to turn left at the same time, which should be the rule due to the four-phase control to be aimed for, the scope of

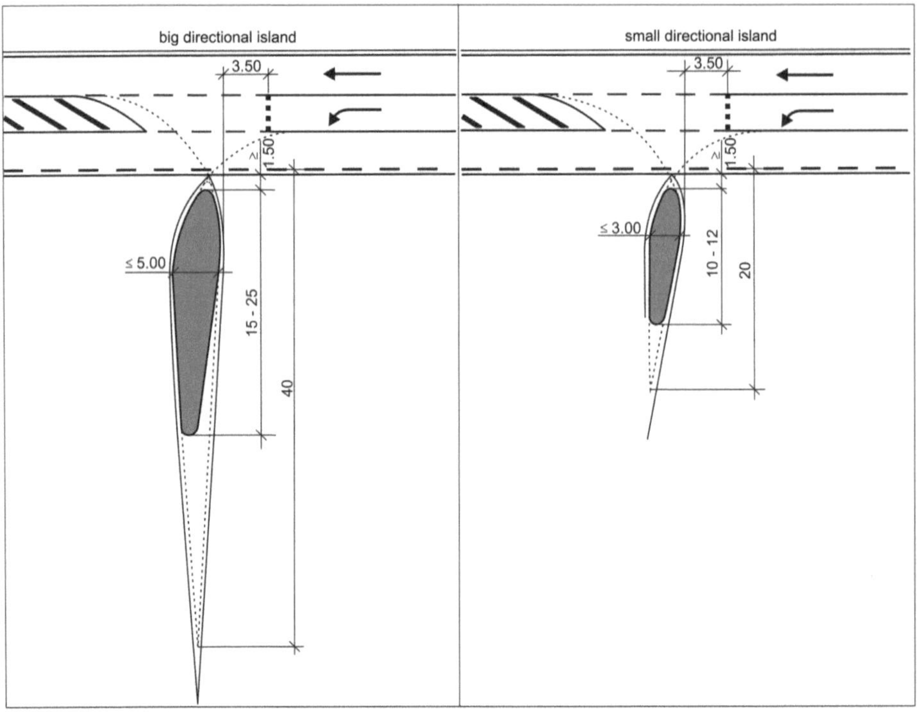

Fig. 10.46 Roadway dividers at intersections and crossings (FGSV 2012a, b, c, d)

movement of the design vehicles must also not overlap. This must also be checked with trailing curves. Here, the directional islands can also be moved slightly away from each other or the directional island offset could be adjusted. The directional islands should be introduced in the minor intersection access by a lane boundary line (sign 295 StVO).

The timely recognition of the roadway divider in the minor intersection access road is particularly important for traffic safety. Therefore, it is necessary to check, for example by means of perspective images, whether the recognisability of the roadway divider and the comprehensibility of the traffic flow are guaranteed. There could be problems with visibility in right-hand curves or on a hilltop. In this case, it must be checked according to the design instructions for large and small directional islands whether the line of sight runs over the road divider. If the test result is negative, the lane divider should be extended.

Construction

The baseline for the construction of directional islands (Figs. 10.47 and 10.48) is usually the axis of the minor intersection access. At crossings and enters of type KE1, in case of additional turning lanes for a left turn in a vehicle in one of the two intersection accesses, instead of the axis, an axis-parallel line must be used as a baseline, which allows a right offset of the two-directional islands of approximately 3 m. Here the crossing angles $\alpha = 80$

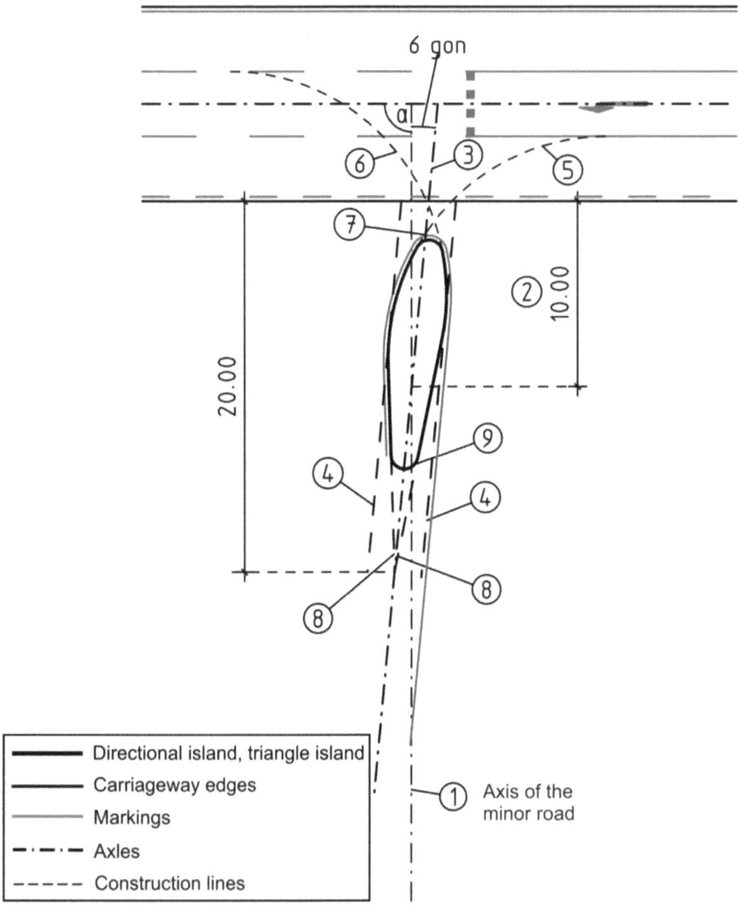

Fig. 10.47 Design instruction of a small directional island at a crossing angle α = 80 to 120 gon (FGSV 2012a, b, c, d)

to 120 gon are shown as an example. The design for smaller or larger crossing angles can be taken from RAL (FGSV 2012a, b, c, d).

Design instructions for a small directional island according to (FGSV 2012a, b, c, d):

1. Definition of the axis of the minor intersection access.
2. Determination of a point on the axis of the minor intersection access road at a distance of 10 m from the edge of the road of the major road.
3. Drawing of a directional island axis rotated by 6 gon to the right against the axis of the minor intersection access through the point determined according to point 2.
4. Construction of two auxiliary lines at a distance of 1.50 m right and left parallel to the directional island axis.
5. Construction of a circular curve for the left turn with R = 12 m. For the left-turning types LA1, LA2 and LA3, this curve touches the left edge of the left-turning lane and

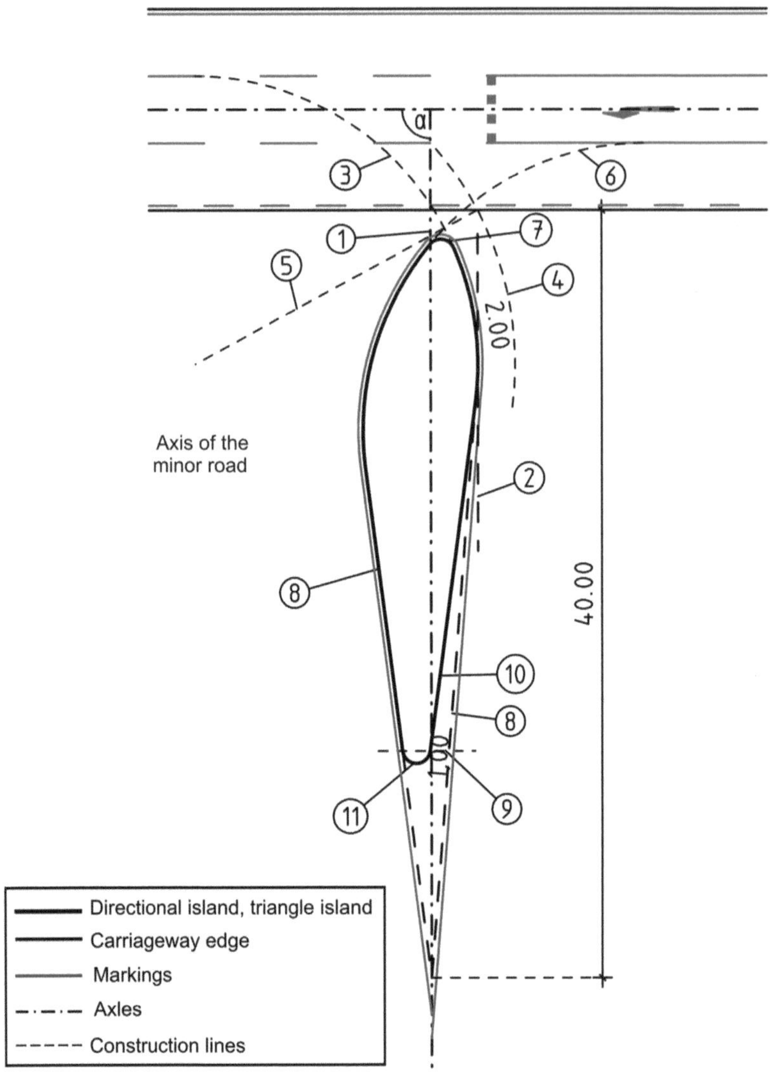

Fig. 10.48 Design instruction of a large directional island at a crossing angle α = 80 to 120 gon (FGSV 2012a, b, c, d)

the left of the auxiliary lines mentioned in point 4. With the left turn type LA4, the curve begins at the axis of the major road. At intersections with access type KE1, it may be necessary to increase the radius up to 15 m to ensure simultaneous left turning.

6. Construction of a circular curve for the left turn with R = 12 m. For the left turn type LA4 the radius can be reduced to up to 10 m. For left turn types LA1, LA2 and LA3, this curve touches the right of the auxiliary lines mentioned under point 4 and the left edge of the lane of the major road into which the turn is to be made. With left turn type

LA4 the arc ends at the axis of the major road. At a crossing angle of $\alpha \leq 100$ gon, the radius may have to be reduced to 8 m to achieve the intended directional island shape.

7. Rounding of the front directional island head between the two curves for the left turn off and the left turn in with R = 0.75 m
8. Fitting of two straight lines so that on the one hand they are tangent to the curve for the left-turning off or the left-turning in the vehicle and on the other hand they jointly intersect the directional island axis at a distance of 20 m from the edge of the higher-level road.
9. Rounding of the rear directional island head with R = 0.75 m.

Design instructions for a large directional island according to (FGSV 2012a, b, c, d):

1. Definition of the axis of the minor intersection access. Determination of the intersection point between the axis of the minor intersection access and the edge of the road of the major road. For intersections with access type KE1, ensure that the axes of the minor intersection accesses are offset to the right according to Table 10.14.
2. Construction of a parallel to the right of the axis of the minor intersection access. The distance between them is 2.50 m for a right-angled intersection ($\alpha = 100$ gon), it decreases by 0.05 m for each gon of absolute deviation of the intersection angle compared to 100 gon. If simultaneous left-hand turns are to be possible at intersections with access type KE1, it may be necessary to reduce the distance of the parallel to about 2 m even at an intersection angle $\alpha = 100$ gon. It must be demonstrated by means of vehicle path that simultaneous left-hand turns are possible (see point 6).
3. Construction of a circular curve for the left turn in. This circular curve touches on the one hand the parallel to the axis of the minor intersection access road constructed under point 2 and on the other hand the left edge of that lane of the major road into which the turn is to be made. At intersections, the radius of the curve for the left-hand turn 2.50 m is larger than the width of the lane of the major road, at crossroads it is to be selected according to Table 10.14 to ensure the simultaneous left-hand turn.
4. Drawing of a curve with a radius 2.00 m larger around the same centre point as the curve drawn for the left-hand turn after point 3.
5. Connection of the centre of the circle with the point at which the curve drawn after point 4, enlarged by 2.00 m, intersects the edge of the higher level road. Determination of the point at which this connecting line intersects the curve drawn after point 3 for the left-hand turn.
6. Construction of a circular curve for the left-turning vehicle, so that it passes through the intersection determined according to point 5 and at the same time touches the left edge of the lane of the major road from which the turn is made. This circular curve and the circular curve drawn according to point 3 form a part of the front directional island head. The radius of this curve for the left-turning driver is usually as large as the radius for the left-turning driver at intersections, but in any case it must be selected so that the directional island width is between 3 and 5 m. At crossroads, the radius for the left-turning vehicle is determined according to Table 10.14. If a check with trailing curves shows that it is not possible to turn left at the same time with the selected radii, a larger

Table 10.14 Deflection radii, bend radii and directional island offsets at crossings for large directional islands

Angle of crossing [gon]	Deflection radius [m]	Turning radius [m]	Right offset at Traffic lights [m]
80	10	18	1.00
100	15	15	3.00
120	20	11	3.00

distance between the curve for the left-turning vehicle drawn according to point 3 and the parallel radius drawn according to point 4 (e.g. on 3 to 4 m) can be used to ensure that the distance between the two curves is sufficiently large for the left-turning vehicle.

Points 4 to 6 are called 2 m construction.

7. Rounding of the front directional island head with a radius R ≥ 0.75 m, so that the front directional island head is at least 2 m and at most 4 m away from the edge of the carriageway of the major road
8. Drawing of two straight lines, so that on the one hand they are tangent to the curves for the left-hand turn or the left-hand turn and on the other hand intersect the axis of the minor intersection access road at a distance of 40 m from the edge of the road of the major road.
9. Fitting a dimension of 2.50 m perpendicular to the axis of the minor intersection access between these two straight lines.
10. Removal of a dimension of 1.00 m from the straight line on the right of the axis to the left. Draw a new straight line from this point so that it touches the curve drawn after point 3 for the left turn.
11. Rounding off the remaining 1.50 m wide rear directional island head with R = 0.75 m.

10.3.8.5 Triangular Islands

Triangular islands are used for quick right turns from the preferred road and hold an area at the intersection that would otherwise not be used during the turning operations. Furthermore, the triangular island limits the area where right turners meet left turners and provides the possibility of clearly marking the right-of-way control. In addition, it is possible to construct traffic lights. They are used with the right-turning types RA1 and RA3, but only with RA3 if there is no footpath and cycle path parallel to the higher-level road. It should also be noted that, based on current research results (Zimmermann et al. 2014), type RA3 should no longer be used, as RA4 is more advantageous for low traffic volumes and intersections should be signalled at higher traffic volumes.

With the right turn type RA1, a bicycle or pedestrian crossing is led over the triangular island. In this case, the island edges that remain next to the crossings should be at least 1.50 m (including the lowering kerb) long. Since out of town cyclists usually dominate and persons with restricted mobility do not or only very rarely occur alone, the kerbs should be lowered to 0 cm. Otherwise, the instructions for the design of crossing points for cyclists and pedestrians of the H BVA (FGSV 2011b) must be observed.

Triangular islands can also be a safety risk as fixed installations. For this reason, recognition should be ensured, for example, by the introductory markings, traffic signs (e.g. signposts (but: observe the fields of vision!)) and light signal transmitters or reflectors. In addition, triangular islands are bordered with light-coloured sloping kerbs.

Construction

Two types are possible when constructing triangular islands. The construction with default edge lengths (Fig. 10.49) is usually used for bicycle and pedestrian belts (right turn type RA1). Without crossings (right turn type RA3) the construction is usually executed without specifying the edge lengths.

Construction of triangle islands with default edge length (Fig. 10.50):

1. Construction of an auxiliary circle around the centre of the curve for the left-turning from the associated directional island construction (M_{LA}) with a radius that is

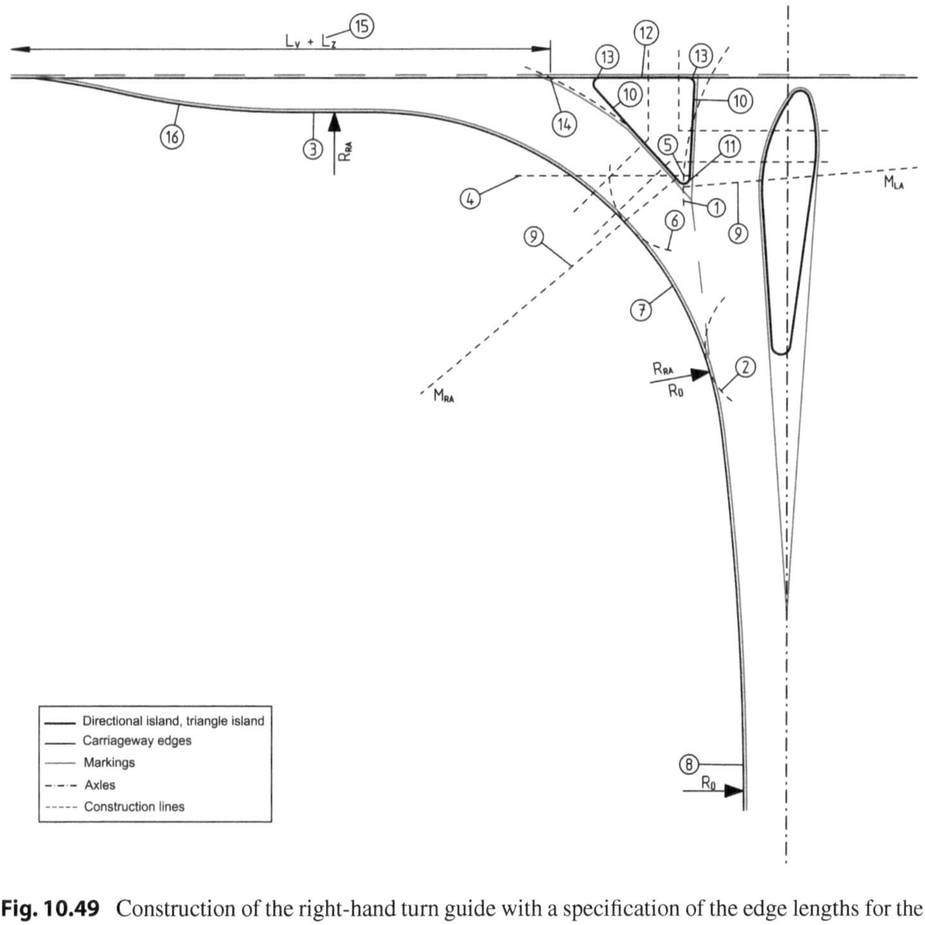

Fig. 10.49 Construction of the right-hand turn guide with a specification of the edge lengths for the triangular island (FGSV 2012a, b, c, d)

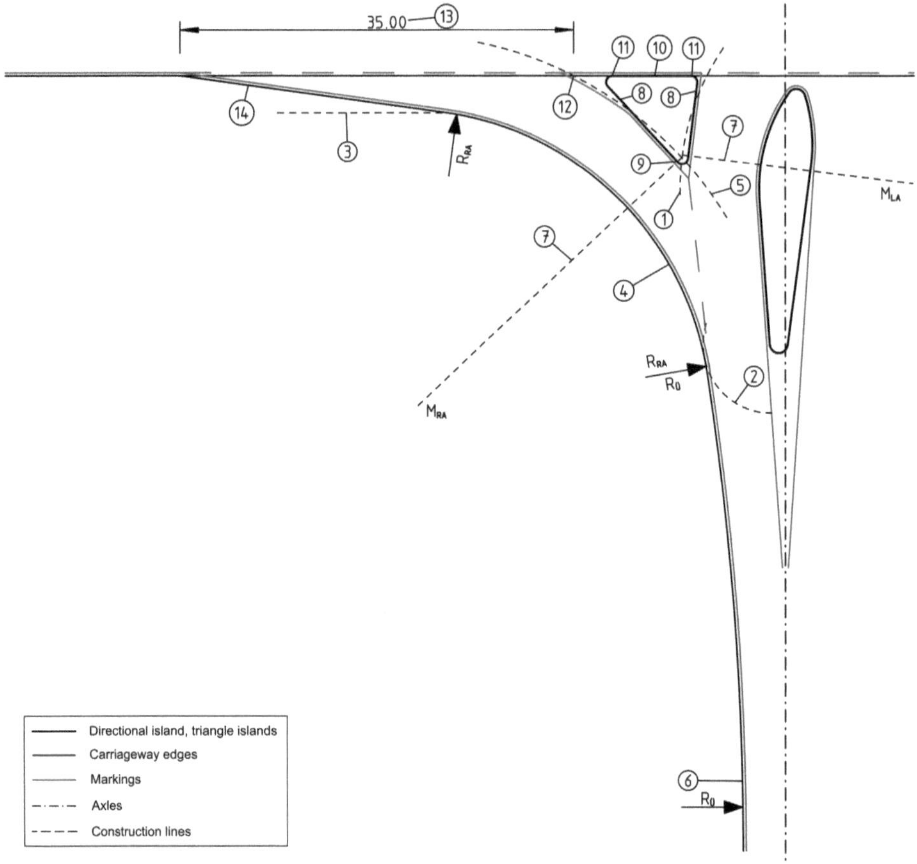

Fig. 10.50 Construction of the right-hand turn guide without specifying the edge lengths for the triangular island (FGSV 2012a, b, c, d)

6.00 m + 0.50 m larger than the corresponding radius of the curve for the left-turning (R_{LA}).

2. Construction of an auxiliary circle around the endpoint of the straight directional island edge, which connects to the arc with R_{LA} (beginning of the rounding radius of the rear directional island head) with a radius R = 5.50 m.
3. Drawing of a parallel to the edge of the major road at a distance of 3.25 m.
4. Drawing a parallel to the edge of the major road at a distance approximately equal to the desired length of the edge of the triangular island opposite the directional island. This edge length of the triangle island is generally to be selected in such a way that crossings across the minor intersection access road are led across both the triangle island and the directional island at a constant distance from the edge of the major road, if possible.

5. Determination of the intersection of this parallel with the auxiliary circle constructed according to point 1 as the centre of the rounding radius (R = 0.50 m) of the rear directional island head.
6. Construction of an auxiliary circle around the center point found according to point 5 with the radius R = 5.50 m + 0.50 m.
7. Fitting or inclusion of a circular curve with a radius to be determined iteratively, which touches the parallel according to point 3 as well as the auxiliary circles according to point 2 and point 6. The determined radius can be rounded if necessary. This requires a new adjustment or calculation.
8. Fitting of a suitable circular arc $R_{ü}$ between the circular arc according to point 7 and the edge of the roadway of the minor intersection access road.
9. Drawing of connecting straight lines between the centre of the rounding of the rear island head of the triangle island and the centres M_{LA} and M_{RA}.
10. Determine the lateral edges of the triangular island as perpendiculars to the two connecting straight lines according to point 9. In the case of intersections, ensure that the side of the triangle facing the directional island is adapted to the intersecting driving relationship. If the boundary conditions result in edge lengths of the triangle island of more than approximately 8 to 10 m, it may be appropriate to design the side of the triangle island facing the right-hand turn as a parallel to the edge of the roadway of the right-hand turn with a distance of 5.50 m.
11. Rounding of the rear island head of the triangle island between the two perpendiculars after point 10 with R = 0.50 m.
12. Determination of the edge of the triangle island at the edge of the roadway of the major road.
13. Rounding of the front island heads of the triangle island created between the perpendiculars after point 10 and the straight line after point 12 with R = 0.50 m
14. Determination of the intersection of the auxiliary circle drawn according to point 5 and the edge of the carriageway of the parent road.
15. Removal of the exit opening from the point determined according to point 14 with the length of the exit opening ($l_v + l_z$) for the right-hand turn lane.
16. Connection of the intersection determined according to point 15 at the edge of the carriageway with the parallel line according to point 3 by introducing distortion at the beginning of the exit opening for the right-hand turn lane.

Construction of triangular islands without specification of the edge lengths:

1. Construction of an auxiliary circle around the centre of the curve for the left-turning from the associated directional island construction (M_{LA}) with a radius that is 6.00 m + 0.50 m larger than the corresponding radius of the curve for the left-turning (R_{LA}).
2. Construction of an auxiliary circle around the endpoint of the straight directional island edge, which is connected to the circular arc with R_{LA} (beginning of the rounding radius of the rear directional island head) with a radius R = 5.50 m.
3. Drawing of a parallel to the edge of the major road at a distance of 3.25 m.

4. Fitting of a curve for the right-hand turn (R_{RA}) tangential to the auxiliary circle according to point 2 and the parallel according to point 3. The curve radius R_{RA} is usually 20 m for crossing angles $\alpha < 80$ gon and 25 m for crossing angles $\alpha \geq 100$.
5. Construction of an auxiliary circle around the centre of the curve for the right-hand turn (M_{RA}) constructed in accordance with point 4 with a radius 5.50 m + 0.50 m greater than the curve for the right-hand turn (R_{RA}) The intersection with the auxiliary circle after point 1 is the centre point for the rounding off (R = 0.50 m) of the rear island head of the triangle island.
6. Adjustment or inclusion of the curve $R_ü$ to the curve according to point 4 and to the edge of the roadway of the minor intersection access.
7. Drawing of connecting straight lines between the centre of the rounding of the rear island head of the triangle island and the centres M_{LA} and M_{RA}.
8. Determine the lateral edges of the triangular island as perpendiculars to the two connecting straight lines according to point 7. In the case of intersections, ensure that the side of the triangle facing the directional island is adapted to the intersecting driving relationship. If the boundary conditions result in edge lengths of the triangle island of more than approximately 8 to 10 m, it may be appropriate to design the side of the triangle island facing the right-hand turn as a parallel to the edge of the roadway of the right-hand turn with a distance of 5.50 m.
9. Rounding of the rear island head of the triangle island between the two perpendiculars after point 8 with R = 0.50 m.
10. Determination of the edge of the triangle island at the edge of the roadway of the major road.
11. Rounding off the island heads of the triangle island between the perpendiculars according to point 8 and the straight line according to point 10 with R = 0.50 m.
12. Determination of the intersection of the auxiliary circle drawn according to point 5 and the edge of the carriageway of the parent road.
13. Remove the exit opening from the point determined in accordance with point 12 using 35 m for the exit wedge or the length of the exit opening ($l_v + l_z$) for a right-hand bend.
14. Connection of the point of intersection determined according to point 13 at the edge of the roadway with the curve RA by a straight line in the case of an exit wedge or with the parallel according to point 3 by introducing distortion at the beginning of the exit opening in the case of a right-hand turn

10.3.8.6 Corner Rounding

The corner rounding should, as far as possible, adapt to the vehicle path and, in particular, enable rapid turning. A three-part circular curve sequence best meets these requirements. The three-part circular curve sequence is characterized by a leading curve R_1, the main curve R_2, which stands for the designation of the curve sequence, and a trailing curve R_3. The sequence of curves must be selected so that vehicles turning off and on usually do not have to use the opposite lane. However, the left turn-off lane may be slightly overpainted.

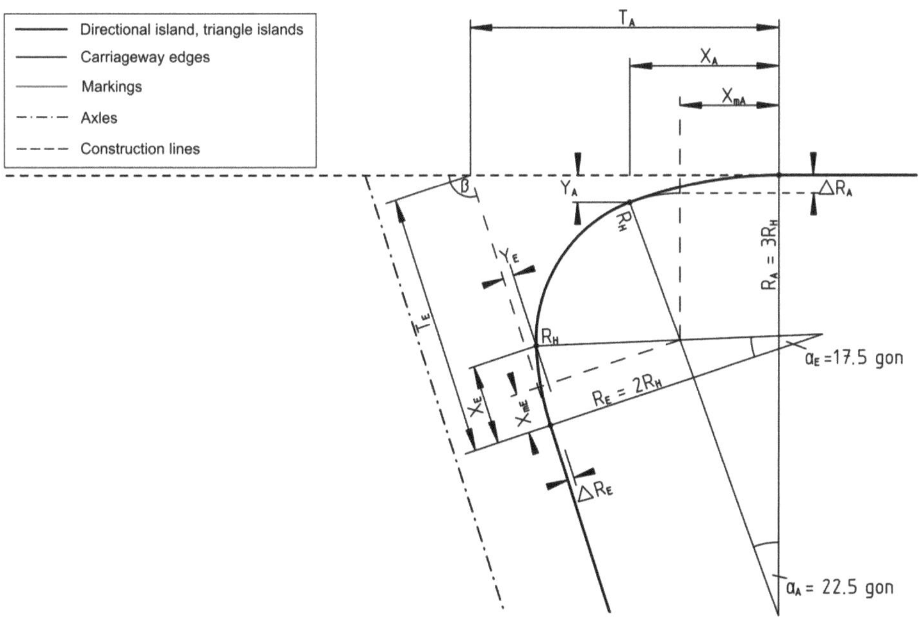

Fig. 10.51 Corner rounding with three-part curve sequence (example: right turn)

This allows the selection of smaller radii and avoids corresponding impairments of the approach view due to the radial positioning of the curves (Fig. 10.51).

$$" R_1 = R_2 \cdot 0.0375 Y_1 = R_2 \cdot 0.0750 X_{m1} = R_2 \cdot 0.2714 X_1 = R_2 \cdot 0.5428$$

$$" R_2 = R_2 \cdot 0.1236 Y_2 = R_2 \cdot 0.1854 X_{m2} = R_2 \cdot 0.6922 X_2 = R_2 \cdot 1.0383$$

$$T_1 = R_2 \cdot \left(0.2714 + 1.0383 \cdot \tan \frac{\beta}{2} + \frac{0.0861}{\sin \beta} \right)$$

$$T_2 = R_2 \cdot \left(0.6922 + 1.1236 \cdot \tan \frac{\beta}{2} - \frac{0.0861}{\sin \beta} \right)$$

The three-part circular curve sequence is constructed with a radius ratio R_1:R_2:R_3 = 2:1:3.
with

R_1: Radius of forward curve [m]
R_2: Main curve radius [m]
R_3: Trailing curve radius [m]

The forward curve R_1 with $\alpha_1 = 17.5$ gon and the outward curve R_3 with $\alpha_3 = 22.5$ gon have a constant central angle independent of the total angle of change of direction. The main curve R_2 is buffered accordingly. With a crossing angle of about 100 gon, RAL (FGSV 2012a) recommends a main curve radius $R_2 = 15$ m for the right-hand turn types RA2, RA4 and RA5 (each with directional islands). For the access types KE1, KE2, KE3, KE4 and KE5 a $R_2 = 12$ m is recommended. For right-hand turn type RA6, a main curve radius

$R_2 = 12$ m is sufficient and for access type KE6 a $R_2 = 10$ m is recommended. In the case of oblique-angled intersections, the sizes must be adjusted if necessary.

The width of the carriageway next to the directional island should be at least 4.50 m. If the value is significantly exceeded, a smaller R_2 should be tested. The passability must then be proved with vehicle paths.

Design Notes for the Right-hand Turn-off

The design of the carriageway edge for right-hand turns is carried out by including a three-part circular curve sequence between the edge of the carriageway of the higher-level road and the right-hand edge of the lower-level road leading away from the intersection. Since this is often located in the widening area next to the directional island, an additional curve back may have to be fitted between the end of the three-part curve sequence and the edge of the roadway of the lower-level street (Fig. 10.52).

1. Construction of an auxiliary circle around the endpoint of the straight directional island edge (at the same time beginning of the rounding radius of the rear directional island head) with a radius R = 4.50 m.
2. Fitting or inclusion of a three-part circular curve sequence for corner rounding between the right edge of the roadway of the major intersection access road and the right edge of the roadway of the minor road leading away from the intersection with the main curve radius of $R_2 = 15$ m.
3. Check that the required minimum distance of 4.50 m between the start of the rounding radius of the rear directional island head and the right-hand edge of the corner rounding is maintained. If this distance is significantly greater than 4.50 m due to local conditions, it should be checked by means of a trailing curve verification whether safe driving is also possible with $R_2 < 15$ m.
4. If the distance according to point 3 falls below the minimum distance of 4.50 m, apply a tangent to the auxiliary circle according to point 1 parallel to the directional island edge. Insert or include an additional curve $R_{ü}$ between this tangent and the right edge of the minor road leading away from the intersection.
5. Fitting or inclusion of a three-part circular curve sequence for corner rounding according to Annex 4 between the right-hand edge of the roadway of the higher-level intersection access and the tangent according to point 4 with the main curve radius of $R_2 = 15$ m.
6. If necessary, iteration of points 4 and 5.

Design Notes for a Right-hand Turn in

At intersections with two separate lanes or at crossroads with only one additional lane, the right-hand edge of the road runs in the area of the directional island over a length of about 40 m, usually parallel to the lane boundary line next to the directional island. The edge of the roadway can then be formed in a straight guide without distortion. For longer required installation areas, the right edge of the roadway must be initiated by a distortion of 70 m. Analogous to the right turn-off, the roadway edge for right turner in is constructed by including a corner rounding in the form of a three-part circular curve sequence between

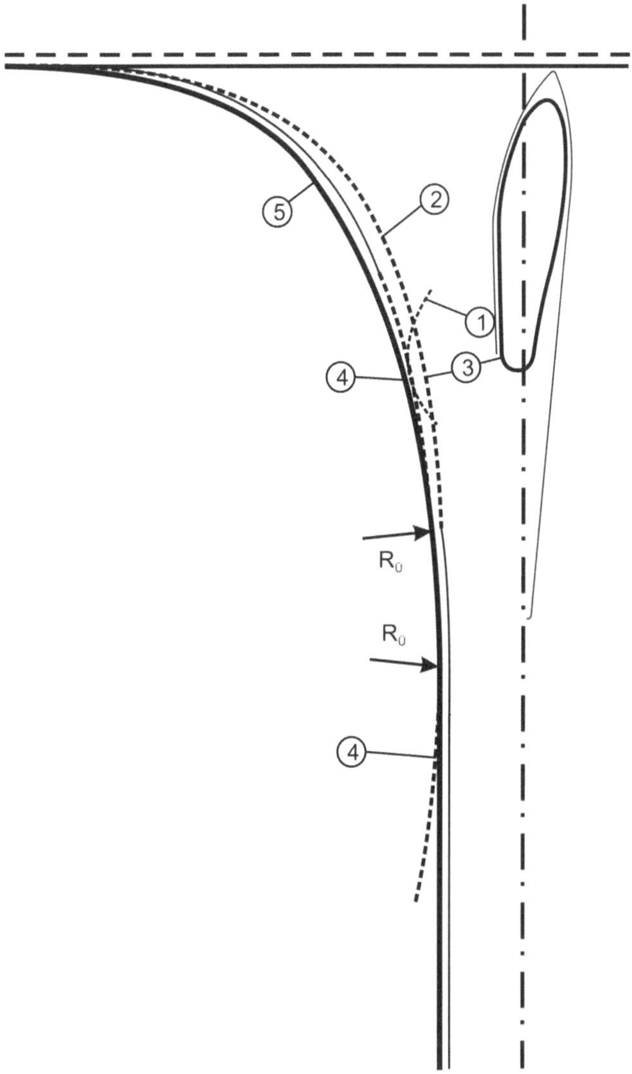

Fig. 10.52 Construction of a three-part circular curve sequence for a right-hand bend without triangular island according to RAL (FGSV 2012a)

the right edge of the minor road leading to the intersection and the right edge of the major road. When constructing the right edge of the roadway, the required number and length of lanes in the installation area must be taken into account (Fig. 10.53).

1. Construction of an auxiliary point on the axis of the minor intersection access. If a single-lane installation area is planned or if a two-lane installation area of more than 40 m length is required according to the HBS (FGSV 2015a), this auxiliary point is located at the intersection of the construction line of the right directional island edge

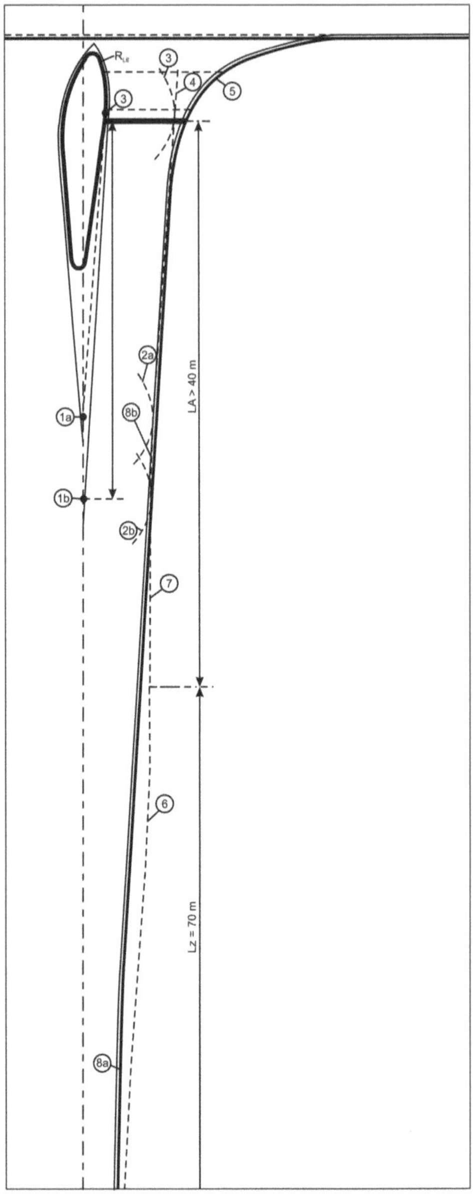

Fig. 10.53 Construction of a three-part circular curve sequence for a right-hand turn in a guide for a single-lane and a two-lane waiting area according to RAL (FGSV 2012a)

with the axis of the minor intersection access. If according to the HBS (FGSV 2015a) a two-lane installation area of up to 40 m length is required, the distance of the auxiliary point from the edge of the carriageway of the higher-level road corresponds to the length of the installation area required according to HBS plus the distance of the stop line from the edge of the carriageway of the higher-level road. However, this auxiliary point must not be closer to the edge of the carriageway of the major road than the

intersection of the construction line of the right directional island edge with the axis of the minor intersection approach.
2. Construction of an auxiliary circle around the auxiliary point determined according to point 1 with a radius R = 4.50 m in the case of a single-lane installation area or the required lane width at the rear directional island head in the case of a dual-lane installation area (sum of the lane widths plus the hard strip widths of 0.50 m at the edge of the lane and at least 0.25 m at the directional island).
3. Construction of an auxiliary circle around the starting point of the construction line of the right directional island edge, which connects to the circle curve with R_{LE}, with a radius R = 4.50 m in the case of a single-lane installation area or with the required lane width in the case of a two-lane installation area (sum of the lane widths plus the hard strip widths of 0.50 m at the edge of the lane and at least 0.25 m at the directional island).
4. Fitting or inclusion of a straight line between the auxiliary circuits according to point 2 and point 3.
5. Fitting or inclusion of a three-part circular curve sequence between the straight line after point 4 and the right-hand edge of the major road.
6. For a two-lane installation area of more than 40 m length, construct a warp with L_Z = 70 m between the right edge of the carriageway of the minor intersection access and a parallel to the axis of the minor intersection access offset by the required carriageway width in the installation area.
7. Extension of this parallel to the axis of the minor intersection access from the distortion in the direction of the intersection.
8. Fitting or inclusion of an additional circular curve $R_{ü}$:
   (a) In the case of a single-lane installation area or the case of a double-lane installation area of up to 40 m Length between the straight line according to point 4 and the right-hand edge of the roadway of the minor intersection access road
   (b) In the case of a two-lane installation area of more than 40 m length between the straight line according to point 4 and the right-hand edge of the roadway widened by distorting according to point 7.

At an intersection, a lane for right turns may be arranged next to the lane for intersecting traffic. In this case, the lane for the crossing traffic is planned analogous to the construction described above. Then, the right edge of the additional lane is added as a parallel to the right edge of the crossing lane. This is started with distortion with a length of 30 m.

10.3.9 Design Elements for Roundabouts on Rural Roads

The design elements refer to single-lane small roundabouts with single-lane entrances and exits. Information on two-lane roundabouts and two-lane access roads can be found in the Technical bulletin for the Design of Roundabouts (Merkblatt für die Anlage von

Kreisverkehren) (FGSV 2008b). In the Netherlands, so-called turbo roundabouts have also been built recently. Good experience has been gained there with these roundabouts. However, the comprehensibility and passability is based on elements that are currently not permitted in Germany under the road traffic regulations. It remains to be seen whether the road traffic regulations will be adapted accordingly. The notes on turbo roundabouts of the FGSV (FGSV 2015b) contain application areas and suggestions for road traffic regulations-compliant signage and marking.

10.3.9.1 Circular Road

The circular roadway is to be designed with a constant radius and constant width depending on the outside diameter as a single-lane roadway. The width is b = 7.50 m for outer diameter D between 35 and 40 m and b = 7.00 m for outer diameter D between 40 and 50 m. Ideally, the circular roadway is horizontal and has a crossfall of 2.5% towards the outside. This system can also be used with a slight longitudinal inclination of the entering intersection arms. If the longitudinal inclination is greater, it may be necessary to design the circular roadway in an inclined plane. In this case, the inclination must not be greater than 6.0%. The same principles apply to the connections of circular entrances and exits to the circular roadway as to the minor intersection accesses at level intersections according to 10.3.2.6.

10.3.9.2 Circle Island

In principle, roundabouts are a very safe form of intersection. Nevertheless, accidents involving crossing the roundabout island are occasionally reported. Therefore, the circle island should be designed in such a way that it can be clearly recognized, but no obstacles should be placed there either. A design of the circular island as a slightly rising hill with vegetation that cannot grow as an obstacle has proved to be positive since on the one hand the visibility is interrupted and on the other hand, vehicles that may be passing over the island are not excessively slowed down by the hill and the vegetation. Inclined kerbs are to be used to border the circular island. Due to the risk of collision, no walls, high kerbs or steeply rising rows of pavement should be arranged.

In recent years, the number of heavy and oversized transports on rural roads has increased, not least because of the ever-larger wind turbines. Such large vehicles are generally not able to drive over roundabouts in normal operation. To ensure passability, the roundabout island may have to be made passable. The required areas are to be verified with tractrix curves. When large vehicles turn off, however, only a small area of the roundabout island remains.

10.3.9.3 Circle Entries and Exits

Circular entrances should and circular exits must be designed as single-lane. To improve visibility and to avoid high entry speeds, the roundabouts should be approached as radially as possible to the traffic circle. For this purpose, the axes of the intersection approach roads must be aligned to the centre of the circle. At all intersection access points, the circular entrances and exits are to be separated by roadway dividers that are as straight as possible (R = 50 m).

Basically, the passability of the circle entrances and circle exits must be checked by means of vehicle paths. RAL recommends that the roadway widths next to the directional island (including the 0.50 m wide hard strips on both sides) should be 4.50 to 5.00 m at the narrowest point in the access road, and 4.75 to 5.50 m at the exit. If the hard strip at the directional island is only 0.25 m wide, the dimensions can be reduced by 0.25 m.

Construction of a Directional Island at Roundabouts

1. Determination of the axis of the roundabout access.
2. Drawing of the outer radius for the roundabout.
3. Construction of two auxiliary lines right and left parallel to the axis of the roundabout access road at a distance of 1.25 m.
4. Construction of an auxiliary line through the intersection of the outer radius of the roundabouts drawn in accordance with point 2 and the axis of the roundabout access perpendicular to this definition of the front head of the directional island. Drawing of parallels to this auxiliary line at a distance of 5.50 m, 8.00 m and 10.00 m (5.50 m-, 8.00 m- and 10.00 m-line); to determine the crossing point for the common footpath and cycle path and the construction of the rear head of the roadway divider. The dimensions are to be checked in each specific case and modified if necessary.
5. Determination of the points of intersection of the axis parallel drawn according to point 3 with the 10.00 m line.
6. Rounding of the rear head of the directional island between the intersection points determined according to point 5 with a radius R = 1.25 m.
7. Drawing of two further auxiliary parallels to the axle parallels constructed according to point 3 at a distance from 50 m.
8. Draw a curve of a circle with a radius of 50 m around the intersection points of the 8 m line with the auxiliary parallels (circle centres) drawn after point 7, so that these are tangent to the axis parallels drawn after point 3.
9. Rounding of the front corners of the directional island by two curves of a radius of R = 0.50 m between the curves drawn according to point 8 and the auxiliary line for the outer radius of the roundabout constructed according to point 4.
10. Construction of the outer edges of the directional island by the curves drawn in accordance with point 8 between the roundings created in accordance with points 6 and 9.
11. Determination of the crossing point for the common footpath and cycle path between the 5.50 m and the 8.00 m line
12. Drawing of one additional auxiliary parallel to the axle parallel drawn in accordance with point 3 at a distance of 4.50 to 5.00 m for the circle approach and at a distance of 4.75 to 5.50 m for the circle exit; thereby determining the edges of the carriageway
13. Construction of circular curves for entrances and exits with radii between 14.00 and 18.00 m, so that they are tangent to the outer radius of the roundabout drawn according to point 2 and the auxiliary parallels according to point 12.

14. Formation of the respective right-hand lane edges of the circular approach and exit through the parallels according to point 12 between the transition to the radii according to point 13 and the rear head of the directional island. From the rear head of the directional island, distorting of the carriageway edges to the dimension of the roadway width of the roundabout access road.

The roadway dividers are bordered by sloping kerbs. At crossing points for cyclists and pedestrians, the directional island must be at least 2.50 m wide (in the middle of the crossing point). Analogous to Sect. 10.3.8.4, it must be checked at circular access points in right-hand curves or on crest curves whether the visibility of the directional island and the comprehensibility of the traffic flow can be improved by extending the directional island.

The corner rounding of circular entrances is formed as a simple curve with an R from 14 to 16 m. Similarly, the circle exit is formed with an R from 16 to 18 m, whereby the curve may be enlarged by up to 30% without a crossing point for cyclists and/or pedestrians.

10.3.9.4 Bypass

A bypass is a right-hand turn lane, where the right turners are guided past a roundabout. A bypass is to be equipped with a diverging lane and a merging lane on rural roads. Diverging and merging lanes are 3.50 m wide (including the interrupted lane boundary marking). The hard shoulder is 0.50 m wide in each case, analogous to the following roadway. The distortion distance l_Z is 20 m for the diverging lane and 30 m for the merging lane strip. The diverging lane should be long enough so that the congestion in the circular entry does not block the diverging lane. The 95% congestion according to the HBS (FGSV 2015a) is decisive for this. The merging lane should be 60 to 80 m long including distorting.

The island remaining between the bypass and the roundabout should be bordered with kerbs and at least 2 m wide. If cyclists and pedestrians have to be led over the bypass, they have to be guided without right of way and the island has to be enlarged accordingly. The lane width of the bypass is 4.50 m plus 0.50 m wide hard strips on both sides. The bypass should be routed with a continuous circular curve without a counter-curve. The passability must be checked with vehicle paths (Fig. 10.54).

10.3.10 Central Islands and Crossing Points

In principle, crossing points for cyclists and pedestrians should be located at intersections. However, occasionally pedestrian and cycle paths that are independent of roads also cross the rural roads. These crossing points represent obstacles in the road space and should therefore be recognisable in good time. The necessary fields of vision for motor vehicle traffic and for crossing cycle and pedestrian traffic must always be kept clear and structural and/or technical safety measures should be planned depending on the volume of motor vehicle traffic, the volume of heavy goods traffic and the volume and composition of cycle and pedestrian traffic. Pedestrians and cyclists must always be taken the right of way (sign

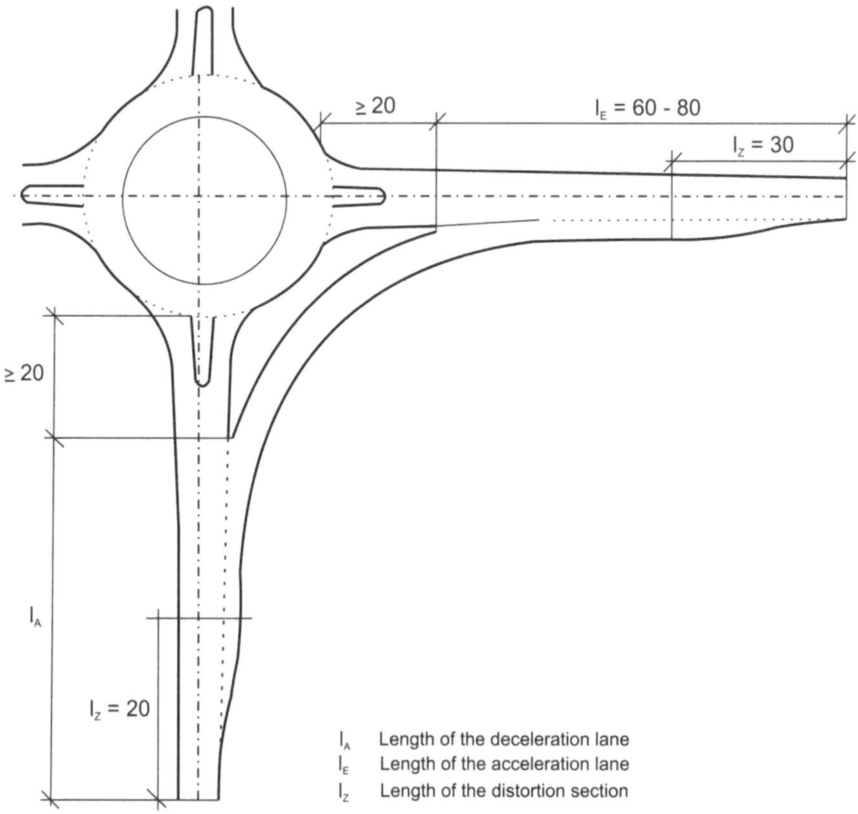

Fig. 10.54 Bypass at the roundabout

205 StVO (give way)) and accordingly, the crossing points must not be marked as crossings and must not be coloured (Figs. 10.55 and 10.56).

Central islands should always be used for protection when a particularly large number of pedestrians and cyclists cross and when there are a particularly large number of people in need of protection (e.g. pupils). Due to the speeds in motor vehicle traffic, all-day visibility is particularly important for motor vehicle traffic. Fixed lighting could improve visibility, but the masts must not be placed on the central island. A central island should be 2.50 m wide and bordered by 0.50 m wide hard strips. The lanes should be at a constant width next to the island, resulting in a widening of 3.50 m with distortion as described in Sect. 14.5 (Fig. 10.54).

Central islands have a particular risk potential, especially for overtaking operations. For this reason, a continuous lane marking (sign 295 StVO) must be placed in good time before the island. Furthermore, sign 138 StVO could be used to indicate the danger area. It also makes sense to limit the maximum permitted speed to i.e. 70 km/h, in order to

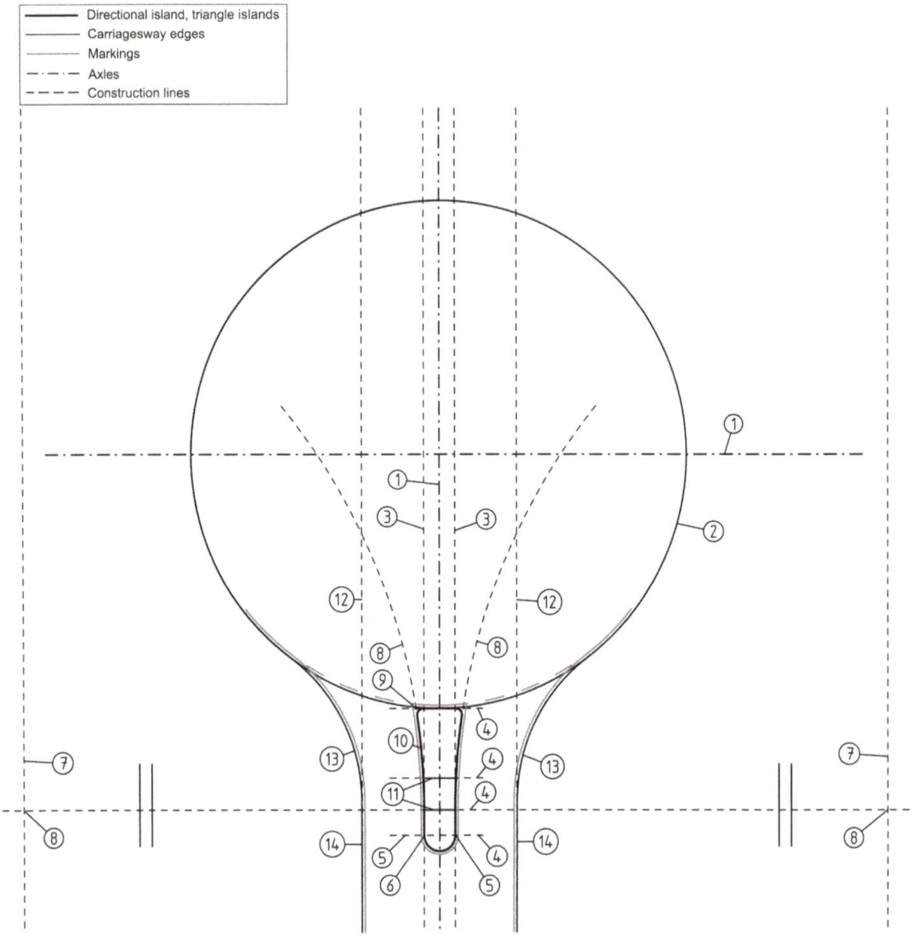

Fig. 10.55 Construction of a directional island at roundabouts according to RAL (FGSV 2012a)

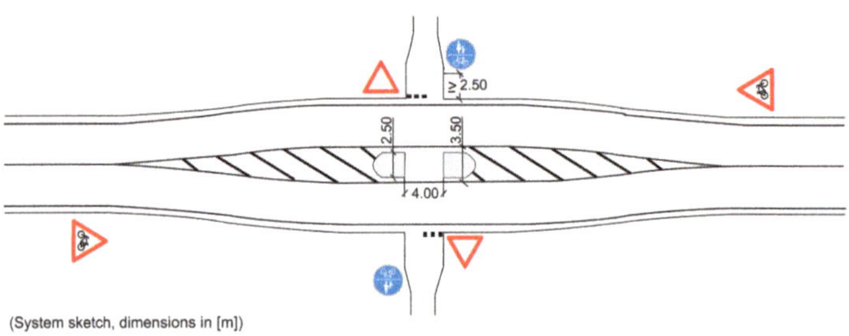

Fig. 10.56 Crossing point for bicycle and pedestrian traffic outside intersections at a road of EKL 3 with a central island as crossing facility

increase attention and make the crossing point safer. On roads with a high traffic volume or road sections with difficult visibility conditions, consideration should be given to setting up a crossing point with traffic lights.

10.3.11 Dimensioning of Intersections on Rural Roads

The dimensioning of grade-separated (sub-) intersections is carried out analogously to the motorways according to the HBS (FGSV 2015a). The dimensioning of at-grade (sub-) intersections on rural roads is also carried out according to the HBS (FGSV 2015a). The decisive quality criterion here is the average waiting time. Due to the different driving characteristics, T-junctions and intersections, as well as roundabouts, are distinguished.

10.3.11.1 Relevant Traffic Volumes

The traffic volumes that are decisive for the design are called design traffic volumes q_B. For dimensioning, the expected traffic volumes should be available for all streams in the intersection, differentiated according to vehicle types light traffic (q_{LT}), trucks and buses (q_{Tr+Bus}) and truck combinations (trucks with trailer and articulated lorry (q_{TrT})). The calculation takes into account the maximum hourly traffic volumes of all flows q_{PCu} in PC-U that occur simultaneously. These are converted into Passenger Car Units PC-U using the factors 1.0 for light traffic (q_{LT}), 1.5 for trucks and buses (q_{Tr+Bus}) and 2.0 for truck combinations (trucks with trailers and articulated lorries). As an approximation, the proportion of heavy goods vehicles (HGV) can be converted to PC-U by a factor of 1.7.

This results in the following conversion:

$$q_{PC\ U} = q_{LT} + q_{Tr+Bus} \cdot 1.5 + q_{TrT} \cdot 2.0.$$

10.3.11.2 Intersection Geometry

The intersection design for the intersections and intersections of single-lane roads considered here is governed by the RAL (FGSV 2012a). The quality of the traffic flow at such at-grade intersections is significantly influenced by the number of lanes in the access roads available for turning traffic.

10.3.11.3 Right of Way, Traffic Streams and their Ranks

The traffic regulations (StVO § 9 (3)(4)) define the classification of traffic streams according to different priority orders. Rank 1 has priority over all other streams. Ranks 2 to 4 are subject to waiting for the respective smaller ranks (Fig. 10.57).

The type of right-of-way control (sign 205 (Give way!) or sign 206 (stop! Give way!)) is important for the capacity of a T-junction or intersection and is taken into account in the calculation of the relevant mainstream load and the basic capacity of the individual traffic streams.

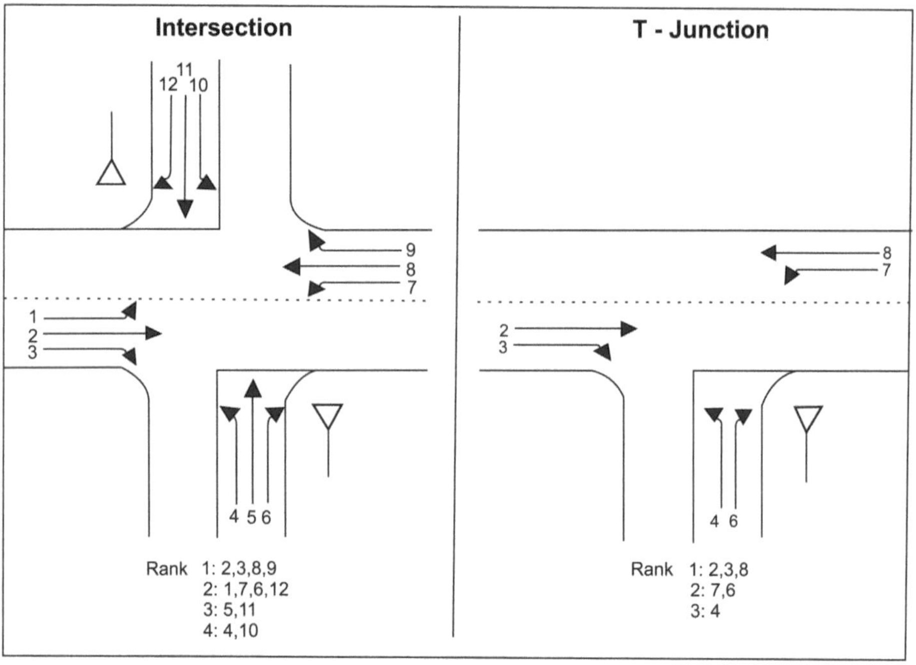

Fig. 10.57 Traffic streams at the intersection without traffic lights

10.3.11.4 Factors of Influence

The capacity of minor intersection accesses or the correlation between mean waiting time and design traffic volume is generally determined for good external conditions (dry roadway, brightness, good lighting conditions).

The different behaviour of waiting drivers depends on the type of vehicle, i.e. essentially on its acceleration capacity and length. This is taken into account by converting the design traffic volumes into PC-U/h according to Sect. 10.3.11.1.

Each minor vehicle stream has to consider different major streams due to the priority hierarchy. For these individual subsidiary streams, this so-called decisive main stream load q_P is made up of different privileged partial streams. Since it is irrelevant to the driver who is obliged to wait whether the usable gap between two privileged vehicles is formed by cars or trucks, the decisive main stream load is given in vehicles per hour [vehicle/h]. The HBS (FGSV 2015a) distinguishes between procedures for T-junctions and intersections.

T-Junctions

If there is no separate lane available for right turners (stream 3), half of the traffic volume of the secondary road (4, 6) is considered to be preferred when calculating the main stream load of the secondary streams ($0.5 \cdot q_3$).

Table 10.15 Determination of the relevant main stream load q_P for intersections; the indices refer to the traffic streams according to Fig. 10.57

Off-line	No.	Decisive main stream load q_P [vehicle/h]
Turn left from the main road	1	$q_8 + q_9^c$
	7	$q_2 + q_3^c$
Right turn from the minor road	6	$q_2^b + 0.5\ q_3^a$
	12	$q_8^b + 0.5\ q_9^a$
End of the minor road	5	$q_2 + 0.5\ q_3^a + q_8^c + q_1 + q_7$
	11	$q_8 + 0.5\ q_9^a + q_2 + q_3^c + q_1 + q_7$
Left turn from the minor road	4	$q_2 + 0.5\ q_3^a + q_8 + 0.5\ q_9^a + q_1 + q_7 + q_{12}^{c,d} + q_{11}^d$
	10	$q_8 + 0.5\ q_9^a + q_2 + 0.5\ q_3^a + q_1 + q_7 + q_6^{c,d} + q_5^d$

[a]If stream 3 or 9 has a separate lane, q_3 or q_9 is not applicable
[b]If the stream has 2 or 8 lanes, the traffic volume on the right lane for q_2 or q_8 shall be used
[c]If stream 3, 9, 6 or 12 is separated by a triangular island with subsequent subordination under the right of way, q_3 q_9, q_6 or q_{12} are not applicable
[d]If a "stop sign" applies to the respective oncoming streams 11 and 12 or 5 and 6, q_{11} and q_{12} or q_5 and q6 are omitted

Intersections

The decisive main stream load for crossings is given in Table 10.15.

The influence of the location of intersections and T-junctions on the capacity is taken into account in the diagrams for determining the basic capacity (FGSV 2015a). Since in conurbations during peak times, local traffic with comparatively short travel distances is predominantly encountered, a distinction is made between intersections

- within agglomerations, and
- outside of agglomerations.

10.3.11.5 Critical and Subsequent Time Gaps

The term limit or critical time gap t_g is used to describe the smallest time gap in a higher-level stream that is assumed by a vehicle requiring maintenance to carry out the intended driving manoeuvre (Table 10.16). The mean time interval between two consecutive departure times of off-stream vehicles is called the subsequent time gap t_f [s] (Table 10.17).

10.3.11.6 Turn-off Device for Right-hand Traffic

If a triangular island or a separate lane is set up for those turning right off the main road at intersections or T-junctions, this will affect the calculation of the relevant main stream loads. This aspect is considered in Table 10.18. Triangle islands or separate lanes for right turns from the main road can also influence the capacity of oncoming left turns. This is taken into account when determining the basic capacity of traffic streams 1 and 7.

Table 10.16 Mean critical time gaps (t_g) for T-junctions and intersections (FGSV 2015a)

		Critical time gap t_g [s]			
		Outside agglomerations		Within agglomerations	
Off-line	No.	With right turn facility in the opposite main road access	Without right turn facility in the opposite main road access	With right turn facility in the opposite main road access	Without right turn facility in the opposite main road access
Turn left from the main road	1 7	6.4	5.9	6.0	5.5
Turn right from the side street	6 12	7.3		6.5	
Crossing from the minor road	5 11	7.0		6.5	
Turn left from the minor road	4 10	7.4		6.6	
Crossing and turning left at one-way streets		6.2		5.6	

Table 10.17 Mean subsequent time gaps for T-junctions and intersections (t_f) (FGSV 2015a, b)

		Subsequent time gap t_r [s]			
		Outside agglomerations		Within agglomerations	
Off-line	No.	Give way!	Stop! Give way!	Give way!	Halt! Stop way!
Turn left from the main road	1 7	2.9		2.6	
Turn right from the minor road	6 12	3.1	3.7	3.1	3.7
Crossing from the minor road	5 11	3.5	4.0	3.5	4.0
Turn left from the minor road	4 10	3.4	3.8	3.4	3.8
Crossing and turning left at one-way streets		3.4	3.8	3.4	3.8

10.3.11.7 Quality Criteria and Levels of Quality of Traffic Flow

The average waiting time w [s] for each partial flow is used as a measure of the quality of traffic flow at intersections without traffic signals (Table 10.19).

The vehicle stream with the worst classification is decisive. This is usually one of the left-hand turn streams 4 or 10 with rank 3 or 4.

Table 10.18 Determination of the decisive main stream load q_P for T-junctions; the indices refer to the traffic streams according to Fig. 10.57 (FGSV 2015a, b)

Minor stream	No.	Decisive main stream load q_P [vehicle/h]
Turn left from the main road	7	$q_2 + q_3{}^{a}$
Right turn from the minor road	6	$q_2{}^{b} + 0.5\ q_3{}^{c}$
Left turn from the minor road	4	$q_2 + 0.5\ q_3{}^{c} + q_7 + q_8$

[a]If stream 3 is separated by a triangular island with subsequent subordination under the right of way, q_3 is not applicable
[b]If there are several lanes available for stream 2, the traffic volume on the right lane for q_2 shall be applied
[c]If stream 3 has a separate lane, not applicable q_3

Table 10.19 Limit values of the average waiting time at intersections without traffic lights (FGSV 2015a, b)

Quality level	A	B	C	D	E	F
Average waiting time, w [s]	≤ 10	≤ 20	≤ 30	≤ 45	> 45	–[a]

[a] Level F is reached when the degree of saturation is greater than 1

The length of the congestion N (Number of vehicles) in the minor access roads as a design criterion becomes significant if the congestion reaches an upstream intersection or is greater than the storage space available on the separate lane.

10.3.11.8 Calculation of Capacity

The capacity at the accesses without right of way is determined in a multi-stage procedure depending on the ranks of the traffic flows.

Basic Capacity

The capacity of traffic flows of rank r = 1 is the capacity of the continuous lanes. This is generally set for the straight lanes with 1800 PC-U/h per lane. The capacity of traffic flows of rank r = 2 is equal to the basic capacity. The capacity of the traffic flows of the rank r = 3 and r = 4 results from the basic capacity and a reduction caused by the right-of-way regulation. This reduction corresponds to the probability of a congestion-free state of the higher-level flows, which nevertheless have to wait in relation to the streams of the first and possibly second and third rank.

The basic capacity G indicates the maximum number of vehicles of a bypass flow in [PC-U/h] that can pass through the intersection by exploiting the time gaps occurring in the major flows (these form the so-called decisive main flow q_p).

$$G_i = \frac{3600}{t_f} \cdot e^{\frac{q_p}{3600}\left(t_g - \frac{t_f}{2}\right)}$$

G_i = Basic capacity of the traffic flow i [car E/h]
q_p = Decisive main stream load [Fz/h]
t_g = Mean critical time gap [s]
t_f = Mean subsequent time gap [s]
The basic capacities can also be taken from the diagrams of the HBS (FGSV 2015a).

Capacity

For the second-ranked streams, i.e. for the left-hand turn off from the main road (streams 1 and 7) and for the right-hand turn in from the minor road (streams 6 and 12), the capacity C_i is equal to the basic capacity G_i.

For the streams of the third rank, i.e. crossing from the minor road (streams 5 and 11) and for the streams of the fourth rank, i.e. for a left turn in from the minor road (streams 4 and 10), it should be noted that they can only flow out if there is no congestion in the streams which are given priority over them (with rank 2 and 3 respectively). The probability of a congestion-free state $p_{o,i}$ in-stream i is

$$p_{0,i} = 1 - \frac{q_i}{C_i},$$

where q_i [PC-U/h] is equal to the existing design traffic volume. The capacity for these minor streams is therefore

$$C_i = p_{0,i} \cdot G_i \quad [\text{Pkw} - \text{E} / \text{h}],$$

with a distinction being made according to the type of crossing:

T-Junction

The capacity of the left turn C_4 (rank 3) is equal to the basic capacity G_4 reduced to the probability $p_{0,7}$ of a congestion-free state in the left turn off stream 7, whereby this only applies if a sufficiently long left turn lane is available and straight-ahead traffic is not obstructed.

$$C_4 = P_{0,7} \cdot G_4$$

Intersection

The capacity of the crossing streams 5 and 11 (rank 3) is equal to the respective basic capacity G_5 and G_{11}, reduced to the independent probabilities $p_X = p_{0,1} \cdot p_{0,7}$ that there is no congestion in both left-turning off stream 1 and left-turning off stream 7 and that the left-turning lanes are sufficiently long.

$$C_5 = p_x \cdot G_5$$
$$C_{11} = p_x \cdot G_{11}$$

If the 95% congestion length N_{95} of the traffic streams 1 and 7 is greater than the length of the respective left-turning lanes or if no separate lanes are available for the left-turning streams, this must be taken into account according to the HBS (FGSV 2015a).

For the fourth rank streams, i.e. for the left turning in streams 4 and 10, the congestion-free conditions at the left turning off (1 and 7) and right turning in (6 and 12) as well as at crossing streams (5 and 11) have to be considered for the determination of the capacity. These probabilities of congestion-free conditions in the second and third rank streams are not independent of each other. Therefore, the two factors $p_{z,5}$ and $p_{z,11}$ are determined with the following equation, which describes these dependencies approximately.

$$p_{z,i} = \frac{1}{1 + \frac{1 - p_x}{p_x} + \frac{1 - p_{0,1}}{p_{0,i}}}$$

with

i: 5 or 11

$P_{0,i}$: Probability of the congestion-free state in the stream i

P_x: Probability of the congestion-free state in streams 1 and 7 = $P_{0,1} \cdot p_{0,7}$

$P_{z,i}$: actual probability of a congestion-free state in streams 1, 7 and 5 or 1, 7 and 11 according to (FGSV 2015a), taking into account the statistical dependencies between the congestion states in streams of rank 2 and 3

The capacity of the left-hand turning in stream (4 and 10) is therefore

$$C_4 = p_{z,11} \cdot p_{0,12} \cdot G_4 \quad [\text{Pkw} - \text{E/h}],$$
$$C_{10} = p_{z,5} \cdot p_{0,6} \cdot G_{10} \quad [\text{Pkw} - \text{E/h}].$$

If the traffic turning right (stream 6 or 12) is separated by a triangular island with subsequent subordination under right of way, $p_{0,6}$ and $p_{0,12}$ are omitted (Table 10.18).

Further details, for example on the consideration of mixed lanes, are contained in the HBS (FGSV 2015a) (Table 10.20).

10.3.11.9 Capacity Reserve, Average Waiting Time, Permissible Traffic Volumes

The value of capacity C represents an unsatisfactory traffic situation associated with generally unwanted waiting times. For the design, therefore, a traffic situation is aimed at that has a capacity reserve R_i [PC-U/h] of the minor stream i compared to the capacity C_i

$$R_i = C_i - q_i \quad \text{or} \quad R_m = C_m - q_m.$$

Here q_i is the existing traffic volume of the minor stream i or mixed stream m. Between the capacity reserve determined according to this equation and the associated quality of the traffic flow, expressed by the criterion mean waiting for time w_i [s] of the minor vehicles, there is a close correlation of the form

Table 10.20 Summary of the calculation of capacities (FGSV 2015a)

Intersection	Stream i	Capacity	Note
	Left turn off 1,7	$C_i = G_i$	
	Right turn in 6, 12	$C_i = G_i$	
T-junction	Left hand turn in 4	$C_4 = p_{o,7} \cdot G_4$	$p_{o,7} = 1 - q_7/C_7$
4-arm intersection	Crossing 5, 11	$C_i = p_X \cdot G_i$	$p_X = p_{o,i} \cdot p_{o,7}$
		$C_4 = p_{z,11} \cdot p_{o,12} \cdot G_4$	$p_{z,i}$ (s. o.)
	Left-hand turn in 4, 10	$C_{10} = p_{z,5} \cdot p_{o,6} \cdot G_{10}$	For i = 5, 11

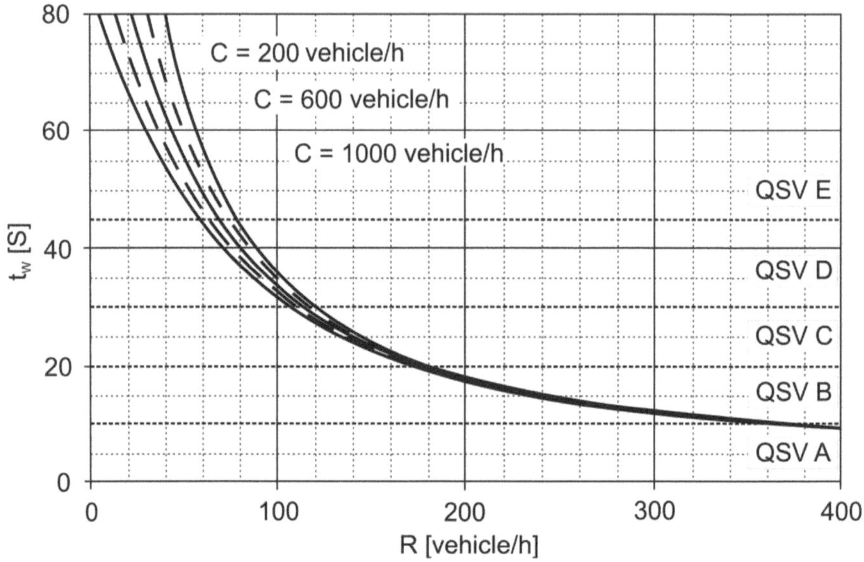

Fig. 10.58 Average waiting time as a function of the capacity reserve R and the capacity C (FGSV 2015a)

$$w_i = f(C_i, R_i).$$

Figure 10.58 shows the dependence of the mean waiting time on the capacity C_i and the available capacity reserve R_i. The curves give approximate values for all streams without right of way as well as mixed streams.

The required capacity reserves and the permissible traffic volumes correspond in each traffic quality level with a corresponding value of the average waiting time. From the curves, in Fig. 10.58 a quality level can be read for each permitted average waiting time. If an upper limit value for the mean waiting time still to be accepted is given, a capacity reserve $R_{erf,i}$ results which is necessary for this quality of traffic flow. This means that the existing traffic volumes must not exceed the permissible traffic volumes $q_{zul,i}$ corresponding to these capacity reserves.

$$q_{zul,i} = C_i - R_{erf,i} \quad [Pkw - E/h]$$

The results of the capacity calculation for the peak hour can be used as a decision criterion for whether an intersection should be controlled with or without a traffic light system. In addition to the reasons for traffic congestion at the intersections investigated here, the decision process must take into account traffic safety aspects and, if necessary, the interests of pedestrians and cyclists.

10.3.11.10 Verification of Queueing Lengths

For the assessment of an intersection without a traffic light system, congestion in the minor traffic streams can also become a decisive criterion. This is the case when the available queueing space is limited. The length of the turn-off lanes should be dimensioned for the 95% congestion length N_{95} (in PC-U) (Fig. 10.59). N_{95} means: In 95% of the time during the considered design interval, the congestion is shorter than N_{95} PC-U. This means that for the remaining 5% of the time the queueing zone is insufficient.

The congestion lengths also need to be checked for the left-hand turn off lanes on the main road, as congestion there leads to obstructions for straight ahead (privileged) vehicles. If, for example, at intersections on fast-moving rural roads, the probability of congestion should only be 1%, the 99% congestion length N_{99} is decisive (Fig. 10.60).

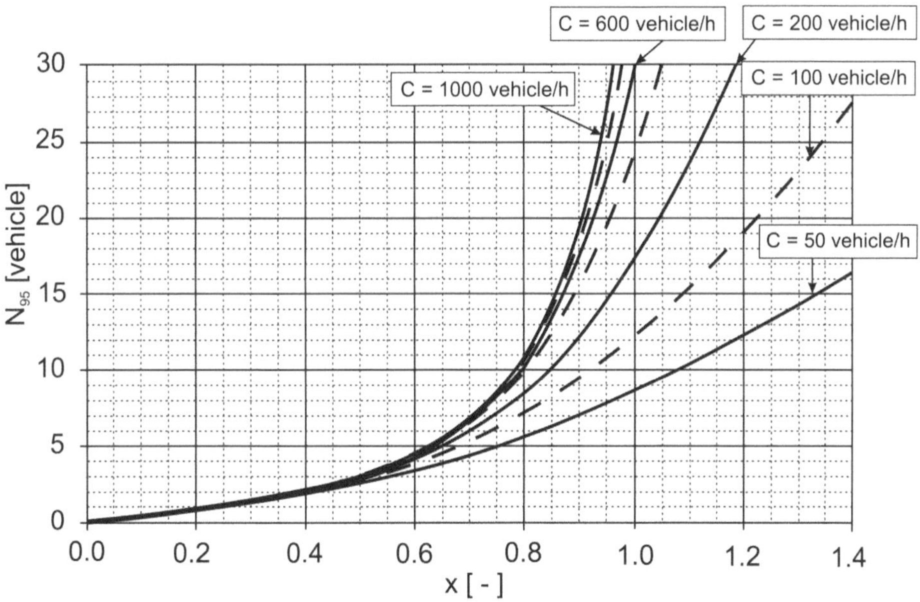

Fig. 10.59 Congestion lengths N_{95} (According to FGSV 2015a)

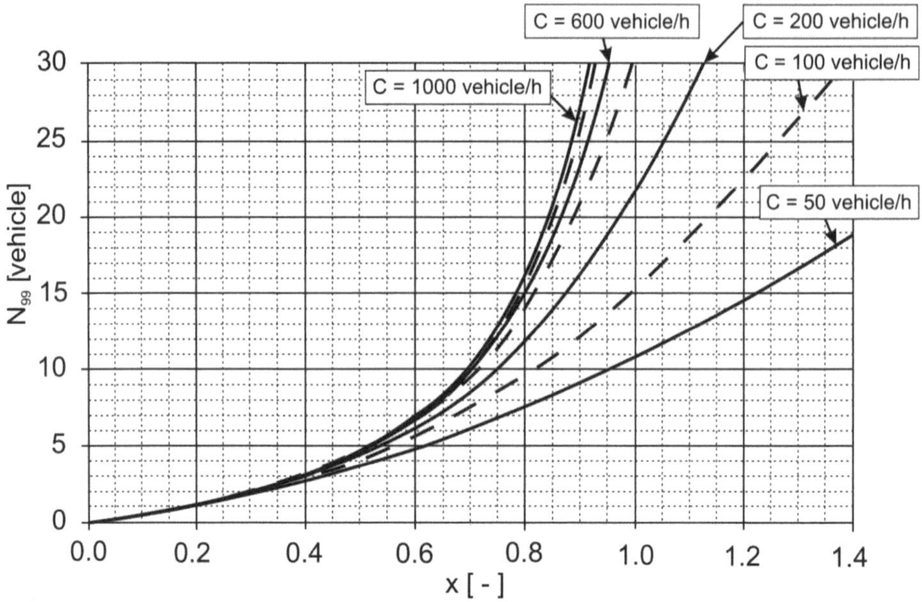

Fig. 10.60 Congestion lengths N_{99} (According to FGSV 2015a)

Table 10.21 Rough procedure for estimating the quality of traffic flow

Capacity reserve R in all minor streams	Rough check of the minimum quality requirement
R > 100 PC-U/h	No traffic lights required
	Sufficient quality (level D) of traffic flow at the intersection without traffic lights
	Average waiting time at 45 s/Fz
	At R > 200 [PC-U] the average waiting time is 20 s/Fz.
	Waiting times would be higher with traffic lights
0 < R < 100 PC-U/h	Traffic lights useful
	Quality of traffic flow critical (level E)
	Longer waiting times without traffic lights than with traffic lights
	If the limit of R > 100 PC-U/h is narrowly missed, more detailed traffic analysis is recommended.
R < 0 PC-U/h	Traffic lights required
	Quality level F, the capacity of the intersection not guaranteed without traffic lights
	Unacceptable waiting times and backwater lengths without traffic lights

10.3.11.11 Assessment of the Quality of Traffic Flow

To assess the quality of traffic flow at intersections without traffic signals, the design method can be applied according to two aspects:

1. Comparison of the waiting times, i.e. the waiting time w_i determined for a design traffic volume q_i should be less than the waiting time $w_D \rightarrow w_i \leq w_D$ targeted for a given quality level (e.g. level D),
2. Comparison of the traffic volumes, i.e. the design traffic volume qi should be smaller than the permitted traffic volume $q_{zuli,D}$ which is defined for a given waiting time at a quality level (here e.g. level D): $q_i < q_{zuli,D}$.

These checks must be carried out for all minor streams i. The approximate procedure shown in Table 10.21 can be used to check compliance with the minimum quality requirements (Level D).

Integrated Consideration of Road Sections and Intersections

11

Summary

The integrated consideration of road sections and intersections has long been a scientific requirement, as the elements influence each other. For example, in Richter (*Entwurfsstandards für Knotenpunkte an Ortsumgehungen,* Typo Druck Verlag, Bonn-Bad Godesberg, 1993), it is shown that on sections of road with intersections with traffic lights, a different driving behaviour is to be expected than on sections of road with intersections without traffic lights. In Steinhauer et al. (*Übergreifende Bewertung von Streckenabschnitten und Knotenpunkten auf Landstraßen* (Bd. 1004). (BMVBS, Hrsg.), Wirtschaftsverlag NW, Bremerhaven, 2008), road sections are evaluated from the point of view of traffic flow and here, too, there was a mutual influence. RAL (FGSV, *Richtlinien für die Anlage von Landstraßen (RAL),* FGSV Verlag, Köln, 2012) takes this fact into account and offers for the individual design classes coordinated cross sections and intersection types with coordinated element sizes.

Therefore, already in the phase of demand planning and line determination, the standard cross sections, the recommended element sizes as well as the intersection types and their operational elements must be taken into account, depending on the importance of the connection and the design class derived from it.

© Springer Fachmedien Wiesbaden GmbH, part of Springer Nature 2023

T. Richter, *Road Planning - Freeways and Country Roads*,
https://doi.org/10.1007/978-3-658-35189-2_11

Design Elements in the Site Plan

12

Summary

The design elements in the site plan essentially determine the driving behaviour and safe usability of the motorways and country roads. While on motorways the elements should be chosen as generously as possible, RAL tries to promote adapted driving behaviour by specifying recommended element sizes.

12.1 State of the Art

The influence of alignment on driving behaviour on rural roads has so far mostly been carried out on a sectoral basis. The route characteristics are determined in the area of alignment by the following features:

- Curves, curvature, crossfall,
- Sight distance,
- Longitudinal inclination, length of the longitudinal inclination.

When taking a holistic view of the design of rural roads by means of standardised route characteristics, as provided for in the RAL (FGSV 2012a) with the self-explanatory road, the findings of the sectoral influences of alignment on driving behaviour described below must therefore also be taken into account.

In numerous studies (e.g. Hiersche et al. 1984; Krebs and Klöckner 1977; Leutzbach and Zoellmer 1989), a direct influence of the curve radius on driving behaviour could be demonstrated. It was found that radii greater than or equal to 500 m have no influence on local speed behaviour (Steierwald and Buck 1992). Especially in the radius range smaller or equal to 250 m, the strong influence of the curve radius is evident. RAS-L-95 (FGSV 1995) also contains a correlation between the curve radius and the speed. Investigations in Switzerland have shown that in both left- and right-hand curves, drivers always keep a

© Springer Fachmedien Wiesbaden GmbH, part of Springer Nature 2023

T. Richter, *Road Planning - Freeways and Country Roads*,
https://doi.org/10.1007/978-3-658-35189-2_12

much greater distance from the edge of the road than from the centre line, and that this distance is largely independent of the width of the road (Spacek 1998).

In addition to the radius of the curve, the size of the angle of change of direction must be considered (Krebs 1973). It was assumed that with direction change angles from 11 to 16 gon, a considerable gain in radius can be achieved by "cutting" curves within the own lane. If the sum of the directional change angles is considered over a distance, the result is the curvature expressed in gon per kilometre. Curvature has also been related to speed in various studies and it has been found that curvature has a dominant influence on driving behaviour (Köppel and Bock 1979). RAS-L-95 (FGSV 1995) contains a diagram for the determination of speed from the curve, in which curves below 100–150 gon/km have no influence on speed. In contrast, other studies argue that the curvature is not sufficient to assess the safety of the route (Lippold 1997). In addition, it is necessary to consider the continuity.

12.2 Basic Principles

Roads as a construction project have a great length extension. The reference line for representation in plans is usually the road axis—an ideal central line, which generally runs in the middle of the carriageway and, in the case of dual carriageways, as the main axis in the middle of the central reserve. The regular division along the road axis with length indications is called stationing.

The aim of the design is to achieve the most optimal and continuous road possible, which is only achieved by combining the three levels of representation—site plan, elevation plan and cross section. These three mapping levels must therefore be designed iteratively in alternating work steps according to the principles of spatial alignment. In the spatial alignment, it has to be checked whether the size and sequence of the spatial design elements allow sufficient recognizability and comprehensibility and whether the entire street space appears calm and visually appealing.

The curvature, ramp and visibility bands usually shown in the elevation plan serve to link these plans (Table 12.1). Although the driver's perspective is the most relevant one for driving behaviour, at the moment the work is usually still done separately in the respective two-dimensional planes. There are first approaches for spatial routing, but they are not yet established in daily design practice. Here the products KorFin (A + S Consult Dresden GmbH) or VIS-All 3D (Software-Service John, Ilmenau) can be mentioned, which allow an integrated three-dimensional design. These are mainly used in the context of demand planning or line determination.

Bridges must be continuously integrated into the spatial carriageway layout, taking into account safety, costs and design considerations. This means that the structures should not be located in areas of small design elements. However, no straight sections should be created for the bridges either, because these can rarely be harmoniously integrated into the carriageway layout.

Table 12.1 Relevant plans in road design

Plan	Content	
Site plan	Layout	Projection onto a horizontal plane
Altitude plan	Site surface	Handling of the terrain at axle height
	Gradient	Processing of the vertical sectional plane of the road
	Curvature band	Schematic representation of the course of the curvature of the site plan
	Ramp band	Schematic representation of the course of the road edge heights in relation to the axle height
	Sight distance band	Specification of the stopping and, if necessary, overtaking sight distances of both directions of travel via the stationing
Cross section	Cross section of the road	Vertical section to the road axis

Design elements in the site plan for the design of the road are

- Straight,
- Curve and
- Transition curve,

in the elevation plan for the design of the gradient

- Straight and
- Curve,

in cross section for the design of the road space

- Vertical and horizontal cross-sectional elements and
- Crossfalls.

By incorporating the road environment, the visual lines are created as a view of the road from the driver's perspective. The processing of the site plan with indication of the elevation points of the terrain results in the course of the terrain along the road axis. The course of the upper edge of the road—the gradient—is formed into this terrain profile. The gradient usually runs along the road axis. Only in justified exceptional cases, does the gradient run along the inner or outer edge of the road (Fig. 12.1).

The elevation plan is always drawn in the same longitudinal scale as the site plan. The elevation plan is the vertical sectional plane of the road as a development. It contains the terrain profile and the gradient. In order to be able to illustrate the elevation clearly, the height scale is usually enlarged in the ratio 1:10 to the longitudinal scale. In very flat terrain, this ratio can be increased to 1:20 to illustrate details important for planning; in topographically strongly structured regions, a fivefold increase is sufficient.

To overcome the height differences in the terrain, road sections with constant longitudinal inclinations are arranged. The points of intersection (tangent intersections) are rounded off by crest and sag curves. Site plan and elevation plan are coordinated with each other under consideration of the spatial alignment. Essential parameters of the design elements of the elevation plan are the longitudinal inclination s [%] and the rounding radius H [m] of crest (H_C) and sag (H_S) curves.

The following routing rules should be followed:

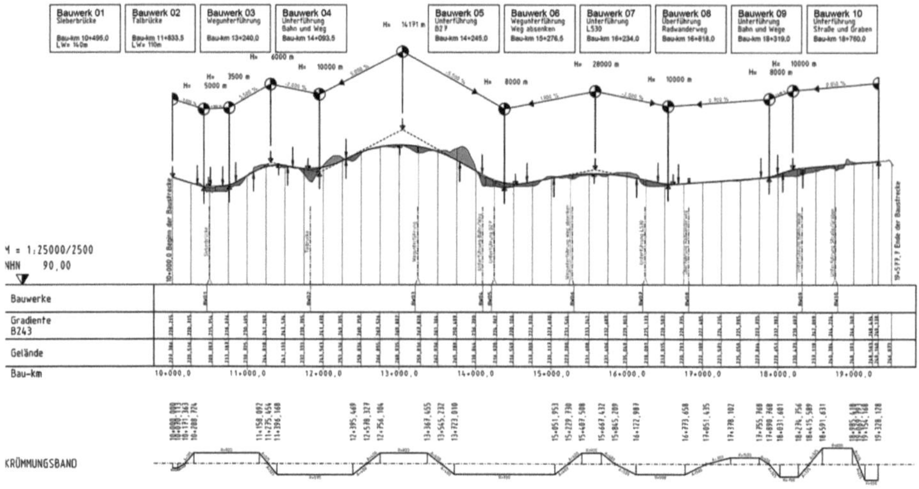

Fig. 12.1 Schematic representation of the elevation plan

- Observe mandatory points from the intersection of existing path and water networks,
- Avoid high dams and deep cuts for reasons of landscape protection
- Do not cross valleys with high dams,
- Optimise valley bridge lengths and abutment heights in comparison to embankment construction-bridge construction,
- Fit bridges into the curved lines of the site and elevation plan (avoid board effect),
- Low longitudinal gradients improve traffic safety, increase capacity and traffic quality, reduce operating costs and road user costs and lower emissions,
- By reducing the gradient, an additional lane can be saved in the gradient area, and by increasing the gradient, an indispensable additional lane can be made more economical,
- Take into account the avoidance of heavy goods vehicles on parallel routes if steep sections are used on motorways with only two lanes per direction,
- In the case of soil types that are unsuitable for construction purposes, the gradients should be in the area at risk of slipping instead of in the cut on a low embankment,
- Route the layout in such a way that a mass balance is generally achieved with a minimum mass sum (transport costs!). Side withdrawals, side deposits or removal of excess soil lead to ecological resistance and high costs (land acquisition, compensation measures through greening and reforestation, transport costs).

12.3 Layout Finding

The site plan of a road in its final state consists of a map base with contour lines, in which, in addition to the constraint points to be observed, the existing road is entered with the essential stakeout dimensions of the site plan elements (straight lines, curves, transition curves).

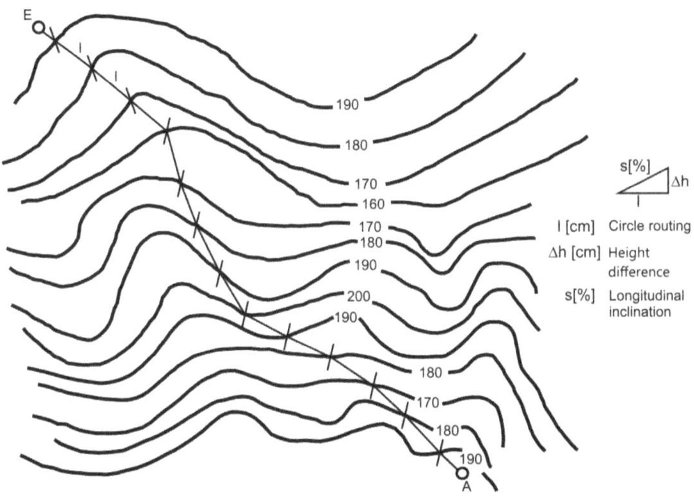

Fig. 12.2 Site plan and circular route

The following routing rules should be followed:

- From a landscape point of view, the route follows relief bends (e.g. valley floor edge) or existing traffic routes.
- On southern slopes, the danger of black ice formation is lower than on northern slopes.
- Soil types unsuitable for construction purposes (marl, alluvial sand) should be avoided by routing the site plan.

As a first point of reference, a freehand line should be designed, in the course of which the above-mentioned points and the recommended radius ranges of the EKL are considered. In order to avoid large cuts and dams in moving terrain from the outset, the site plan should be aligned with the so-called circular route (Fig. 12.2). Whether valleys can be extended or whether they can be crossed with a bridge is to be decided according to the importance of the road (minimum radii, maximum longitudinal inclination, …) and the ecological boundary conditions.

12.4 Straight Lines

In the historical overview (see Chap. 1), the straight line was the ideal route. It offered the best routing to allow marching columns and carriages the shortest connection between two places. The straight line was also the dominant design element in the first "car-only" roads in Germany between 1930 and 1950.

Despite these supposedly great advantages, many accidents had to be noted after the completion of the first sections of road, most of which occurred on the straights that were considered safe. They were often caused by hitting vehicles that were stopping or by the vehicle leaving the road. As it turned out, these accidents were mostly due to driver fatigue.

Today it is known that long straights with constant longitudinal inclinations have even more disadvantages besides their monotony. They have the disadvantage that they

- make it difficult to estimate the distances and speeds of oncoming and following vehicles,
- tempt the driver to high speeds,
- increase the risk of glare from oncoming vehicles at night,
- the constant crossfall forces the driver to make permanent steering corrections for drainage reasons,
- may fatigue the driver, and
- are difficult to adapt to the structure of the landscape in hilly terrain.

When building new roads outside built-up areas, longer straight lines with constant longitudinal inclinations should therefore be avoided and replaced by curves with very large radii. The straight line can nevertheless be advantageous as a design element

- for special topographic conditions, e.g. in the plain and in wide valleys,
- in the area of junctions to improve clarity,
- to create overtaking sight distances on two-lane roads, especially in sag curves and when overtaking lanes cannot be created, and
- where the route is based on railway lines, canals and borders.

In the cases mentioned, straight lines with constant longitudinal inclination in their length max L_G [m] should be limited to 1500 m on country roads and to 2000 m on motorways. On country roads, straight lines between curves curved in the same direction should be avoided. If this is not possible, their length min L_G [m] in EKL 1 to 3 should be at least 600 m and in EKL 4 at least 400 m in order to ensure the continuity of the optical guidance. In the case of roads within built-up areas, the use of the straight line is generally unproblematic, since the line layout is usually influenced by urban planning criteria and the disadvantages of excessively long straight lines hardly arise.

12.5 Circular Curves

In the past, the circular arc was mainly used to connect two straight sections when unavoidable changes of direction had to be made. In today's road design, it is the most frequently used routing element. Mathematically, the circle curve is the simplest and most regular curve. It has the same radius at every point and therefore the same curvature. The constant radius of the circle leads to a constant curvature value, just like the straight line (Fig. 12.3).

In the case of country roads, the largest possible radii should be selected for the circular curves, taking into account the respective topography, in order to achieve short construction lengths and travel distances, large stopping distances and as far as possible sufficient overtaking opportunities, as well as a smooth driving style.

On the other hand, the radii should only be large enough to harmonize in size and sequence with the structure of the terrain and the landscape elements, to be in harmony

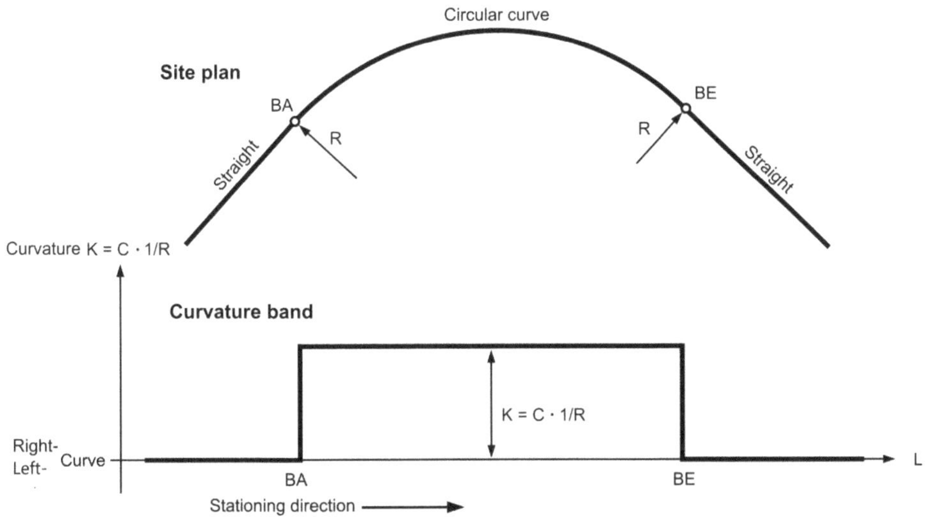

Fig. 12.3 Straight lines and circular curves in the site plan and curvature band

with the terrain-related elevation plan elements in size, sequence and spatial effect, and to allow for a uniform driving style appropriate to the design class. The appropriate driving style is of particular importance for RAL, as the design classes are based on a planning speed that is to be supported by the elements to be selected.

Above all, the landscape protection concerns to be observed when upgrading existing rural roads and the urban planning constraints that exist when building new roads, converting existing roads or upgrading existing urban roads require careful consideration in each individual case when determining radii between the requirements of adjacent land use and harmonious alignment.

The radii of successive curves should be balanced for reasons of traffic safety on country roads (relational alignment). For the radius size of adjacent curves, the relations are given in Fig. 12.4. In design classes EKL 1 and EKL 2, the radius sequences should be in the good range. In design class EKL 3, the good range should be aimed at, but the usable range is sufficient. In design class EKL 4, the usable range is generally sufficient.

The minimum curve radius (min R) was derived in the predecessor guidelines to RAL (FGSV 2012a) driving dynamics as a function of the design speed V_e, the degree of utilization n of the radial frictional connection and the limit values of the crossfall max q (Chap. 8).

The RAL (FGSV 2012a) does not contain any minimum radii directly dependent on speed. Rather, radius ranges are recommended depending on the design class and the planning speeds on which these are based (Fig. 12.5).

In many cases, especially in flat terrain or with straight-line spatial structures, larger radii are advisable. They can be used if it is taken into account that, especially in design

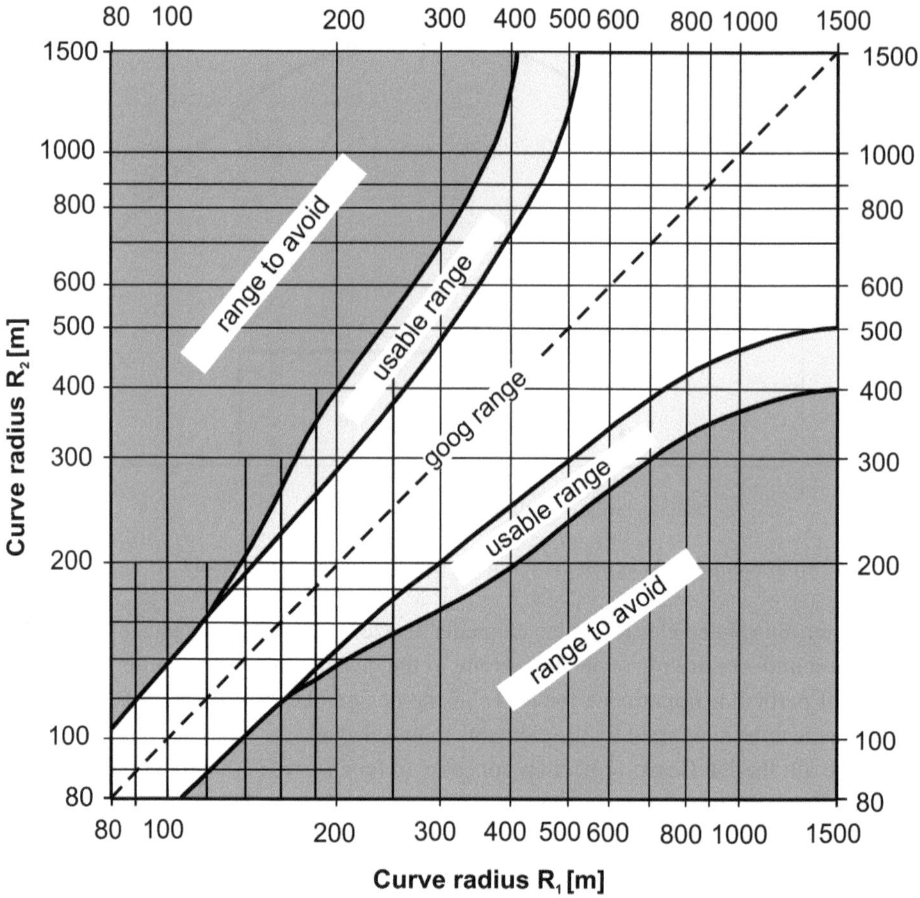

Fig. 12.4 Relationship of consecutive radii (FGSV 2012a)

Design class	Radius ranges R [m]	Minimum length of circular curve min L [m]
EKL 1	≥ 500	70
EKL 2	400 – 900	60
EKL 3	300 – 600	50
EKL 4	200 – 400	40

Fig. 12.5 Recommended radius ranges and minimum lengths of curves according to RAL (FGSV 2012a)

classes 3 and 4, higher speeds than the planning speeds are driven and sight distances can arise that provoke overtaking manoeuvres in sections of the road where the sight distances are insufficient. Furthermore, in the sense of the self-explanatory road, a uniform characteristic should be achieved over as many sections of the same design class as possible. Individual generous route sections would undermine this endeavour.

In other cases, radii smaller than the recommended ones may be appropriate. Examples are a moving topography or adaptation to spatial structures such as areas requiring special protection or a large number of fragmented settlements. Since the smaller radii are assigned higher accident characteristics according to the literature (e.g. (Hiersche et al. 1984)), deviations from the recommended radius ranges must always be justified, since the deviations are critically questioned in the safety audit at the latest. The recommended radii can be undercut by up to 15% in EKL 2 to 4, whereby the ratio of the successive radii must then be in the good range.

A further boundary condition for the selection of the curve radius is the recognizability as a single element. Therefore, minimum lengths are defined for the curves.

In addition to the absolute minimum radii, the permissible minimum radii following straight lines must also be observed, since a high speed level is reached in longer straight lines and small curve radii are not expected to follow. The radii following straight lines (Fig. 12.6) should be in the good range in design classes EKL 1 to 3, otherwise the usable

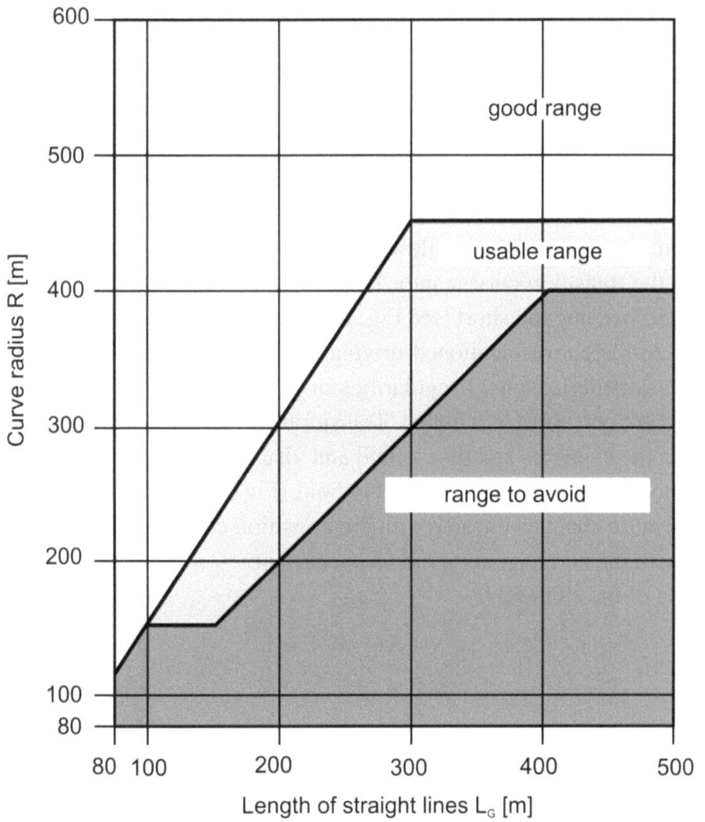

Fig. 12.6 Permissible radius range in connection with straight lines (FGSV 2012a)

range is sufficient. Short straight lines with a length $L_G < 300$ m act like longer clothoids, therefore the two adjacent radii must lie within the radius relation (Fig. 12.4).

Occasionally, the good or usable range according to Fig. 12.4 cannot be maintained due to constraint points. In such cases, it is particularly important to ensure that the sequence of radii can be identified. This should be supported by flanking traffic engineering measures. Direction signs in curves (sign 625 StVO), danger signs in curves (sign 103 StVO), suitable planting, protective devices with reflecting elements or the restriction of the permissible maximum speed may be suitable for this purpose.

12.6 Transition Curve

The third design element in the site plan is the transition curve. This arrangement is necessary for reasons of driving dynamics and aesthetics.

Theoretically, the steering wheel angle at the transition from straight line to curve would suddenly change. This would result in the transverse jerk increasing to a dangerous level. In practice, a change in curvature requires the steering wheel to be moved or turned. Since this movement takes place while driving, there is a constant change in the curvature value of the line driven. This is possible with a transition curve. The gradual change in the curvature of the transition curve results in a steady line and thus enables a continuous change in the radial acceleration occurring at constant speed. This also gives a finite value to the transverse jerk. In the case of very large radii, the transverse jerk can be compensated for within the lane. In principle, an attempt should always be made to use a transition curve. For radii with $R \geq 1000$ m (following straight lines) and for radii $R \geq 2000$ m with a reverse line, the transition curve can be dispensed with on country roads if it cannot be fitted or the radii become too short (see Fig. 12.5).

In addition to the above-mentioned driving dynamics reasons, the transition curve is also used for aesthetic reasons. Irregularities or kinks in the lines are perceived by the human eye and are perceived as a defect. Transition curves cause a gradual change of curvature between the elements and thus a fluid and visually appealing alignment.

The transition curve is designed as a clothoid (Fig. 12.7). With this transition curve shape, the curvature changes linearly with the transition curve length. The product of any curve length with the corresponding radius is constant.

The formula of the clothoid is

$$A^2 = R \cdot L$$

with:

A: Clothoid parameter [m],

R: Radius at end of clothoid [m],

L: Length of the clothoid up to the radius R [m].

With the clothoid, the curvature changes linearly with the curve length. As a result, the driver—assuming a constant speed—can turn his steering wheel at a constant angular

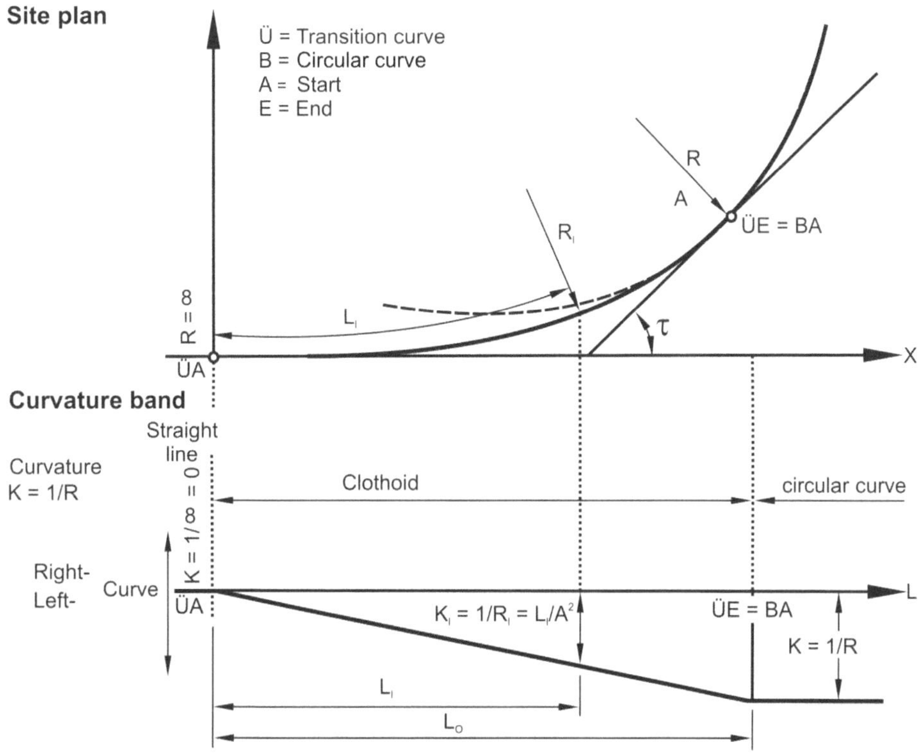

Fig. 12.7 Site plan and curvature line of the clothoid

speed. All clothoids are geometrically similar and can be traced back to the unit clothoid with the parameter A = 1. The parameter A serves as an enlargement or reduction factor comparable to the radius R, so that an infinite number of clothoid transition curves can result. While with the parameter, all length values also change linearly proportional, all angles and ratio values remain unchanged at the same forming point. These characteristic forming points are called characteristic points (Fig. 12.8). One of the most important identification points is the identification point 1, which marks the point on the clothoid where R = L = A. The tangent angle at this point is τ [rad] = 0.5 or τ [gon] = 31.8310. The angle τ between tangent and main tangent has a different value at each point and increases with increasing length for the same parameter. The following relationship applies:

$$\tau = L/(2\cdot R)[\text{rad}] \quad \text{or} \quad \tau = \left(L/(2\cdot R)\right)\cdot(200/\tau) \quad [\text{gon}].$$

The coordinates of any clothoid point P_n are difficult to calculate. For the practical design work, therefore, extensive tables were created from which all essential determination values for certain clothoid parameters A can be taken or calculated using the table values of the unit clothoid A (Fig. 12.9). Clothoid rulers are also available for practical design work

Identification points	τ [gon]	τ [rad]	A [m]		R [m]		L [m]	
1	31.83	0.50	1.00 R	1.00 L	1.00 A	1.00 L	1.00 A	1.00 R
1,5	14.16	0.22	0.67 R	1.50 L	1.50 A	2.25 L	0.67 A	0.45 R
2	7.96	0.13	0.50 R	2.00 L	2.00 A	4.00 L	0.50 A	0.25 R
3	3.54	0.06	0.33 R	3.00 L	3.00 A	9.00 L	0.33 A	0.11 R
4	1.99	0.03	0.25 R	4.00 L	4.00 A	16.00 L	0.25 A	0.06 R
5	1.27	0.02	0.20 R	5.00 L	5.00 A	25.00 L	0.20 A	0.04 R
6	0.89	0.01	0.17 R	6.00 L	6.00 A	36.00 L	0.17 A	0.03 R
R/A	$100/r^2$	$1/2r^2$	R/r	r*L	r*A	r^2*L	A/r	R/r^2

Fig. 12.8 Characteristic of clothoid identification points

R = 450										
A	L	τ^g	τ^0	ΔR	X_M	X	Y	T_K	T_L	A
140	43,556	3,0809	2 46 22	0,176	21,776	43,545	0,703	14,522	29,040	140
150	50,000	3,5368	3 10 59	0,232	24,997	49,985	0,926	16,671	33,339	150
160	56,889	4,0241	3 37 18	0,300	28,441	56,866	1,198	18,970	37,934	160
170	64,222	4,5428	4 05 19	0,382	32,106	64,190	1,527	21,418	42,826	170
175	68,056	4,8140	4 19 57	0,429	34,021	68,017	1,715	22,697	45,384	175
180	72,000	5,0930	4 35 01	0,480	35,992	71,954	1,919	24,015	48,016	180
190	80,222	5,6746	5 06 26	0,596	40,100	80,159	2,382	26,761	53,504	190
200	88,889	6,2876	5 39 32	0,731	44,430	88,802	2,924	29,657	59,290	200
225	112,500	7,9578	7 09 43	1,171	56,221	112,324	4,682	37,556	75,062	225
250	138,889	9,8244	8 50 31	1,785	69,389	138,559	7,132	46,402	92,709	250
275	168,056	11,8875	10 41 56	2,612	83,930	167,471	10,434	56,205	112,242	275
300	200,000	14,1471	12 43 57	3,697	99,836	199,015	14,763	66,982	133,680	300
325	234,222	16,6032	14 56 34	5,089	117,095	233,131	20,306	78,751	157,043	325
350	272,222	19,2558	17 19 49	6,839	135,697	269,742	27,267	91,539	182,359	350
375	312,500	22,1048	19 53 40	9,003	155,624	308,753	35,859	105,378	209,664	375
400	355,556	25,1504	22 38 07	11,640	176,857	350,046	46,303	120,309	239,004	400
450	450,000	31,8310	28 38 52	18,584	223,138	438,879	73,671	153,666	304,025	450
500	555,556	39,2975	35 22 04	28,192	274,287	534,757	111,238	192,180	378,043	500
550	672,222	47,5500	42 47 42	41,017	329,957	635,676	160,812	236,705	461,985	550
600	800,000	56,5884	50 55 46	57,615	389,693	739,060	223,991	288,510	557,220	600

Fig. 12.9 Section of the representation of the tabulated determinants of simple clothoids for R = 450 m (Kasper et al. 1954)

(Fig. 12.10). These rulers consist of a convex and a concave curve part and have the shape of an S-curve. The rulers are labeled for the scale 1:1000 as standard (conversion possible).

In order for the transition curve to appear visually, it must make a change of direction of at least $\tau = 3.5$ gon from the zero point (identification point r = 3, R/A = 3). For safety reasons (recognizability of the main curve, the curvature of the main curve is greater than the continuous curvature of the clothoid), $\tau = 31{,}8$ gon is recommended as upper limit (identification r = 1, R/A = 1). This applies to roads of all category groups:

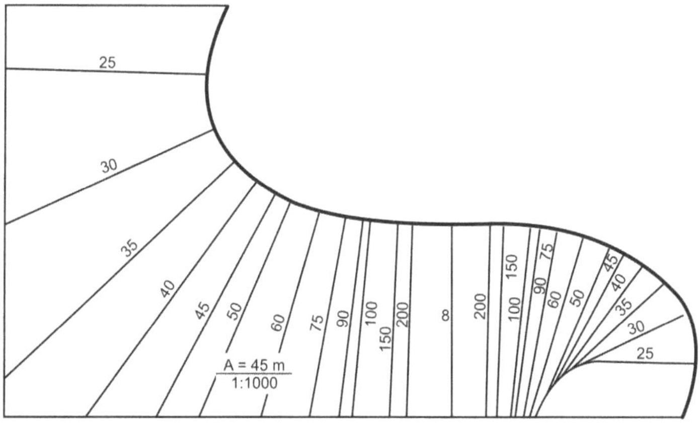

Fig. 12.10 Example of a clothoid ruler (A = 45 m)

$$\frac{R}{3} \leq A \leq R$$

with:
A: Clothoid parameter [m],
R: Radius at end of clothoid [m].
For small radii (R < 150 m), clothoid transition curves with A = R should be used if possible. This prevents too small design elements. In principle, clothoid lengths l < 50 m and clothoid parameters A < 100 m should be avoided. For larger radii, clothoid parameters in the direction A = R/3 should be used if possible, so that the clothoids do not become too long in relation to the radii (Fig. 12.11).

The clothoid parameter transition curve must be selected at least large enough to accommodate the ramping of the roadway edges within the transition curve without exceeding the permissible ramp inclinations max Δ s. The required clothoid length for ramping and twisting is available in most cases, as the other conditions are usually more extensive. If it becomes necessary to calculate the minimum clothoid parameter from the ramping and twisting conditions, the following formula must be applied:

$$A_{min} = \sqrt{\frac{a \cdot (q_e \pm q_a)}{\Delta s_{max}}} \cdot R_e \quad [m]$$

A_{min}: Minimum clothoid parameters [m],
R: Radius at end of transition curve [m],
a: Distance of the edge of the roadway from the axis of rotation [m],
Δs_{max}: Maximum ramp gradient [%],
q_a: Crossfall at the beginning of the transition curve [%],
q_e: Crossfall at the end of the transition curve [%].

R	250	275	300	325	350	375	400	450	500	R
	76.67	67.55	60.39	54.67	50.02	46.18	42.96	37.89	34.11	Circle distance D
250	28.29	26.19	24.34	22.70	21.26	19.98	18.83	16.88	15.28	Auxiliary size ε in new degree
	0.00	10.39	19.29	26.98	33.65	39.49	44.64	53.28	60.24	Auxiliary size E
	67.55	59.04	52.41	47.14	42.87	39.37	36.44	31.87	28.47	
275	26.19	24.20	22.46	20.92	19.57	18.37	17.31	15.49	14.02	
	10.39	0.00	8.90	16.56	23.22	29.05	34.17	42.78	49.71	
	60.39	52.41	46.22	41.33	37.39	34.17	31.49	27.31	24.24	
300	24.34	22.46	20.81	19.37	18.10	16.98	15.99	14.30	12.93	
	19.29	8.90	0.00	7.66	14.31	20.13	25.25	33.84	40.75	
	54.67	47.14	41.33	36.77	33.10	30.11	27.63	23.79	20.98	
325	22.70	20.92	19.37	18.01	16.82	15.78	14.84	13.27	11.99	
	26.98	16.56	7.66	0.00	6.65	12.46	17.57	26.15	33.05	
	50.02	42.87	37.39	33.10	29.66	26.87	24.57	21.01	18.42	
350	21.26	19.57	18.10	16.82	15.71	14.72	13.84	12.37	11.17	
	33.65	23.22	14.31	6.65	0.00	5.81	10.92	19.49	26.39	
	46.18	39.37	34.17	30.11	26.87	24.25	22.10	18.78	16.37	
375	19.98	18.37	16.98	15.78	14.72	13.79	12.97	11.58	10.45	
	39.49	29.05	20.13	12.46	5.81	0.00	5.11	13.68	20.57	
	42.96	36.44	31.49	27.63	24.57	22.10	20.06	16.95	14.70	
400	18.83	17.31	15.99	14.84	13.84	12.97	12.19	10.88	9.82	
	44.64	34.17	25.25	17.57	10.92	5.11	0.00	8.57	15.46	
	37.89	31.87	27.31	23.79	21.01	18.78	16.95	14.17	12.16	
450	16.88	15.49	14.30	13.27	12.37	11.58	10.88	9.70	8.75	
	53.28	42.78	33.84	26.15	19.49	13.68	8.57	0.00	6.89	
	34.11	28.47	24.24	20.98	18.42	16.37	14.70	12.16	10.36	
500	15.28	14.02	12.93	11.99	11.17	10.45	9.82	8.75	7.89	
	60.24	49.71	40.45	33.05	26.39	20.57	15.46	6.89	0.00	

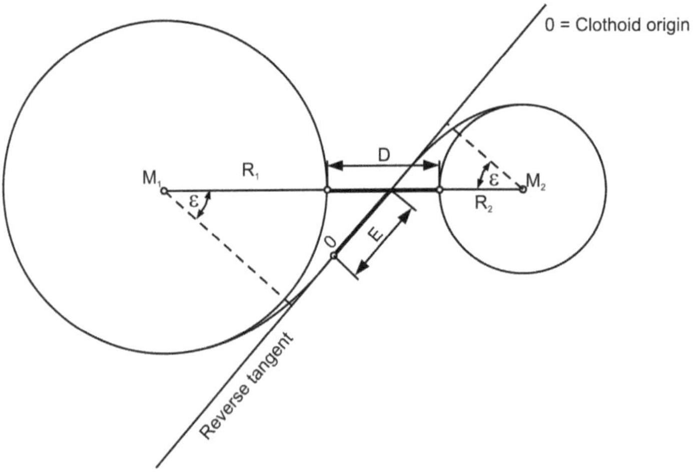

Fig. 12.11 Example of a clothoid table for a reversible clothoid (Kasper et al. 1954)

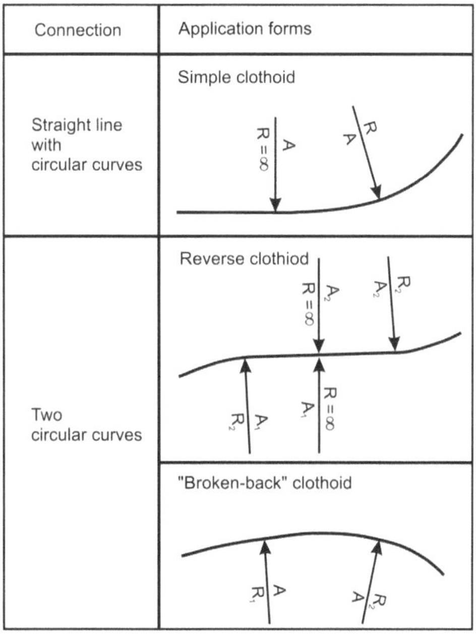

Fig. 12.12 Application possibilities of the clothoid

The common transition curve shapes are shown in Fig. 12.12. The simple clothoid provides the transition between the straight line and the circular curve. Basically, the smaller the following radius, the larger the clothoid parameter should be. Thus, the transition becomes longer and optically "softer". If the requirements for the sequence of radii cannot be met, e.g. after longer straight lines, the following circular curve must be pointed out as early as possible. In this case, the selection of a smaller clothoid parameter (A = R/3) may be appropriate as an exception. This makes the transition appear optically "harder", the driver is made aware of the following curve and is prompted to reduce speed, provided that he can recognise the optically harder transition in good time. The turning clothoid consists of two simple clothoids, which meet at their zero point and are curved in opposite directions. If the line includes the two adjacent curves, a reverse clothoid is formed. For each of the two clothoid sections, the conditions of the simple clothoid apply. In the interest of a harmonic line and an even ramping, both clothoid sections should have approximately the same parameters.

A flat curve is a circular curve with a small angle of change of direction ($\gamma \leq 10$ gon). At such flat curves, a clothoid can be omitted, if they are at least 300 m at motorways and at least 200 m (EKL 1 and 2) or 150 m (EKL 3) or 100 m (EKL 4) at country roads.

12.7 Site Plan Curves

The design elements straight line, circular curve and transition curve can be combined to form different layout curves. These are shown in Fig. 12.13.

Designation of the site plan curve Image	Element sequence Curvature band	Conditions of use	Evaluation
Compound curve 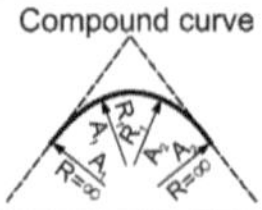	$R=\infty - A_1 - R_1 - A_2 - R=\infty$ 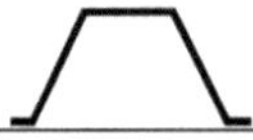	- As symmetrical as possible ($A_1 \approx A_2$) - In case of asymmetrical formation, the ratio $A_1 : A_2$ must be ≤ 1.5.	very good
Reverse line 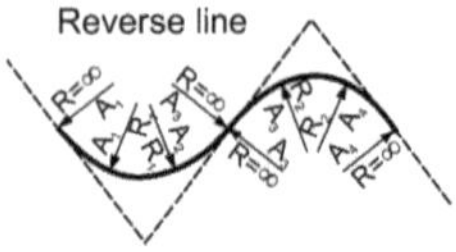	$R=\infty - A_1 - R_1 - A_2 - A_3 - R_2 - A_4 - R=\infty$ 	- Both arms of the reverse clothoid should have the same parameters ($A_2 \approx A_3$). - In case of asymmetrical formation, the ratio $A_2 : A_3$ should be ≤ 1.5. - The length of the straight line between the two arms of the reverse clothoid may not exceed $L_Z \leq 0{,}08 \cdot (A_2 + A_3)$. Otherwise, both curves are considered independent compound curves.	very good
Line with "broken-back" clothoid 	$R=\infty - A_1 - R_1 - A_2 - R_2 - A_3 - R=\infty$ 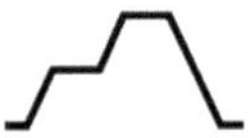	- The circular curves lie one inside the other, are of different sizes, must not touch and are not concentric to each other. - The sequence of radii must be coordinated according to Fig. 12.5. - The angle of change of direction of the "broken-Back" clothoid should be at least $\tau \geq 3.5$ gon.	good (check operating conditions)
Flat circular curve 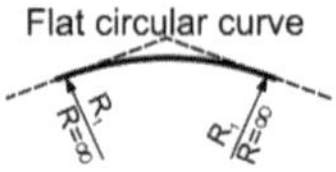	$R=\infty - R_1 - R=\infty$ 	- They are only permitted for small changes in direction ($\gamma < 10$ gon) and/or large radii. - The curve lengths should be 200 m in EKL 1 and EKL 2, 150 m in EKL 3 and 100 m in of EKL 4. - The cross slope should remain constant over a length of 60 m in the middle of the circular curve.	Satisfactory (check operating conditions)
Vertex clothoid 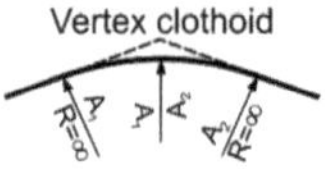	$R=\infty - A_1 - A_2 - R=\infty$ 	- They are only permitted for small changes in direction ($\gamma < 10$ gon) and/or large radii. - The curve lengths should be 200 m in EKL 1 and EKL 2, 150 m in EKL 3 and 100 m in of EKL 4. - The cross slope should remain constant over a length of 60 m in the middle of the circular curve.	Satisfactory (check operating conditions)

Fig. 12.13 Summary of the possible site plan curves (FGSV 2012a)

13 Design Elements in the Elevation Plan

Summary

The elevation plan shows the road in elevation. The reference value is the axis of the road, which is usually located in the center of the road. For carriageways separated by center reserves, an elevation plan with an axis in the middle of the carriageway (without hard shoulder) is usually created for each carriageway separately. This simplifies the representation of the required torsions and allows separate gradients for the optimization of the crossfall conditions in the central reservation, especially in bridge structures.

The elevation plan shows the gradient, the axis of the road in the elevation plan, and the terrain heights at axle height. These are either interpolated from the contour lines of the topographic maps or automatically generated from a digital terrain model. The more moving the terrain is and the more broken edges (water bodies, slope edges) there are, the more densely the terrain heights must be determined. This is especially true for intersecting roads or railway lines, which must be included in the design survey in a sufficient length. In the case of an alignment in the regional planning procedure, the draft survey is generally carried out in a corridor from 30 to 50 m on both sides of the road in order to leave sufficient scope for alignment optimisation within the framework of the design planning.

Taking into account the structural, geological and topographical boundary conditions as well as the limit values for longitudinal inclinations, crest and sag curves, a bending line is first laid in the terrain profile constructed from the site plan, which is later gradually replaced by longitudinal inclinations and crest and sag curves. It should be noted, however, that for reasons of harmonious spatial alignment, too many changes of inclination should be avoided (fluttering see Chapter 16: Spatial alignment). Furthermore, the sight distances resulting in combination with the site plan alignment must be taken into account, which should not suggest areas that are avoidably suitable

© Springer Fachmedien Wiesbaden GmbH, part of Springer Nature 2023

T. Richter, *Road Planning - Freeways and Country Roads*,
https://doi.org/10.1007/978-3-658-35189-2_13

for overtaking, but which are not suitable for overtaking after all. If such areas do occur, they should be marked by appropriate overtaking bans (see Chapter 15 Sight distances).

In addition, the elevation plan receives at least one curvature band, one slope band (ramp band) and one sight distance. Furthermore, the elements of drainage such as trenches, troughs and culverts are usually shown.

13.1 State of the Art

There are no current research results in Germany on the elements of the altitude plan. In earlier studies (Hiersche et al. 1984; Krebs and Klöckner 1977; Leutzbach and Zoellmer 1989; Durth and Levin 1991), the accident rate as a function of longitudinal inclination and crest radius was considered in particular. Mean longitudinal inclinations between ±2% have no influence on the speed behaviour and the occurrence of accidents. The range will currently have increased due to the significantly higher motorized vehicles. However, too large longitudinal inclinations ($s > 4\%$) increase the speed difference between vehicles and increase the risk of accidents. In addition, steep gradients increase the braking distances required.

Small crest curve radii in the range of the stop sight distance or below can no longer guarantee a safe stop in front of a suddenly appearing obstacle. Nevertheless, no accident black spots can be observed at the small crests that frequently occur in the existing network (Durth and Levin 1991). This is mainly due to the fact that suddenly appearing obstacles are extremely rare and can usually be avoided so that no accident happens. Nevertheless, such small crest curve radii should be avoided for safety and comfort reasons.

Somewhat more clearly, a safety risk was identified at sight distance levels between half and full overtaking sight distance. This should be avoided by choosing a suitable combination of site plan elements and the crest curve radius. It should be noted, however, that there are no jumps in visibility. This means that in the sequence of large sight distance (> 600 m) and small sight distance ranges just above the holding sight distance range, there are always sight distance ranges between half and full overtaking sight distance range. Here it should be examined whether the risk can be reduced by imposing overtaking bans (if necessary with the addition "agricultural traffic may be overtaken").

13.2 Longitudinal Inclinations

In principle, low longitudinal gradients ($s \leq 4\%$) should be aimed for, as they improve traffic safety and the quality of traffic flow. Low longitudinal gradients also reduce operating costs and emissions. Longitudinal gradients $s > 6\%$ are to be avoided despite higher limit values in the EKL 3 and 4. In the case of moving terrain, these advantages must be

Table 13.1 Maximum longitudinal inclination according to RAA (FGSV 2008a) and RAL. (FGSV 2012a)

Design class	EKA 1 A	EKA 1 B	EKA 2	EKA 3
Max s [%]	4.0	4.5	4.5	6.0
Design class	EKL 1	EKL 2	EKL 3	EKL 4
Max s [%]	4.5	5.5	6.5	8.0

weighed against better adaptation to the terrain, less interference with the surroundings and lower construction costs.

On the other hand, larger longitudinal inclinations avoid zones with weak drainage in torsional areas (see Sect. 14.3). In torsion areas, a longitudinal inclination of $s \geq 1.0\%$ (better 1.5%) must be maintained. If this is not possible, the minimum longitudinal inclination (in exceptional cases to be justified) must be $s = 0.7\%$. In flat terrain, the requirements are often difficult to meet or gradients contradicting the natural course of the terrain must be arranged, which can disturb the landscape. Remedial measures are described in Sect. 14.3.

Depending on the design class, the maximum longitudinal inclinations given in Table 13.1 should not be exceeded.

In order to prevent slipping on winter ice, the maximum slope must not exceed 10%, even if the maximum longitudinal slope is exceeded in exceptional cases due to the existing topography. In areas of at-grate intersections, small longitudinal gradients ($s \leq 4\%$) are to be aimed for due to acceleration and deceleration processes and the adaptation of the road surfaces; longitudinal gradients $s > 6\%$ are to be avoided.

In water protection areas, for example, roadways can also be limited by kerbs. In this case, the longitudinal inclination on board should be at least $s = 0.5\%$, otherwise pendulum gutters or other drainage measures must be arranged. In the case of long bridges (width $\geq$ 100 m) and in tunnels, a minimum longitudinal inclination of $s = 0.7\%$ must be maintained to ensure drainage. In the case of short bridges, especially at intersections or railway underpasses, this cannot always be guaranteed and is often not sensible for economic and design reasons. In these cases, drainage must also be ensured by other drainage measures (e.g. dense sequence of drains or box gutters).

13.3 Crest and Sag Curves

The intersections of the tangents with different longitudinal inclinations are rounded off with crests and sags. Crests are convex roundings and can be distinguished into true crests with changes in inclination from uphill to downhill and fake crests with changes in inclination without change in uphill or downhill. Sag curves are correspondingly concave roundings.

Legend for Fig. 13.1

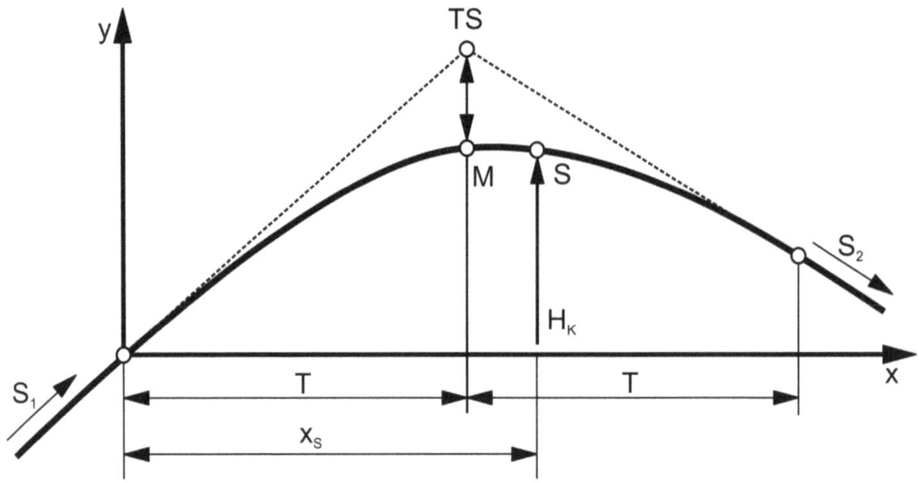

Fig. 13.1 Model for crest and sag curves. (FGSV 2012a)

H [m]	Crest and sag curve radius (radius of curvature of the apex of the square parabola)
T [m]	Tangent length of the curve $T = \frac{H}{2} \cdot \frac{s_2 - s_1}{100}$
s_1, s_2 [%]	Longitudinal inclinations of the two tangents
s(x) [%]	Longitudinal inclination of the gradient at any point within the curve $s(x) = s_1 + \frac{x}{H} \cdot 100$
y[x] [m]	Ordinate at any point within the curve $y(x) = \frac{s_1}{100} \cdot x + \frac{x^2}{2 \cdot H}$
x_s [m]	Abscissa of the vertex of the rounding $x_s = -\frac{s_1}{100} \cdot H$
f [m]	Stitch dimension from the tangent intersection point to the rounding off curve $f = \frac{T^2}{2 \cdot H} = \frac{T}{4} \cdot \frac{s_2 - s_1}{100} = \frac{H}{8} \cdot \left(\frac{s_2 - s_1}{100}\right)^2$
M	Center of rounding
S	Vertex (corresponds to the high point for crests and the low point for sag curves)
TS	Tangent intersection point

Sign rule:

Longitudinal inclination s as gradient [%].	Positive (+)
Longitudinal inclination s as gradient [%].	Negative (−)
Sag curve radius H_W [m]	Positive (+)
Crest curve radius H_K [m]	Negative (−)

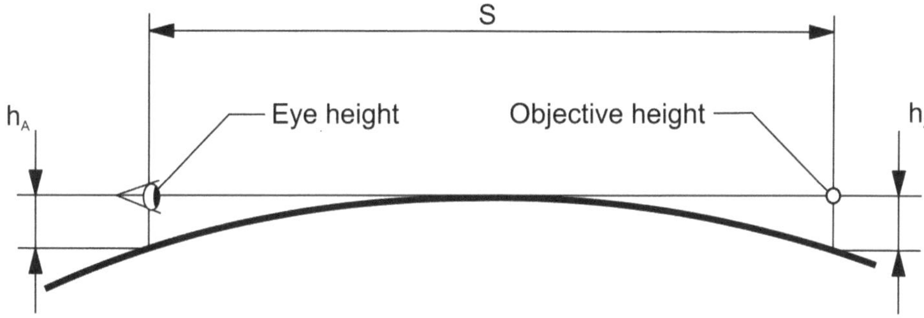

Fig. 13.2 Relationship between sight distance and tip radius. (FGSV 2012a)

In principle, the crest and sag curve radii are selected according to the topography in such a way that they result in a balanced spatial line in combination with the elements of the site plan (see Chap. 16). Furthermore, the required stopsight distance must be ensured.

The rounding of the crests and sag curves is done by circular curves (Fig. 13.1). They are approximately calculated with square parabolas.

The minimum required tip radius results from the requirement that an obstacle with a target point height of 1.00 m (generally a stationary car, from which about 0.50 m is visible) is detected in good time so that a following vehicle can still stop in front of this obstacle when the road is wet (Fig. 13.2). In the RAA (FGSV 2008a), the target point height is specified as 0.50 m. However, a uniform target point height of 1.00 m should be used, especially since at high speeds on motorways smaller obstacles cannot be detected at a great distance and obstacles other than the ends of traffic jams are rare. The eye point height is also set at 1.00 m.

S [m]	Sight distance
h_A [m]	Height of the eye point (1.0 m)
h_Z [m]	Height of the target point (1.0 m)

$$\min H_K = \frac{S_h^2}{2\cdot\left(\sqrt{h_A}+\sqrt{h_Z}\right)^2} = \frac{S_h^2}{8}$$

The RAA contains minimum tip radii while the RAL contains recommended values from which there may be a deviation of up to 15% in exceptional cases, which must be justified. All values are based on experience and ensure that the required holding ranges are given in the area of straight domes. However, this does not release the designer from the need to provide proof of sight distance. Dome radii that result in sight distance ranges between half and full overtaking sight distance encourage critical overtaking and should therefore be avoided on single-lane two-lane roads with guidance. If, due to constraining points, such dips cannot be avoided, overtaking using the lane for oncoming traffic should be prevented (Table 13.2).

Table 13.2 Top and trough radii on motorways and country roads. (FGSV 2008a, 2012a)

Design class	Minimum tip radius min H_K [m]	Minimum tub radius min H_W [m]	Minimum tangent lengths min T [m]
EKA 1 A	13.000	8800	150 (120[a])
EKA 1 B	10.000	5700	120
EKA 2	5000	4000	100
EKA 3	3000	2600	100
Design class	Recommended tip radius H_K [m]	Recommended tub radius H_W [m]	Minimum tangent lengths min T [m]
EKL 1	8000	4000	100
EKL 2	6000	3500	85
EKL 3	5000	3000	70
EKL 4	3000	2000	55

[a]Exception value for conversion and removal

If the recommended values are deviated from downwards, the spatial line management, i.e. the coordination of the site and elevation plan elements according to size and allocation, is of particular importance. The overlapping of small trough radii with small curve radii is to be avoided, because this causes problems with the recognizability of the design elements, whereby the required holding sight distance must always be maintained. In the case of moving topography, the tip radius must be selected to be larger than the trough radius, whereby the lower limit $H_W \geq 0.5\ H_K$ must be observed. In the case of flat topography, the trough radius must be selected to ensure better integration into the landscape and a visually harmonious course of the carriageway, and must be significantly larger than the tip radius. However, the minimum longitudinal inclinations in the twisting areas must be observed.

Design Elements in Cross Section 14

Summary

The cross-sectional components (carriageway, hard shoulders, verges), the drainage facilities, the transitions to the adjacent terrain (embankments or retaining walls) and the structural design of the road body are shown in cross sections (Fig. 14.1). Due to the cross-sectional representation, the road axis must be relocated in the layout or elevation plan if embankments, cuts or structures are too large.

14.1 State of the Art

In some research studies, a dependence of the V_{85} on the road width in connection with the curvature and the radius of the curve was proven (Lippold 1997). In the classes 6.10 to 7.00 and 7.10 to 8.00 m, the road width has only a small influence on the driving speed, because the linear relationship between the curvature and the V_{85} was very similar in both width classes. This was confirmed by the determination procedure for the V_{85} in the RAS-L (FGSV 1995).

In Sweden, there was a study on EKL 4 similar roads that showed that 6-m wide roads without centre markings are driven 6 km/h slower than with centre markings (Ihs 2006). The slowest driving speeds were found to be without centre markings and only poorly visible edge markings.

At the TU Dresden, the influence of road planting and road space design on the driving behaviour on off-road roads was investigated on the basis of real drives and drives in a driving simulator (Lippold et al. 2009). The influence on speed and lane behaviour could be proven and an orientation visual range was recommended, which corresponds to 1.3 times the minimum stopping sight distance.

© Springer Fachmedien Wiesbaden GmbH, part of Springer Nature 2023

T. Richter, *Road Planning - Freeways and Country Roads*,
https://doi.org/10.1007/978-3-658-35189-2_14

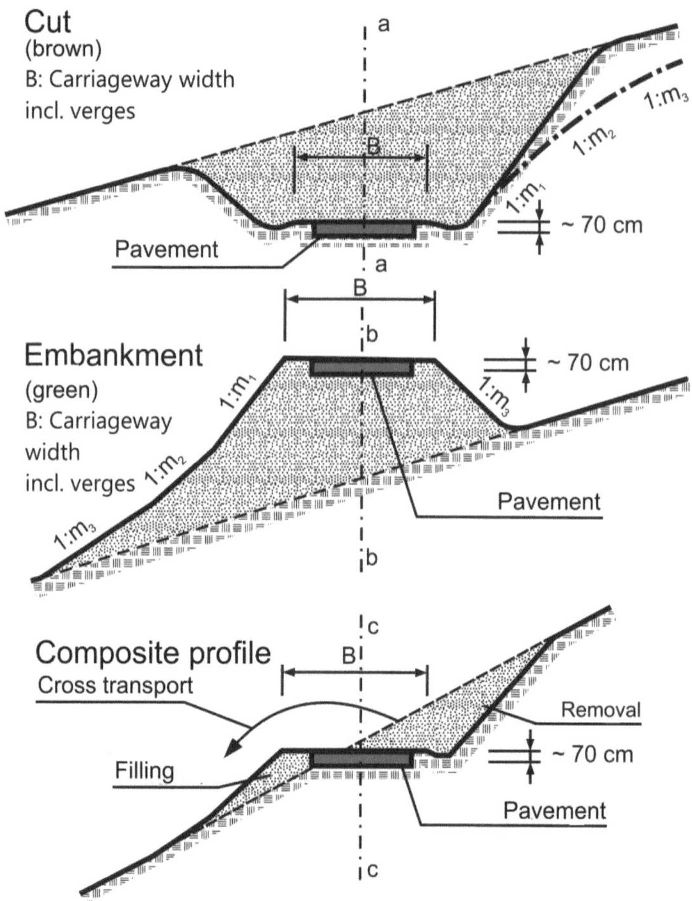

Fig. 14.1 Cross sections in cuts, embankment and composite profile (colours: representation in the elevation plan)

14.2 Crossfall

In principle, a crossfall of 2.5% is required for drainage. In the superelevation areas, where the crossfall changes and has a 0-passage, drainage must be ensured via a sufficient longitudinal inclination (Sect. 13.2). Verges are inclined outwards at 12% if the rainwater from the roadway is drained via this, otherwise at 6%.

14.2.1 Crossfall in the Straight Line

The crossfall of the carriageway in the straight line necessary for drainage must be designed according to Fig. 14.2. The roadway includes the lanes and the hard shoulders.

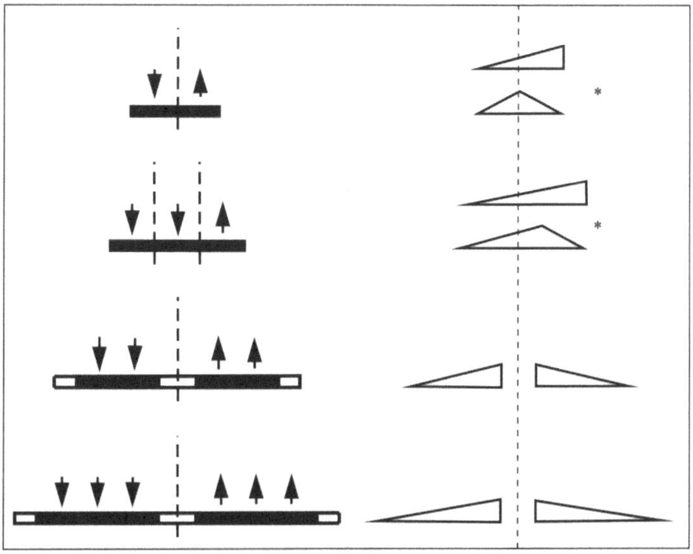

* When upgrading existing roads in exceptional cases

Fig. 14.2 Crossfall forms in the straight line

The minimum and standard crossfall of the carriageway in the straight line for all category groups is min q = 2.5%. Single carriageway roads with two or three lanes are usually arranged with a one-sided crossfall in the straight line. In exceptional cases, a roof profile may be economical for existing roads.

14.2.2 Crossfall in Circular Curves

The initial relationship for determining the relationship between curve radius and curve crossfall is the context of forces on the vehicle in the curve (Chap. 8).

For reasons of driving dynamics, curves are generally to be designed with a crossfall to the inside of the curve. The maximum crossfall max q is 6.0% on motorways according to RAA (in exceptional cases 7.0% to illustrate narrow circular curves) and 7.0% on country roads according to RAL. On bridge structures, the crossfall should be limited to 5.0% for manufacturing reasons. For drainage reasons, the minimum crossfall in a circular curve is equal to the minimum crossfall in the straight line q = 2.5%.

For reasons of manufacturing and driving dynamics, additional lanes, merging and diverging lanes and paved hard shoulders have the same crossfall as the roadway in curves in direction and size. The crossfall depending on the curve radius are shown in Figs. 14.3 and 14.4. The crossfall determined should be rounded up to 0.5%.

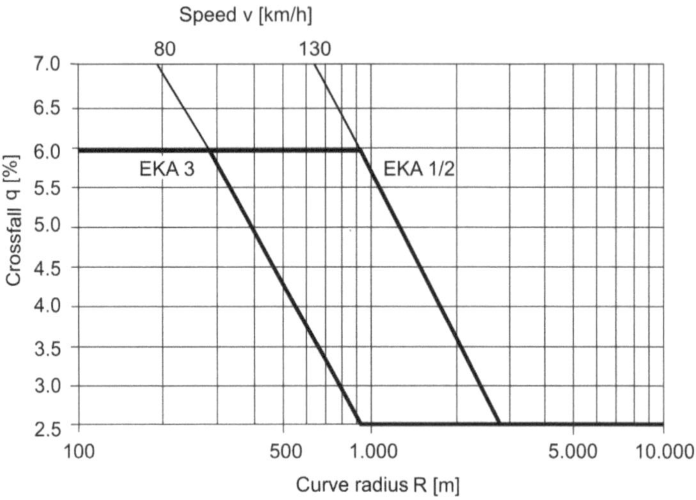

Fig. 14.3 Crossfalls on motorways as a function of the curve radii. (FGSV 2008a)

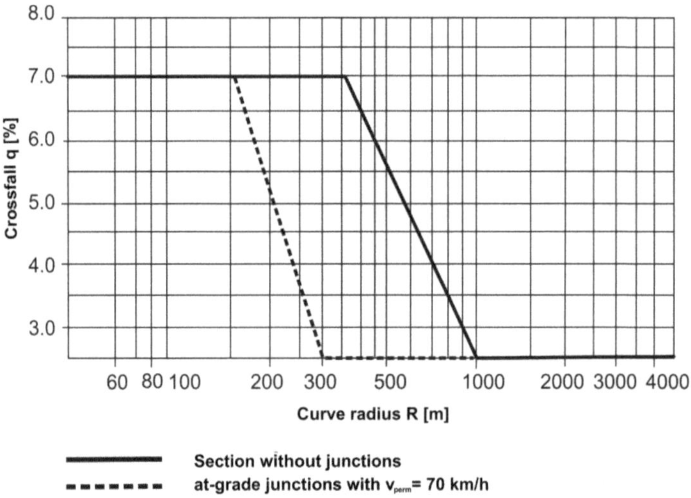

Fig. 14.4 Crossfalls on country roads as a function of the curve radii. (FGSV 2012a)

In the case of very large radii, the portion of the crossfall is not necessary to absorb the radial centrifugal forces, since sufficient radial force is available between the roadway and the tyres. It is therefore possible to use a crossfall of −2.5% to the outside of the curve to avoid zones with weak drainage in areas with insufficient longitudinal inclination or to avoid expensive central reservation drainage on two-lane roads. On country roads, this is generally possible at curve radii R > 3000 m. On motorways, the application depends on the design class (Table 14.1). Due to the significantly lower adhesion coefficients in wet

Table 14.1 Minimum radii for a crossfall to the outside of the curve according to the RAA. (FGSV 2008a)

Design class	min R [m]	Perm. $V_{wetness}$ [km/h]
EKA 1 A	4000	–
EKA 1 B	3200	120
EKA 2	1900	100
EKA 3	1050	80

conditions, the maximum permissible speed in wet conditions must be limited according to Table 14.1. A change of direction of the transverse gradient must be excluded in a curve sequence in the same direction.

14.3 Superelevation and Superelevation Development

When the radius and/or the direction of curvature is changed, the roadway crossfall changes in parallel. The change takes place within the transition curve (clothoid), regardless of the reference line of rotation and of which crossfall form (one-sided or roof-form crossfall) is used. Here, the edges of the roadway are ramped and the road surface is rotated. The road axis (usually the centre of the road) remains unchanged during the superelevation. An overlapping of the ramp onto the straight line or the curve of a circle is to be avoided.

If a transition curve is omitted for very large radii, the superelevation is done half in both elements. The change of the crossfall is usually done by turning the road surface around the axis of the carriageway for one-lane roads (Fig. 14.5) and around the axis of the directional carriageways for two-lane roads (Fig. 14.6).

In exceptional cases, single carriageway roads can be turned around the edge of the road. Also in exceptional cases, e.g. with a narrow central reservation (to avoid large crossfalls in the central reservation), when creating at-grade intersections in curves or when crossing the central reservation (i.e. for construction sites), dual carriageways can be turned around the edges of the central reservation or around the road axis. Alternatively, separate gradients can be used for the directional carriageways. A further argument for separate gradients is the restriction of sight distance in tight curves. The mode of operation of the safety barriers (separate or combined), which may limit the height difference of the road edges on the central reservation, must always be taken into account.

If two curves of a circle with the same direction of curvature meet directly ("basket curve", which should be avoided at first), the entire ramp should be carried out within the curve with the larger radius. On buildings, torsion should be avoided if possible for technical reasons and because of the greater risk of icing.

The relative grade of ramp Δs is the difference between the longitudinal inclination of the edge of the carriageway (edge of the paved surface) and the axis of rotation.

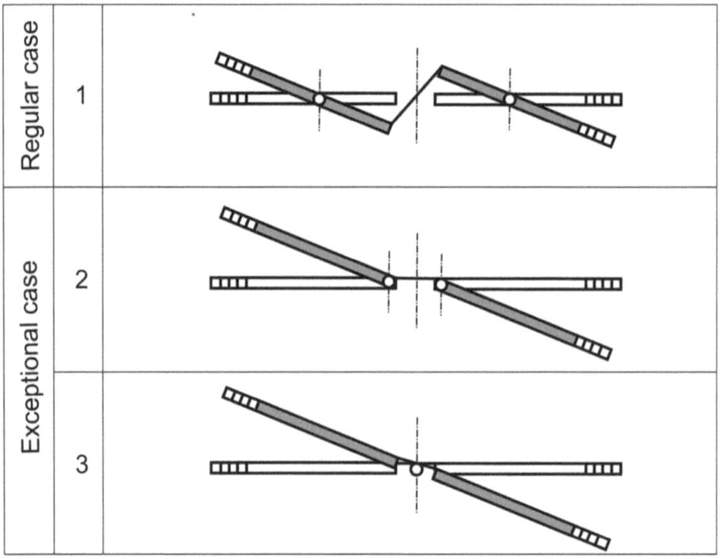

Fig. 14.5 Axis of rotation of the carriageway in superelevation sections of country roads. (FGSV 2012a)

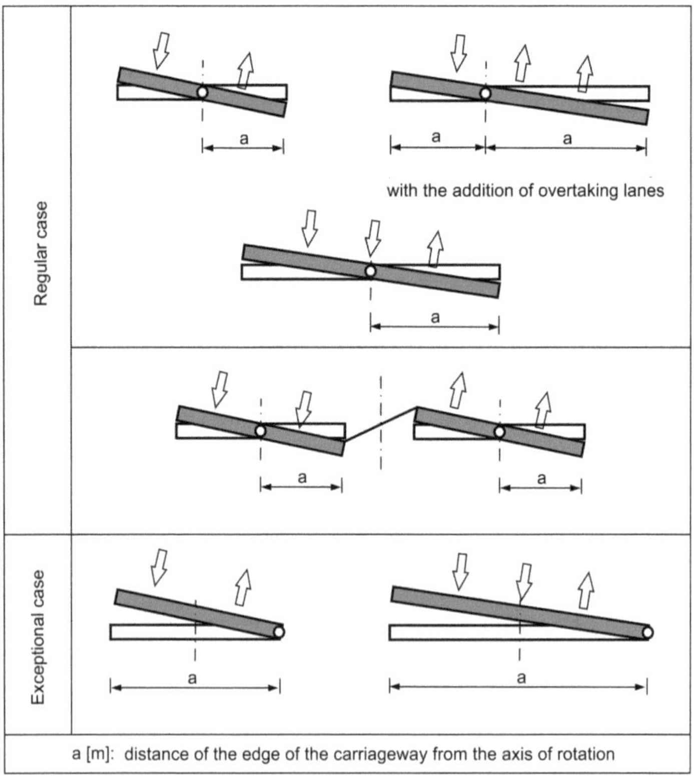

Fig. 14.6 Axis of rotation of the directional carriageway in superelevation sections of motorways. (FGSV 2008a)

$$s = \frac{q_e - q_a}{L_V} \cdot a$$

with:
Δs: Relative grade of ramp [%],
q_e: Crossfall of the roadway at the end of the superelevation section [%],
q_a: Crossfall of the roadway at the beginning of the superelevation section [%], (insert q_a negative if opposite to q_e),
L_V: Length of the superelevation section [m],
a: Distance of the edge of the roadway from the axis of rotation [m].

Within the transition section, a rapid increase in the crossfall, which can have an unfavourable effect on driving dynamics and appearance, should be avoided. Therefore, the maximum ramp relative grade max Δs should not exceed the values in Table 14.2. In case that due to an eccentric position of the rotation axis min Δs is greater than max Δs, min Δs must be applied.

The minimum length of the superelevation distances min L_V is determined taking into account the maximum relative grade of the ramp max Δs and the distance a of the edge of the road from the axis of rotation:

$$\min L_V = \frac{q_e - q_a}{\max \quad s} \cdot a.$$

It is not necessary to calculate for a rounding off of the kinks that occur at the beginning and end of the superelevation section, since the height differences are usually in the order of magnitude of the manufacturing tolerances.

In the superelevation areas, the ramp relative grade in the range + min q by 0 to −min q must not fall below the ramp minimum relative grade min Δs in Table 14.2. This is to ensure that the section of the road that is unfavourable from a drainage point of view is as short as possible. If this is still the case, a split superelevation must be arranged (Fig. 14.9), i.e., the range from +min q to −min q must be ramped up with min Δs. In addition, the longitudinal inclination and minimum ramp relative grade must be coordinated to ensure adequate drainage. On the section of the transition curve still available, the remaining ramps are completed until the crossfall required at the beginning of the curve is reached. If it is not possible to ensure sufficient longitudinal inclination due to special constraints for the area of the clothoid turning point, the zero point of the crossfall can be shifted by the length $L = 0.1 \cdot A$ in relation to the clothoid turning point (L, A [m]). This also applies to the sequence straight line - clothoid - curve. On roads with kerbs, it may be useful to perform the superelevation by turning around the inner edge of the curve of the carriageway, if otherwise, e.g. by superimposing s and Δs, the minimum longitudinal inclination of the drain gutter min $s_r = 0.5\%$ would be undercut.

In detail, the basic forms of roadway superelevation are shown in Figs. 14.7 and 14.8. The superelevation of all additional paved lanes or hard shoulders adjoining the carriageway is also performed along the length of the transition curve or the corresponding transition section of the carriageway. No special limit values apply to the ramp relative grade that occur in this case. The relative grade of the unbroadened carriageway is decisive for the relative grade with simultaneous widening or widening of the carriageway. No special limit values apply to the ramp gradient of the carriageway edge in the transition area.

Figure 14.9 shows the example of a long superelevation section with a split superelevation, which must always be arranged when min Δs is underrun.

The ramp relative grade Δs must not be greater than the longitudinal gradient, so that the direction of the longitudinal gradient (up or downhill) is the same in the road axis and at the edge of the road. In order to ensure drainage at the edge of the carriageway, RAL (FGSV 2012a) stipulates that the difference between the longitudinal gradient and the ramp relative grade (longitudinal gradient at the edge of the carriageway s_{Fbr}) will be

$$s_{Fbr} = s - \quad s \geq 0,2\%$$

s [%] = Longitudinal inclination
Δs [%] = Ramp relative grade

In the RAA (FGSV 2008a), a demand is made for $s_{Fbr} = \geq 0.5\%$ and only in exceptional cases 0.2%. This exceptional case should be dispensed with, as the fixed hard shoulders are not included in the calculation and no longitudinal inclination opposite to the longitudinal inclination of the axle should occur at their edges.

According to the RAS-Ew (FGSV 2005b), there is an additional requirement that the oblique inclination p, the resultant of longitudinal inclination and crossfall must be at least 0.5%.

In exceptional cases, which must be justified, if, for example, there is a zone of weak drainage in the area of the superelevation to be regularly provided for, the change in crossfall can also be carried out on a straight line. This requires a steering movement within the straight line, which is unfavourable in terms of driving dynamics. If the carriageway is simultaneously ramped and widened, the unwidened carriageway is used for the calculation of the ramp gradient. In the transition areas of city bypasses or widened sections of

Table 14.2 Limits of the ramp relative grade. (FGSV 2008a, FGSV 2012a)

		RAA: max Δs [%] for		
Design class	max Δs [%]	a < 4.00 m	a ≥ 4.00 m	min Δs [%] for q ≤ 2.5%
ECA 1, ECA 2		0.225 · a	0.9	0.1[a]
EKA 3		0.25 · a	1.0	
EKL 1, EKL 2	0.8			
EKL 3	1.0			
EKL 4	1.5			

[a][m]: distance of the edge of the road from the axis of rotation

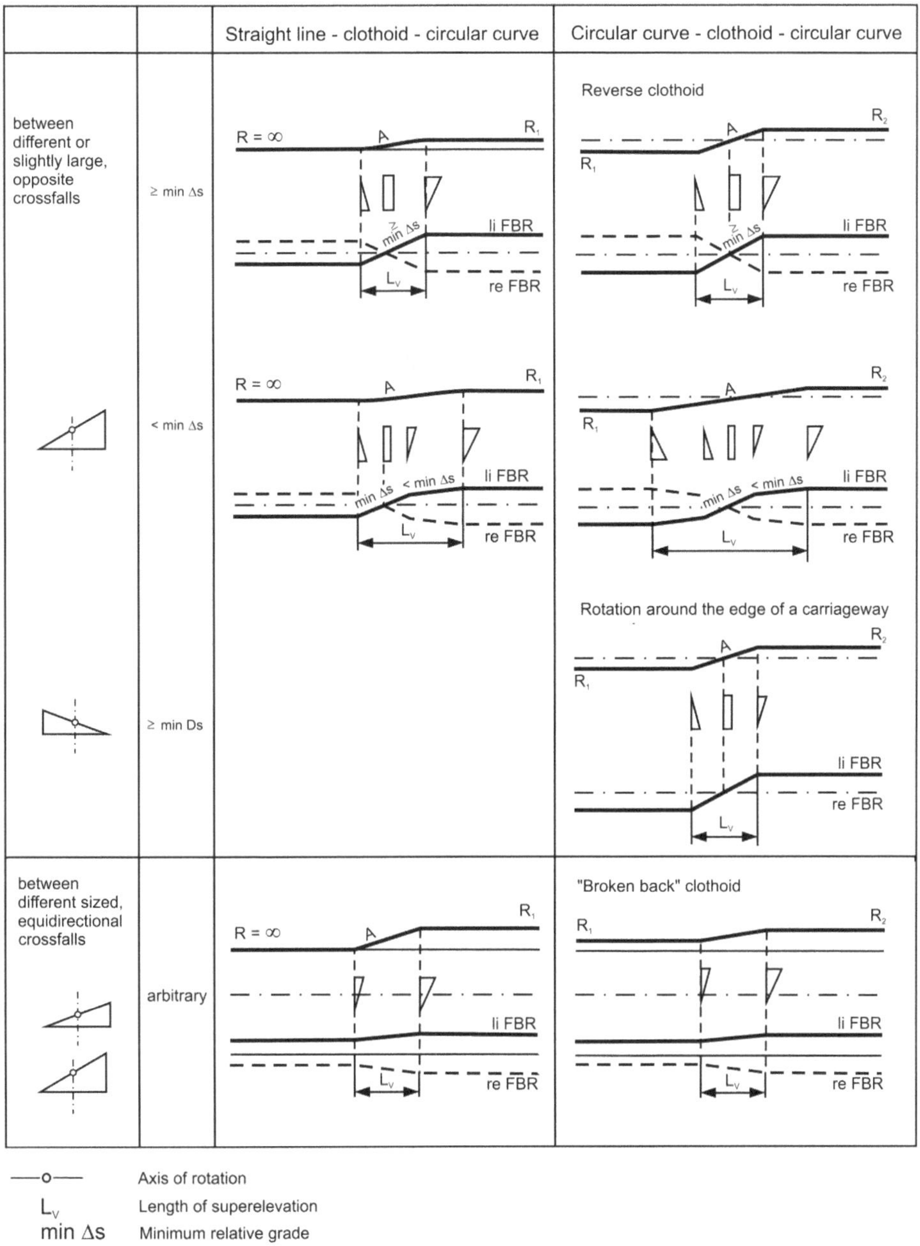

Fig. 14.7 Forms of the roadway superelevation, here: one-sided crossfall. (FGSV 2012a)

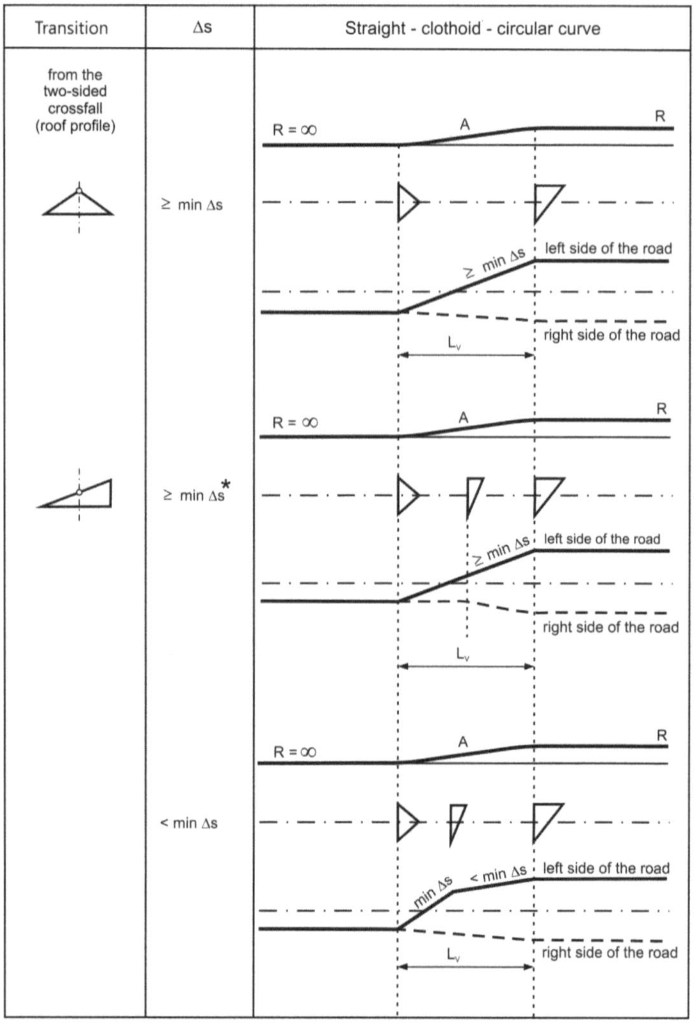

*	structural advantages: shorter ridge in the roadway axis, but longer area with min q in the transition curve
o	Axis of rotation
L_V	Length of superelevation development
min Δs	Minimum relative grade

Fig. 14.8 Forms of the roadway superelevation, here: roof profile. (FGSV 2012a, b, c, d)

road to existing roads, a transition from a one-sided inclination to a roof profile (two-sided crossfall) may be necessary. These are to be designed analogously to Fig. 14.8.

An alternative to avoid a zone with weak drainage is the installation of a diagonal superelevation. A study at the Technical University of Dresden (Lippold et al. 2011) has shown that there are no problems with the traffic flow and traffic safety at diagonal

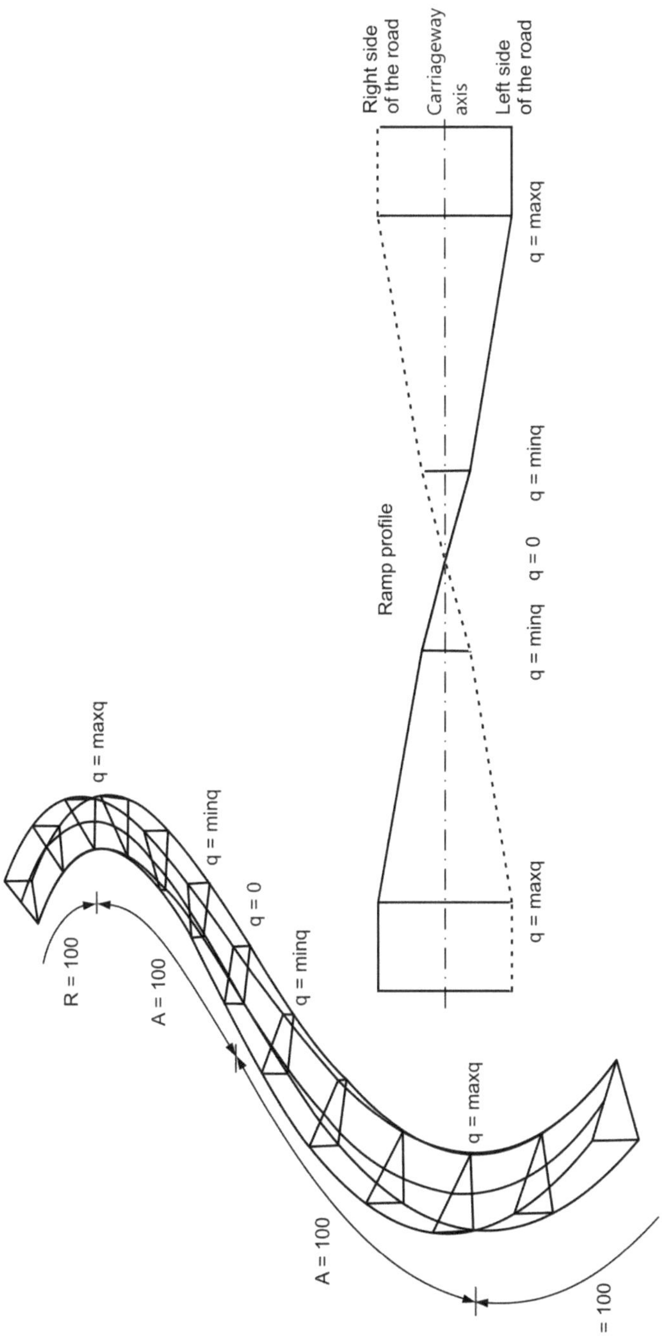

Fig. 14.9 Example of a long torsion range with split superelevation. (FGSV 2012a)

superelevations on motorways. Diagonal superelevations are more complex in construction than classical superelevations and require manual installation in the gusset areas. However, no durability problems have been reported. For this reason, diagonal superelevations are recommended as a good alternative in zones with poor drainage.

The RAA (FGSV 2008a) currently defines a critical water film thickness of 2 mm and points out that this could be exceeded on three- or four-lane carriageways despite compliance with the standard values of the alignment. As countermeasures, increases in longitudinal inclination (often not applicable due to constraint points), diagonal superelevations (cautiously recommended in the RAA, but based on the investigations of the TU Dresden (Lippold et al. 2011)), crossfall to the outside of curves with large radii (always to be aimed for due to the possible abandonment of central reservation drainage), open-pore asphalt (cost- and maintenance-intensive) and transverse box gutters (calculation method PlanUS or PSRM for federal measures available from the Federal Highway Research Institute (Bundesanstalt für Straßenwesen) or the state road construction authorities). However, maintaining a water film thickness of 2 mm requires relatively complex measures even at higher longitudinal gradients, so that an increase in the limit values is under discussion. Further results are expected shortly from the Technical University of Dresden and the University of Stuttgart within the framework of a research project of the Federal Highway Research Institute (FE 02.0321/2010) on the subject of "Comparison and evaluation of constructional solutions of low-discharge zones in torsional areas".

14.4 Widening of Carriageway in Tight Curves

When driving a curve, the rear wheels of a vehicle describe a narrower curve than the front wheels. As a result, the width of the roadway is required to be larger by the amount i in the curve than in the straight line. This applies to curves with radii $R < 200$ m. If the calculated roadway widening is less than 0.50 m, it can be omitted. The widening of the roadway is carried out over the entire length of the curve at the inner edge of the curve. The widening is distorted linearly within the clothoid.

$$i = \frac{100}{R}$$

i [m] = Widening of the roadway,
R [m] = radius of the curve to be widened.

If the linear distortion area is longer than the clothoid, the distortion is still linear with uniform inclusion of the adjacent elements. Often, very tight distortions result in an optically unsatisfactory solution. Here it should be checked whether a separate alignment of the inner edge of the carriageway is more appropriate.

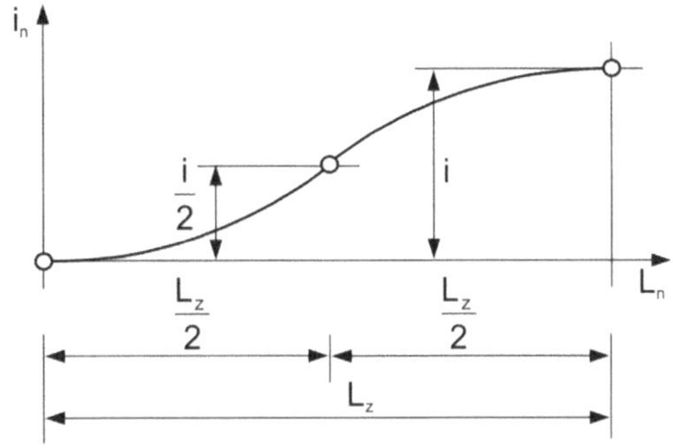

Fig. 14.10 Distortion of the roadway edges during road widening with two square parabolas without an intermediate straight line

$$i_n = \frac{2 \cdot i \cdot L_n^2}{L_z^2} \quad \text{for} \quad 0 \leq L_n \leq \frac{L_z}{2}$$

$$i_n = i - \frac{2 \cdot (L_z - L_n)^2}{L_z^2} \quad \text{for} \quad \frac{L_z}{2} \leq L_n \leq L_z$$

i [m] Road widening
L_z [m] Length of extention
i_n [m] Road widening at the point under consideration
L_n [m] Length of the extention at the point under consideration

14.5 Road Widening

When changing the cross section, when changing the central reservation width, for the installation of a lane divider, an additional lane or an merging/diverging lane, the continuous lanes must be distorted according to the changed cross section. In order to achieve an optically satisfactory guidance of the continuous lanes, the distortion is to take place in the area of small radii ($R < 300$ m) at the inner edge of the curve, in the area of a stretched line on both sides of the road axis. If widening is necessary for an overtaking lane, the widening of the lane is carried out completely on the side of the overtaking lane.

If possible, the edges of the carriageway shall be aligned independently of the road axis or shall be distorted with two square parabolas in the form of S-curves according to Fig. 14.10 and Table 14.3. If this does not result in a visually satisfactory solution or if there are counter-curves in tight radii, the distortion can be extended or the edges of the roadway can be aligned separately. If the width of the central reservation is changed, the edges of the road are also traced separately. The required lengths L_z are given in Table 14.4.

14.6 Special Features of the Alignment in the Area of Bridges and Tunnels

In order to keep bridges as short as possible and to avoid oblique bridges, the angle of intersection with the intersecting road or water body should be between 80 and 120 gon. In principle, it is advantageous if bridges are located in a section of co-extended alignment

Table 14.3 Unit distortion for expansion with two square parabolas

$a=\frac{L_n}{L_z}$	e_n	Δe_n	$a=\frac{L_n}{L_z}$	e_n	Δe_n
0.00	0.000		0.50	0.500	
		0.005			0.095
0.05	0.005		0.55	0.595	
		0.015			0.085
0.10	0.020		0.60	0.680	
		0.025			0.075
0.15	0.045		0.65	0.755	
		0.035			0.065
0.20	0.080		0.70	0.820	
		0.045			0.055
0.25	0.125		0.75	0.875	
		0.055			0.045
0.30	0.180		0.80	0.920	
		0.065			0.035
0.35	0.245		0.85	0.955	
		0.075			0.025
0.40	0.320		0.90	0.980	
		0.085			0.015
0.45	0.405		0.95	0.995	
		0.095			0.005
0.50	0.500		1.00	1.000	

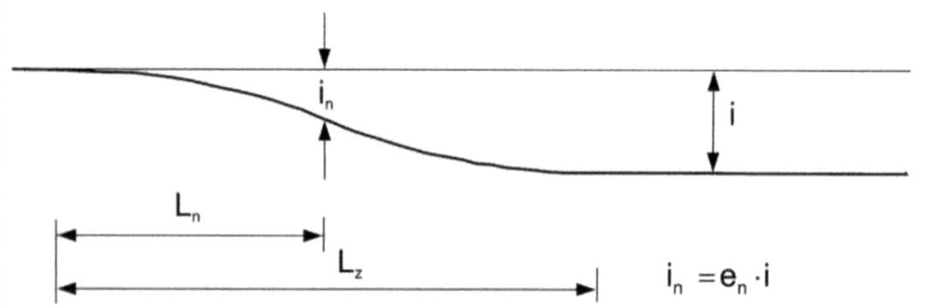

Table 14.4 Distortion lengths as a function of roadway widening and design class

Road widening i [m]	≤ 1.5	≤ 2.5	≤ 3.5	> 3.5
EKL 1 and 2	80	100	120	170
EKL 3	60	80	100	140
EKL 4	50	60	70	–

or a straight line in order to keep construction costs low or to allow special construction methods (incremental launching method).

On the other hand, the bridge structures should fit harmoniously into the general layout, so the position in a circular curve or in a crest or sag curve is optically more favourable. Here a compromise must be found between the construction costs and the harmonious lines. The size of the curve should be chosen so that the crossfall on the bridge structure is limited to 5.0% (on motorways of the EKA 1 and EKA 2 R > 1200 m and on country roads R > 550 m). This simplifies the construction of the bridge and allows the use of mastic asphalt. The use of a transition curve in the area of the bridge is unfavourable, but structurally feasible. A change in the crossfall should be avoided on the bridge.

In principle, a constant longitudinal inclination with a minimum value of s = 0.7% is advantageous. This should be aimed for in particular on long bridges (> 100 m). In the case of short bridges, for example at junctions or underpasses of waterways, this is often not advisable because otherwise artificial high or low points would have to be created, which contradict the course of the terrain, whereby low points should also be avoided for safety reasons. At such high and low points, drainage must be ensured by small distances between road gullies or by special constructions.

In the area of tunnels, the most generous possible site plan and elevation plan elements should be used. Here, max s = 3.0% should be used. In long tunnels (L > 500 m), max s = 2.5% should be aimed for.

Sight Distance

15

Summary

The German Road Traffic Regulations (BMVBS Straßenverkehrsordnung. Dortmund: Verkehrsblatt-Verlag, 2013) stipulate that the driver must choose his speed according to the sight distance conditions. If this legal standard were to be observed, the existing sight distance would be a good way of influencing speed behaviour. However, measurements and observations of the relationship between sight distance and speed show that existing sight distance has only a limited effect on the choice of speed. The following reasons can be given for this:

- Distances and speeds are difficult to estimate, especially while driving.
- At lower speeds, the actual braking distances are underestimated. On fast-traffic highways and motorways, speeds are underestimated.
- Emergency braking before sudden and unexpectedly occurring obstacles are rare events. The driver has no feeling for the forces involved.

As a consequence, the road design is based on the assumption that sight distance has a significant influence on traffic safety, performance and driving comfort and that it must be guaranteed as a design parameter. In order to achieve this, sight distance analyses are required already at the design stage. By reducing the driving space to only one dimension, the geometric sight distance is obtained. The minimum requirement is to check that the geometric sight distance (sight distance plan) in the driving space corresponds to the stopping sight distance required by the driving dynamics depending on the design class.

Basically, the most important design principle for the alignment of motorways and country roads is that a vehicle travelling alone can stop safely at any time when the road is wet and clean in front of a suddenly appearing obstacle. Since the required stopping

© Springer Fachmedien Wiesbaden GmbH, part of Springer Nature 2023

T. Richter, *Road Planning - Freeways and Country Roads*,
https://doi.org/10.1007/978-3-658-35189-2_15

distances are strongly dependent on the speed driven, design class-dependent values must be set for the speeds driven. On motorways, the recommended speed of 130 km/h is used for this purpose, unless speed limits are known in advance (e.g. on regional or urban motorways). In this case, these can be used. On country roads, the required stopping distance is based on the planning speed.

15.1 State of the Art

In a study of spatial alignment, based on the results of driving behaviour and accident events, limit and threshold values were defined, especially for hidden curve starts and visual distortions (Zimmermann et al. 2007a, b). An overtaking sight distance model was developed that considers the full overtaking sight distance according to RAS-L-95 (FGSV 1995) for the entire length of the overtaking path. Furthermore, a reduced overtaking sight distance, which allows a safe termination of an overtaking process that has started, and a marking sight distance, which can be used as a basis for marking lane boundary lines, were developed. At present (status 12/15), a research project to verify these overtaking models (FE 02.336) is underway at the TU Dresden, Chair for Design of Road Traffic Facilities on behalf of the Federal Highway Research Institute. It is to be expected that the value given in the RAL of 600 m for the safer execution of an overtaking procedure will be largely confirmed.

Fluctuating sight distance creates an excessive demand on the driver (Durth 1974). In contrast, a constant range of vision causes an adjustment of the driving behaviour.

A study on the so-called orientation sight distance came to the conclusion that the recommended sight distances should be in the range of 150 to 250 m on tural roads, depending on the design class and the desired traffic quality (Lippold et al. 2007). With sight distances below 150 m, obstacles on the road at 100 km/h can only be overcome with difficulty. It was also found that while low sight distance reduces speeds, it also increases the stress on drivers on country roads when sight distance drops below 200 m, causing faster driver fatigue. Therefore, the further course of the road should be sufficiently visible so that the driver can adapt his driving behaviour to the characteristics of the road without excessive demands or by using abrupt driving manoeuvres.

15.2 Required Stop Sight Distance

The required stopping sight distance s_h is defined as the distance required by a driver driving at a given speed (permissible V, recommended speed on motorways or planning speed on country roads) on a wet, clean roadway to bring his vehicle to a stop in front of an unexpected obstacle of a defined height on the carriageway (Fig. 15.1). It is composed of the distance travelled during the reaction and impact time (s_1) and the pure stopping

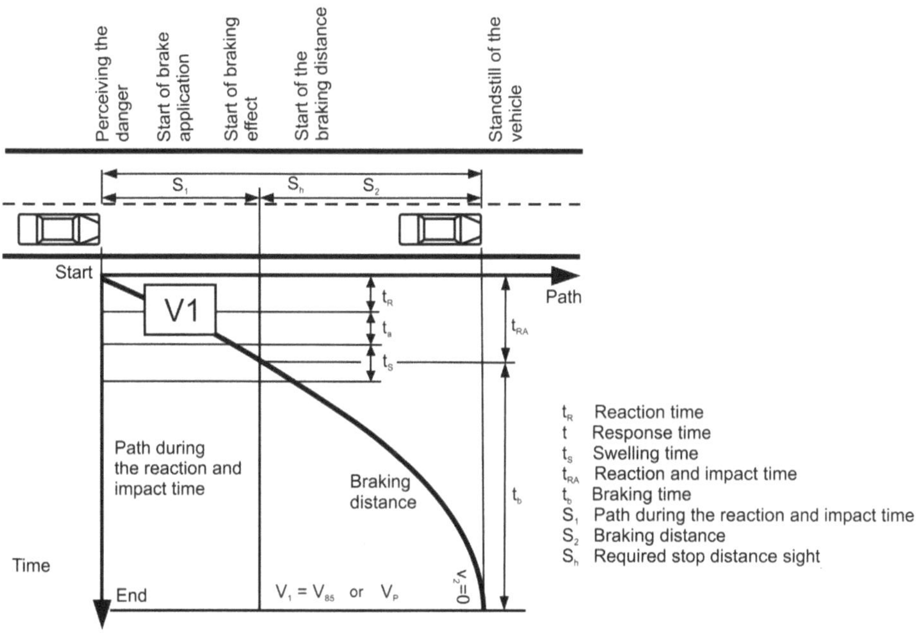

Fig. 15.1 Model for determining the required stop sight distance

distance (s_2) ($s_h = s_1 + s_2$). The reaction and effect duration is set at $t_R = 2$ seconds. This results in:

$$s_1 = \frac{V_0}{3{,}6} \cdot t_R.$$

The pure braking distance depends on the initial speed V_0, the tangential coefficient of friction f_T (V), neglecting the air resistance of the vehicle. For a braking process up to a stop, the following results

$$s_2 = \frac{\left(\frac{V_0}{3{,}6}\right)^2}{2 \cdot g \cdot \left(f_T \pm \frac{s}{100}\right)} = \frac{\left(\frac{V_0}{3{,}6}\right)^2}{2 \cdot \left(a \pm g \cdot \frac{s}{100}\right)}$$

s_1 [m] = Distance during the reaction and effect duration,
V_o [km/h] = initial speed,
t_R [s] = Reaction and action time,
s_2 [m] = braking distance,
g [m/s^2] = Acceleration of gravity,

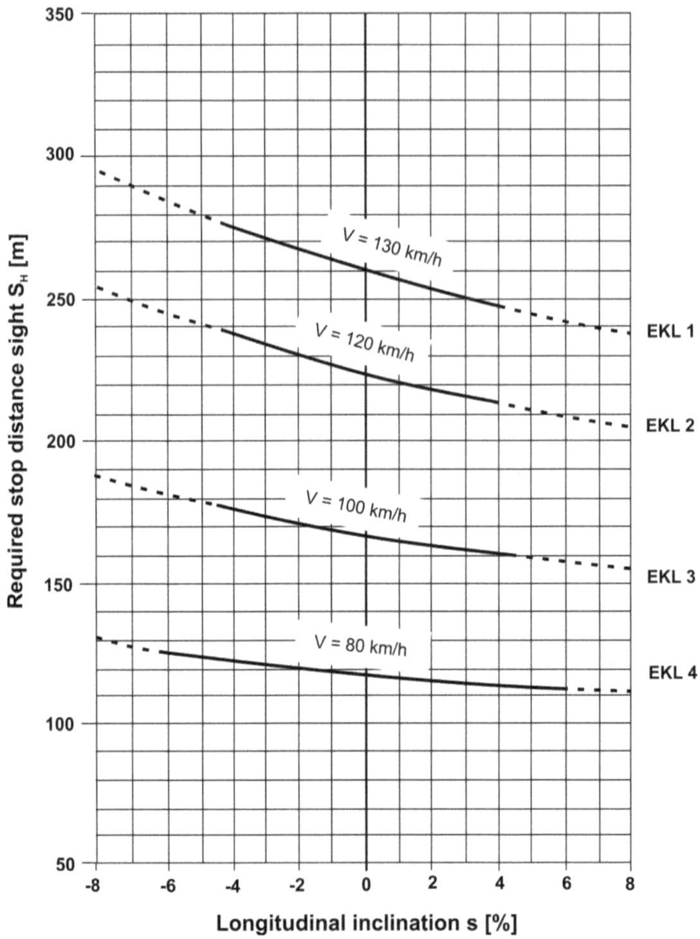

Fig. 15.2 Required stop sight distance s_h on motorways. (FGSV 2008a)

f_T [−] = tangential coefficient of friction a/g (Sect. 8.3),
a [m/s²] = transmittable wet braking deceleration (without ABS, medium value),
s [−] = longitudinal inclination.

The required stop sight distance can be found in Figs. 15.2 and 15.3. Average longitudinal inclinations are to be assumed and determined section by section.

15.3 Orientation Sight Distance

Low sight distance in the area of the stop sight distance requires increased attention from drivers and should therefore be limited to a few route sections (Lippold et al. 2007). It is advantageous if, on the predominant sections of the road section, the existing sight

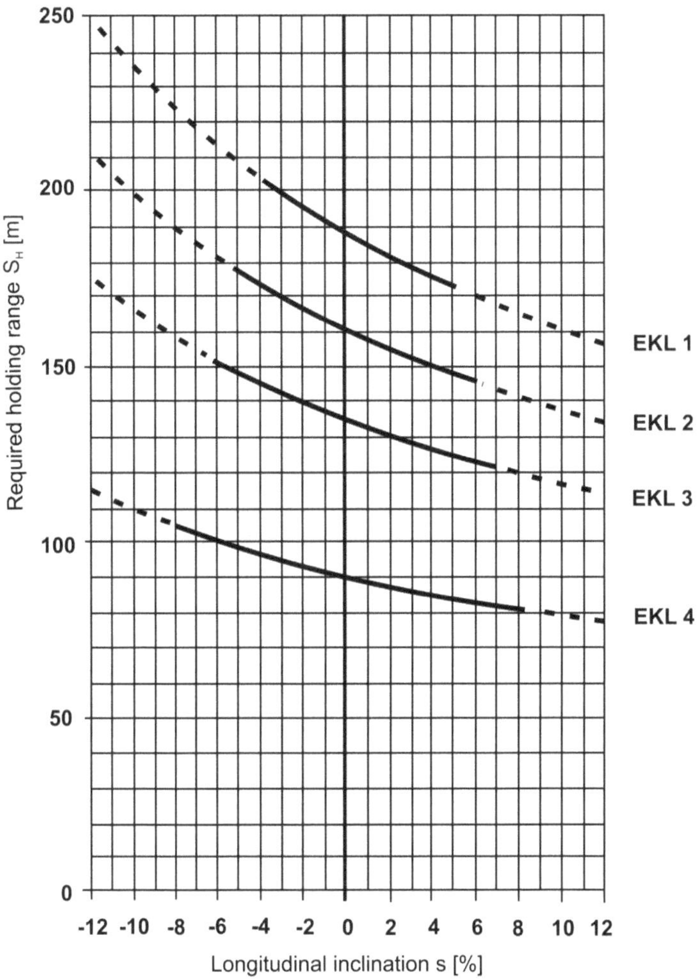

Fig. 15.3 Required stop sight distance s_h on country roads. (FGSV 2012a)

distance is at least 30% greater than the required stop sight distance, especially where greater attention is required (e.g. at intersections).

15.4 Required Overtaking Sight Distance

Required overtaking sight distance $s_ü$ is defined as the distance required to safely execute an overtaking manoeuvre (whereby overtaking is an increased risk even if sight distance is sufficient). It is composed of the path of the overtaking person, the path of the oncoming person during the overtaking process and the safety distance between these two at the end of the overtaking process (Fig. 15.4).

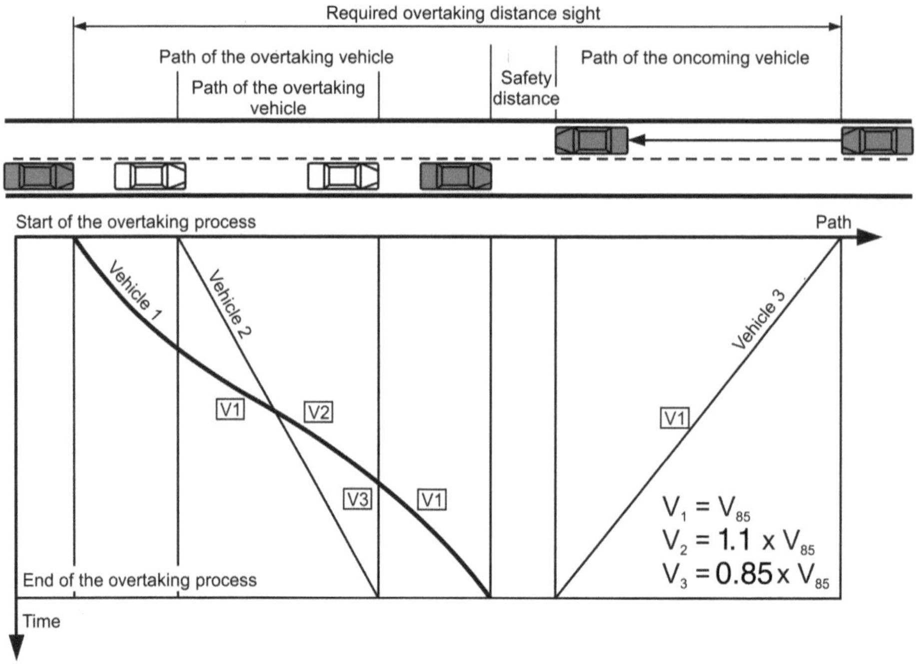

Fig. 15.4 Model for determining the overtaking sight distance. (FGSV 1995)

The absolute size of the required overtaking sight distance depends on the assumed speeds V_1, V_2, V_3. In the RAS-L (FGSV 1995), the speed of the person overtaking was assumed to be $1.1 \cdot V_{85}$. This posed the legal problem that with a V_{85} generally available at 100 km/h, the model assumed a driven speed above the permissible maximum speed on country roads.

In the combination of different speeds, a minimum sight distance of 600 m can be derived, which is required for an overtaking manoeuvre. For example, 325-m sight distance is required to overtake an agricultural vehicle. In principle, it is advantageous for the traffic flow if there are many sections of road with sufficient overtaking sight distance to satisfy the overtaking needs of faster moving vehicles.

It is problematic, however, that despite sufficient overtaking sight distance, a large number of serious overtaking accidents can be observed on country roads. It is also problematic that many drivers overtake even when sight distance is poor and in some cases are not deterred by bans on overtaking (Richter and Ruhl 2014).

In EKL 1 and EKL 2, where mainly high average travel speeds are to be achieved, and as a rule, longer distances are covered, the need for overtaking is to be satisfied to a large extent in the traffic-secured overtaking sections (overtaking lanes). A release of the two-lane road sections in the EKL 2 in areas with sufficient overtaking visual distances ($s_{ü}$ > 600 m) with a double broken line should only be considered in exceptional cases where overtaking distances are insufficient.

Also in EKL 3, overtaking should not be taken as a design principle, as overtaking is not absolutely necessary due to the average distances and the desired average speeds. Therefore, no minimum road sections with overtaking sight distances are required. Rather, it should be checked whether, in the case of an increased need to overtake and few sections of road suitable for overtaking, overtaking lanes can also be arranged at the EKL 3.

Sections of the route where sight distance is between 300 and 600 m are problematic, because overtaking opportunities are suggested there, which are not available after all. Such sight distances cannot be avoided, however, as the existing sight distances usually gradually increases and decreases. In the above-mentioned study on overtaking accidents (Richter and Ruhl 2014), it was found that overtaking was also carried out in sections with even lower sight distance than 300 and that these overtaking manoeuvres are particularly risky. Therefore, it could be considered to prohibit overtaking in all road sections with sight distances less than 600 m. On sections of road with a higher volume of agricultural traffic, consideration could be given to allow such vehicles to overtake by means of an additional sign (at sight distance > 325 m). However, this currently contradicts the principle of the StVO that traffic restrictions should only be imposed in areas where the potential hazard clearly exceeds the general risk. In this case, a less restrictive application of the StVO would be desirable due to traffic safety.

15.5 Existing Sight Distance

The actual sight distance of a road is determined by the lines of sight in the site plan and elevation plan, the cross section and the road environment. When determining the existing sight distances, an eye point height and a target point height of 1.00 m is assumed. The reference point for determining the stop sight distances is the centre of the right lane. On roads of the EKL 4, the centre of the lane is the reference point. For determining the stop sight distances in left-hand curves of directional carriageways, the eye point and the target point are located in the middle of the left lanes. For overtaking sight distances, the eye point is located in the centre of the right-hand lane and the target point in the centre of the left-hand lane.

In principle, the existing sight distance is determined spatially. Since the two-dimensional sight distances determined in the site plan and elevation plan are usually smaller than the spatial sight distances, it is sufficient in the preliminary planning or if the required sight distances are clearly exceeded, to determine them separately according to the site plan and the elevation plan. For the spatial determination of the sight distance, it is recommended to use a road design software, because the graphical determination is relatively complex. When using the software, particular attention should be paid to the correct consideration of vertical sight distance obstacles (planting, safety barriers, ...) and the results should always be critically examined.

In the site plan, the existing sight distance must first be determined in those street sections where cuts > 1 m (eye level $h_{Z} = 1.00$ m) occurs, because only there are visual impairments due to cuts. In all other sections, the existing sight distance at known visual obstructions (bridges, plantings, safety barriers, …) must be determined.

The determination can be done using the circle segment formula:

$$s = 2\sqrt{2rh - h^2} \quad \text{and} \quad h = r - \sqrt{r^2 - \left(\frac{s}{2}\right)^2}$$

s [m] = Sight distance,
r [m] = Radius,
h [m] = Segment height = visual dimension = distance between the visual axis of the eye and the visual obstacle at a height of 1.00 m each (Fig. 15.5).

On single-carriageway two-lane roads, it can be assumed in a simplified way that the distance between the eye's visual axis and the visual obstacle is approximately 7 m (1/2 lane 1.75 m, hard shoulder 0.50 m, verge 1.50 m, trough 2.00 m, slope width at 1.00 m height at n = 1:1.5 corresponds to 1.50 m). This results in a sight distance of

$$s = 2\sqrt{14r - 49}.$$

For example, R = 300 m results in a sight distance of about 130 m. The required stop sight distance for an EKL 3 (134 m at s = 0.0%) is thus just barely met. Either the distance (at 8.00 m results in a sight distance of about 138 m) or the radius (at R = 350 m results in a sight distance of about 139 m) has to be increased.

The existing sight distance in the elevation plan is determined in the same way as the calculation of the minimum crest curve radii using the following formula:

$$\min H_K = \frac{S_h^2}{2 \cdot \left(\sqrt{h_A} + \sqrt{h_Z}\right)^2} = \frac{S_h^2}{8} \rightarrow s_h = \sqrt{8H_K}.$$

The existing sight distance levels are compared with the required sight distance levels by means of sight distance plan, which must be set up separately for both directions of travel.

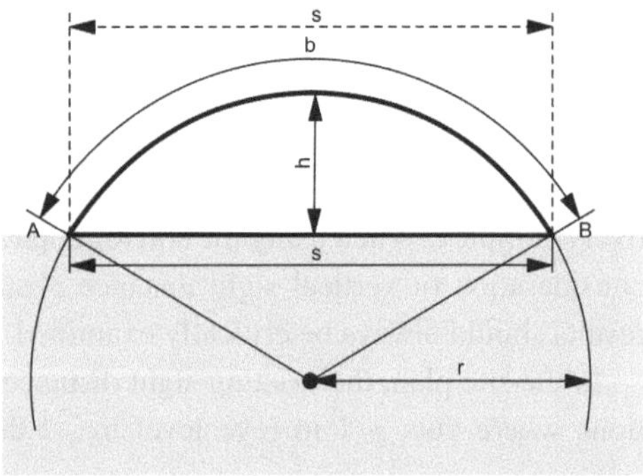

Fig. 15.5 Circle segment formula

In detail, the following requirements must be met:

- For safety reasons, there must be unrestricted stop sight distance on all roads along the entire length of the road. If stop sight distance cannot be guaranteed, a speed limit in the wet should be considered.
- The change in sight distance (decrease and increase) should be balanced and adapted to the situation. The decrease in sight distance should be gradual.
- In the field of vision to be kept clear, all obstacles impairing sight distance (e.g. embankments, planting, noise protection equipment, etc.) should be avoided up to the height of the visual beam. Rows of trees in a loose formation as well as individual trees and bushes may remain in the field of vision to be kept clear if their obstruction of vision is insignificant (very short in time and space) and if they serve the visual traffic guidance or the design of the driving area.

On carriageways with central reservations, the safety barrier systems are often the visual obstacle. On the basis of the simplified geometric model shown in Fig. 15.6, it is

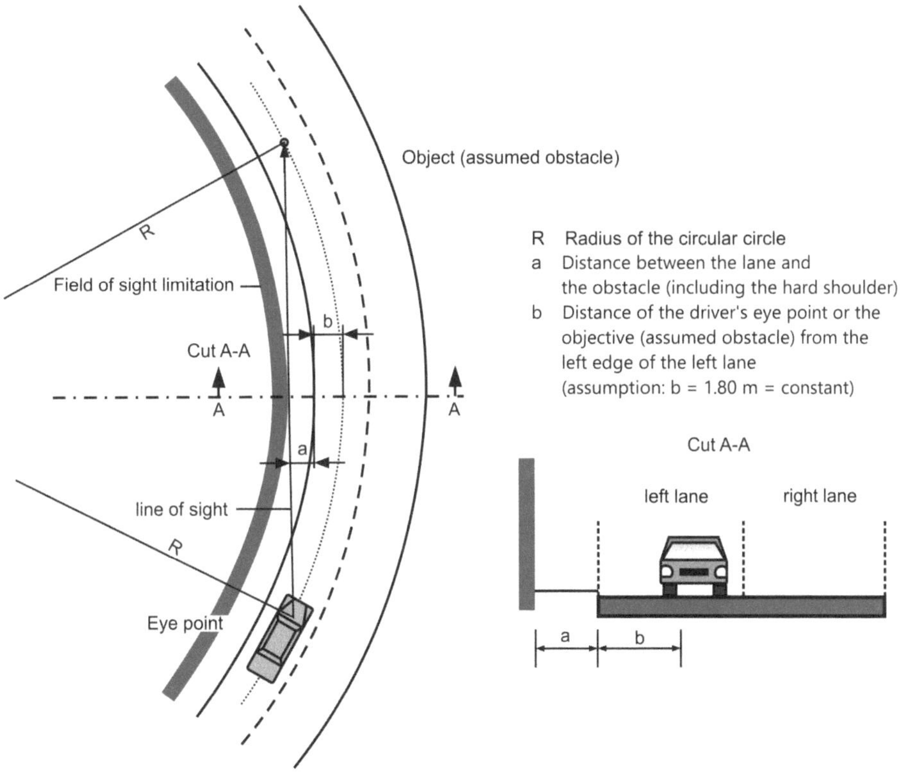

Fig. 15.6 Geometric model for determining the existing sight distance on directional carriageways in left-hand curves. (FGSV 2008a)

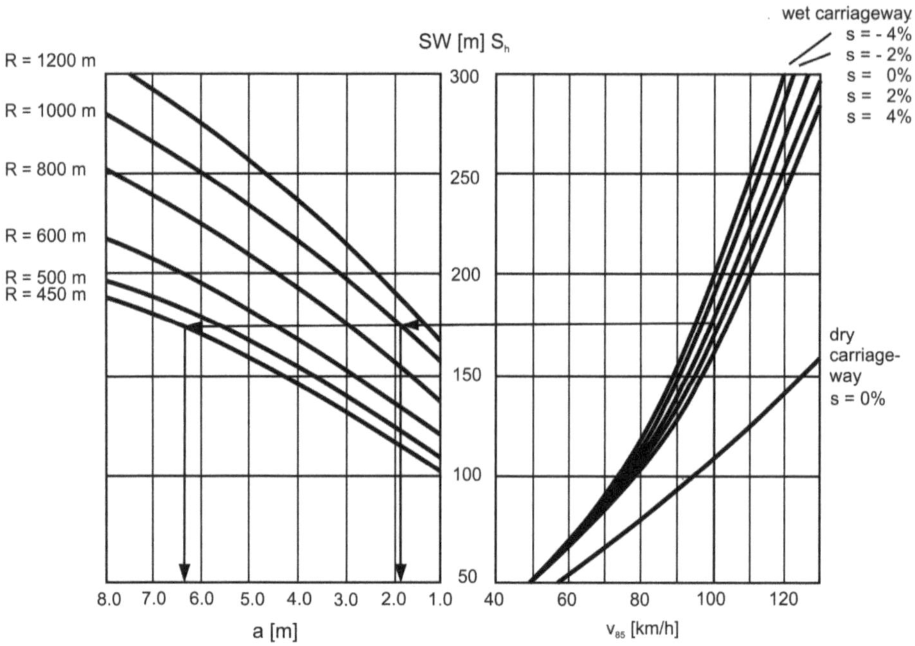

Fig. 15.7 Required stop sight distance and distances between the left edge of the inner lane of a carriageway and visibility obstacles in the central reservation. (FGSV 2008a)

necessary to check whether the required stop sight distance is maintained in a left-hand curve and with vehicle restraint systems higher than 0.90 m or with appropriate vegetation (Fig. 15.7). A distance of the eye point from the edge of the left-hand lane of 1.80 m is assumed. In the case of crest curves, a three-dimensional model shall be used to check whether there is an obstruction to sight distance depending on the radius, the radius of the crest curve and the height of the safety barrier systems.

If there is an obstacle to visibility, it shall be checked whether the stop sight distance can be maintained by larger crests curves, larger radii or lower safety barrier systems. Otherwise, speed restrictions may only have to be imposed in wet conditions.

Particularly at high speeds, it should be noted that in darkness and bad weather conditions, the existing sight distance is limited not only by the road space but also by the range of the headlamps.

Spatial Alignment 16

Summary

The alignment of a road is three-dimensional and must therefore always be viewed spatially. However, the processing and presentation of the road design is mainly done separately in the site and elevation plan and in the cross section. The overlapping of the horizontal and vertical design elements, including the cross-sections, creates spatial elements. These spatial elements and the spatial element sequences resulting from their combination can be illustrated in the form of perspective images or models. The driver's perspective is the only useful perspective representation for traffic assessment of a road.

The driver orients himself to the spatial course of the road and adjusts his driving behaviour accordingly. The spatial alignment thus has a significant influence on road safety. The spatial course of the road is determined by the carriageway and the adjacent side area. Timely recognition, comprehensibility, clarity and unambiguousness of the routing are important for good orientation and an appropriate choice of speed based on this. Orientation is decisively supported by the road equipment such as markings and signs as well as by guiding devices such as delineators and traffic guidance systems. Furthermore, it is possible to improve orientation and to integrate the road harmoniously into the surrounding landscape by appropriate design of the surroundings through terrain modelling or planting. Especially the course at crest curves and in non-visible curve areas is to be clarified with vertical elements. However, a sufficient safety distance must be ensured.

16.1 State of the Art

Current research on spatial alignment is available in particular from the universities of Dresden and Karlsruhe. Weise et al. (2002) have developed a practicable method for the consideration of the spatial alignment of out-of-town streets. The findings have been

© Springer Fachmedien Wiesbaden GmbH, part of Springer Nature 2023
T. Richter, *Road Planning - Freeways and Country Roads*,
https://doi.org/10.1007/978-3-658-35189-2_16

incorporated in the development of the RAL. The same applies to the contribution of KIT (Zimmermann et al. 2007a, b), which deals with the control and evaluation of the spatial alignment of out-of-town streets on the basis of quantitative parameters. In this work, spatial element sequences were evaluated and overtaking visibility distance models were verified. One result of further development of previous work by Zimmermann was the QuaSi-Band, which describes the quality of visibility conditions.

Lippold et al. (2009) have investigated the influence of roadside vegetation on driving behaviour and road safety. Basically, a positive influence of planting on driving behaviour and pre-orientation was found. They also recommended sight distances greater than the minimum required stopping sight distances, which then resulted in the RAL in the recommendation of the orientation sight distance. This also confirmed earlier work specifically on the orientation visibility range by Lippold et al. (2007). The approach of the research work was the question of whether, due to the shorter braking distances of modern passenger cars, smaller required stopping distances will be sufficient in future. As a result, reductions are possible, which were also implemented to a moderate extent within the framework of the RAL, but for psychological and physiological reasons, larger visibility ranges (orientation sight distance = 1.3 fold of required stop sight distance) make sense.

16.2 Site Plan Elements

Straight lines should only be used if they can be well adapted to the terrain or existing structures (e.g. railway lines or canals). The unfavourable impression in moving topography caused by the rigidity of a layout line can be somewhat mitigated by a sag curve with a large rounding radius. As a rule, a more favourable option is a site plan curve with a very large radius. In both cases, it is then possible to assess the distance of vehicles and to better estimate the speed of oncoming vehicles.

Short curves with small tangent lengths between straight lines appear from the driver's perspective as a kink, which can only be avoided by increasing the radii of the curves. The radii should be all the larger, the smaller the change in axis direction. The size of successive design elements in the site plan is determined by the given radius relations and relations between transition and circular curves (Sects. 12.5 and 12.6). The safety of the traffic flow on a winding route with many curves per unit of route is not significantly affected by a smaller curve. The higher curvature in EKL 3 and 4 associated with a lower planning speed is not equivalent to an increased risk of accidents. It is crucial that discontinuities (Fig. 16.1) are avoided. For this reason, especially in the area of small and medium curve radii, the relations of adjacent radii according to Sect. 12.5 must be observed.

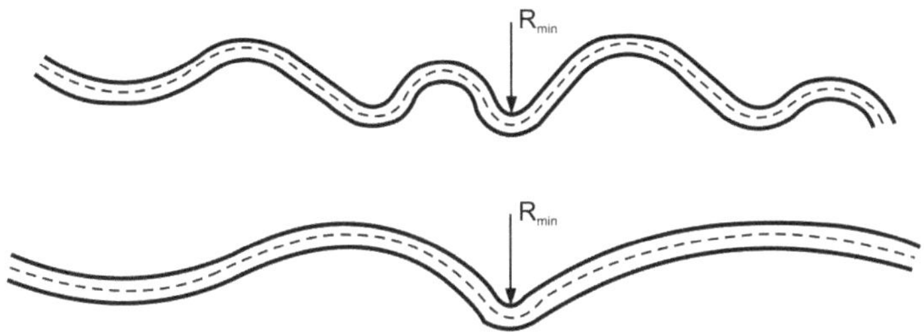

Fig. 16.1 Continuous and non-continuous road. (FGSV 2012a)

16.3 Height Plan Elements

The straight line in the elevation plan is a route with constant longitudinal inclination. It does not pose any major problems with regard to the spatial guidance of a road if it is adapted to the topography. A short straight line between two successive sag curves can simulate a crest curve and should be avoided. Similarly, a short straight line between two successive overlooked crests can simulate a sag curve. These two phenomena must be taken into account in the course of the line, especially in the case of bridges (board effect).

The sag curve is an element with good optical guidance. However, in order to achieve harmonious lines, short sag curves should be avoided between long sections with constant longitudinal inclination (kink effect). It is irrelevant whether the route in the site plan is straight or curved. In sag curves, however, there is the problem of optical distortion (see Sect. 16.4). The spatial layout of the line in the crest area has a decisive effect on the visibility conditions. Crests with small radii restrict visibility.

The element sequence in the elevation plan should largely follow the shape of the terrain. In hilly terrain, the crest curve radius should be larger than the sag curve radius to facilitate the visibility dimensioning (Fig. 16.2). On the other hand, in the case of small differences in height (up to 10.0 m) and on level terrain, the radius of the sag curve should be larger than that of the crest curves to ensure a visually satisfactory roadway (while maintaining the stop sight distance).

Multiple changes in inclination due to short crests and sag curves in a section that can be overlooked should be replaced by a more rapid gradient. Flat cuts and embankments can also be accepted, as these can be adapted to the landscape by means of appropriate greening.

16.4 Overlay of Site Plan and Elevation Plan

The overlapping of sag curves and curvatures of the site plan can lead to misjudgements by the driver. In particular, sag curves in site plan curves give the driver a more generous line than it actually is. Therefore, the relationship between curve radii and sag curve radii

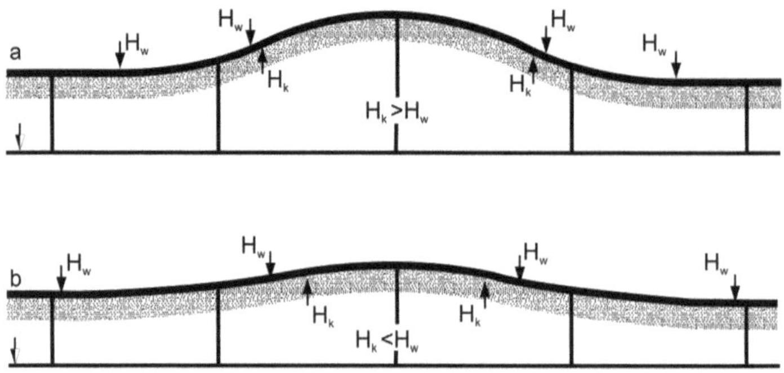

Fig. 16.2 Ratio of crest curve radius (H_k) to sag curve radius (H_w). (FGSV 2012a)

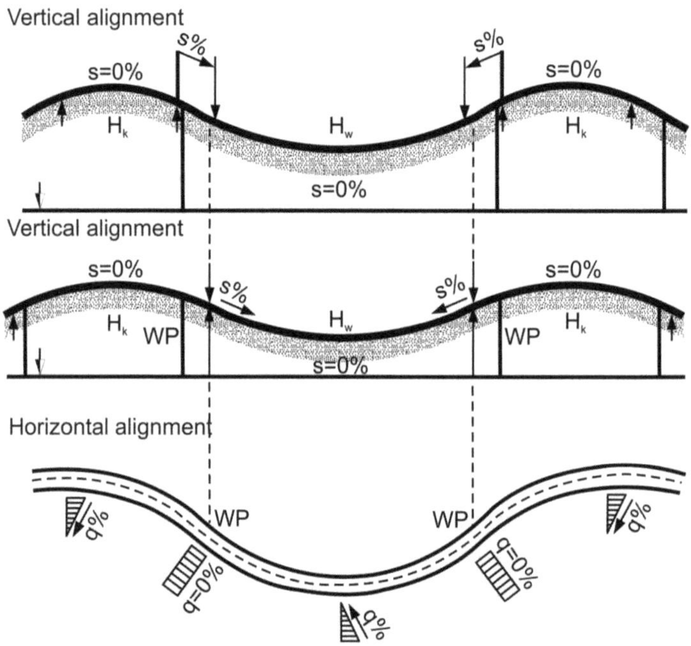

Fig. 16.3 Assignment of the elements in the site and elevation plans

must be coordinated. Experience has shown that the ratio site plan curve radius/sag curve radius should be as small as possible, not larger than about 1/5 to 1/10. The flatter the terrain, the larger the crest and sag curve radii should be in relation to the site plan curve radii.

An optically, drainage-technically and driving dynamically advantageous alignment of the road is generally guaranteed if the turning points of the curvature in the site and elevation plan are approximately at the same place (see Fig. 16.3).

This rule assumes an equal number of turning points in the site and elevation plan. Depending on the topography and other route boundary conditions, the number of turning points can vary considerably. It is particularly unfavourable if the number of turning points in the elevation plan is greater than in the site plan. If a different number cannot be avoided, the turning points of one plane should not coincide with tangent intersections of the other plane but should be as close as possible. The surplus turning points then fall in between. The longitudinal section plan element should be enclosed by the site plan element.

An exception are isolated narrow curves in sag curves. They can appear larger than they actually are due to the "sag curve effect". In order to avoid optical illusion by overlapping of sag curve and site plan curve, the beginning of the curve and the beginning of the sag curve should be pulled apart as far as possible. Either the start of the sag curve must be moved far enough into the site plan curve so that the sharpness of the curve is clearly visible, or the sag curve is moved forward so that the site plan curve only begins at its end.

For reasons of recognisability and clarity, junctions should be located in sag curves wherever possible. If this is not possible for both roads because of the topographical conditions, at least one, if possible the one with the secondary traffic significance should be designed as a sag curve. In the first place, therefore, is the view of the minor road to the point where right of way must be granted, and in the second place is the stop sight distance of the driver with the higher speed on the major road. The recognisability of intersections can be increased by suitable means (planting, guidance systems, signage). However, if the junction is located in the sag curve, special attention must always be paid to drainage.

Good visual guidance is a prerequisite for sufficient safety and traffic quality of a road system. It is achieved when the image of the road appears "calm" and when its course is sufficiently clear and can therefore be grasped and clearly understood in time. Optical guidance is initially achieved by imaging the carriageway and becomes all the clearer the better the visually sensitive carriageway edges and carriageway separation (e.g. by marking) are marked. Special importance is attached to edge guidance in warped areas in the case of carriageway widening. Bridge structures should be integrated into the alignment of the road. To avoid the board effect of a straight bridge between two sag curve, the sag curve radii should be particularly large and have long roundings.

In principle, good optical guidance is achieved by choosing generous elements such as the radius and length of the curves as well as the radius and tangents of the crest and sag curves. A disadvantage, especially on roads of EKL 3 and EKL 4, is that a too generous choice of elements does not result in the desired driving behaviour. For example, the large speed differences between cars and heavy goods vehicles build up overtaking pressure, which can lead to unsafe driving manoeuvres.

16.5 Standard Spatial Elements

The overlapping of horizontal and vertical design elements creates spatial elements and spatial element sequences. The sequence of these spatial elements leads to a satisfactory spatial alignment when standard spatial elements are used. Standard spatial elements (SREs) are created when a site plan element (e.g. curve) and a height plan element (e.g.

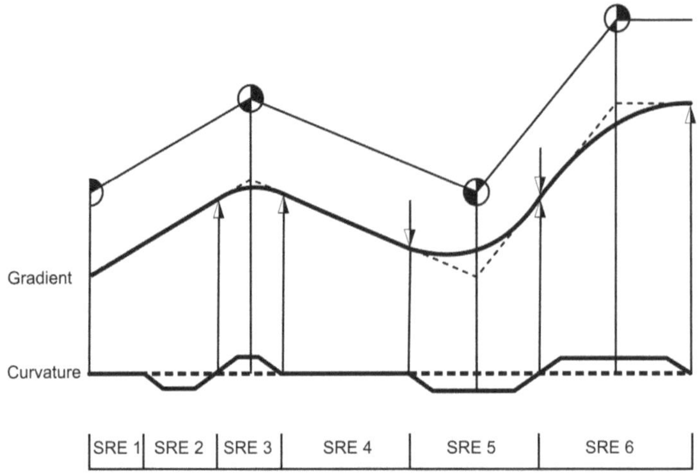

Fig. 16.4 Division of a road section into standard spatial elements (SREs). (FGSV 2012a, b, c, d)

crest curve) overlap, i.e. the beginning and the end are approximately one above the other. Straight lines in the site plan and in the elevation plan correspond to curves with $R = \infty$ or crests with $H = \infty$. When superimposing, the start and end points may be shifted up to about 20% of the length of the site plan element (Fig. 16.4).

Six different SREs can be derived from the superposition of the site plan elements straight line and circular curve with the elevation plan elements straight line, crest and sag curve. These elements are shown in Fig. 16.5 with their perspective images and spatial evaluation.

Very important for road safety is the timely recognition of the element sequences. This is especially true for the beginning of a curve. When overlapping the elements of a site plan and crest curve, the recognizability of the beginning of the curve must be ensured. This means that the curve must have a minimum directional change angle of 3.5 gon in the field of vision in front of the crest. Otherwise, the start of the curve must be shifted before the start of the crest curve according to Table 16.1. Approximations of a straight line and an opposite curve result in almost identical values.

When a circular curve is superimposed on a sag curve with a ratio $R/H > 1{:}10$, the circular curve may appear optically larger than it actually is and may be driven on at excessive speeds. In order to avoid this so-called critical strains, the beginning of the curve should also be before the beginning of the sag curve in this case, i.e. a displacement is only permitted if the elevation plan element is enclosed by the site plan element (see Sect. 16.4).

16.6 Deficits in Spatial Alignment

Deficits in the spatial alignment can (but need not) arise if no spatial elements are created and the sequence of spatial elements does not correspond to the above principles. The deficits in terms of traffic safety must be assessed differently (Table 16.2).

Horizontal/Vertical alignment	Perspective image	Evaluation
Straight line with constant gradient		
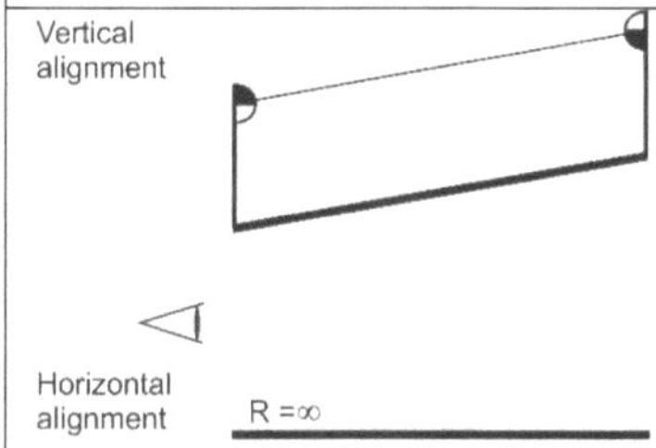		- works with large length monotone - works in the landscape frequently rigid - favors glare effects - suitable for flat landscapes - good for sight on intersections
Straight sag curve		
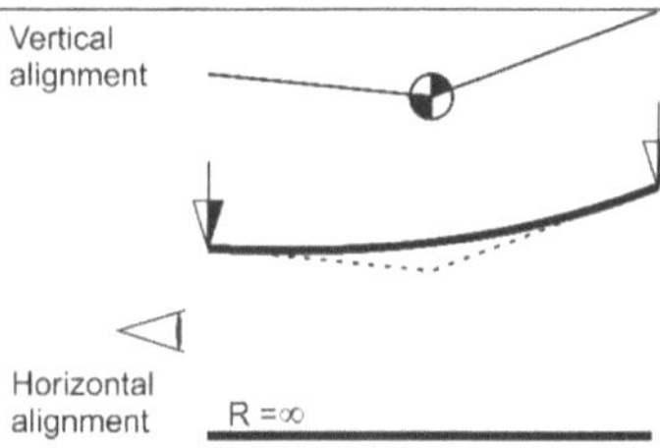		- guarantees good visibility and a good optical guidance - with large Radius one alternative to flat straight lines - suitable for intersections - convenient for overhauls
Straight crest		
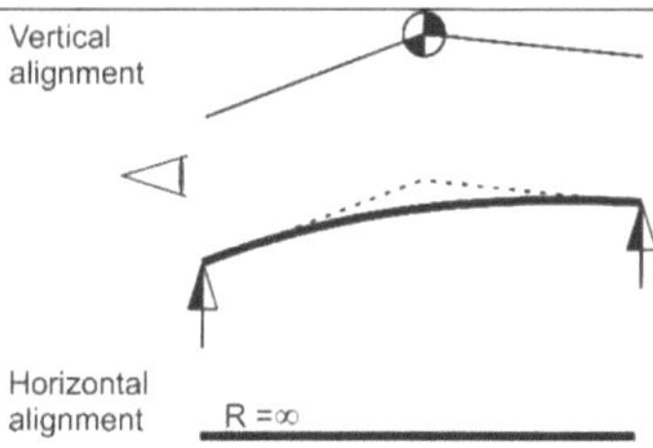		- restricts the view - worsens the optical guidance - not suitable for intersections

Fig. 16.5 SREs by superimposing site plan and longitudinal section plan elements

Curve with constant longitudinal inclination

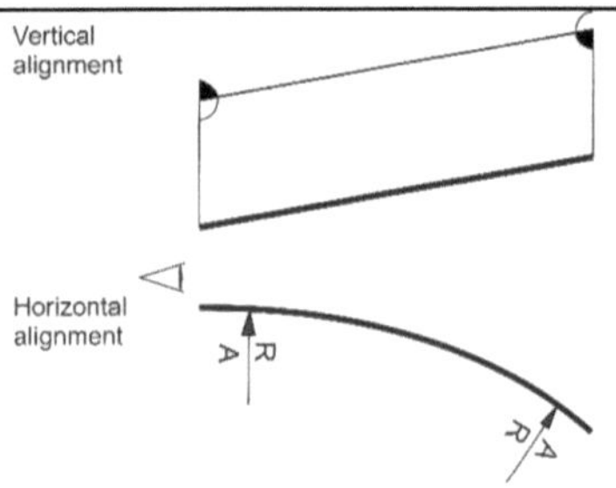

- with sufficient visibility unproblematic
- enables a good adaptation to the environment
- is suitable for creation of overtaking lanes and transition areas

curved sag curve

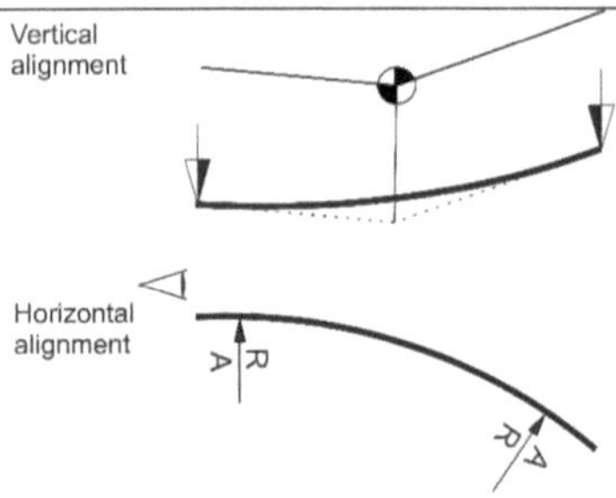

- is with sufficient big design elements unproblematic
- tends to lead to high speeds
- enables a good adaptation to the environment
- improves the optical guidance
- is critical for overlay with small horizontal radii

curved crest

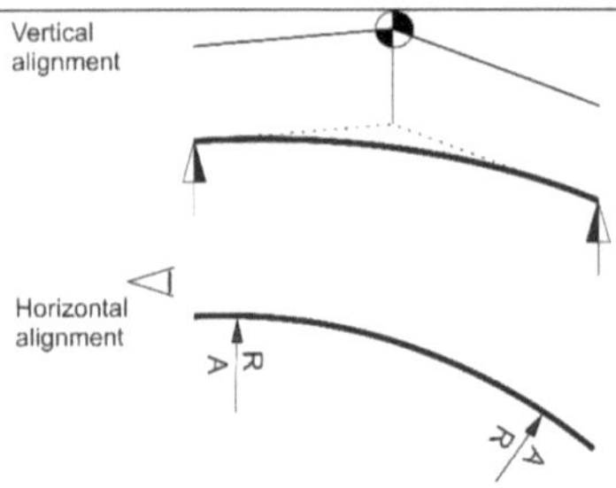

- is with sufficient big design elements unproblematic
- improves the optical guidance
- tends to lead to low speeds
- is for intersections not suitable

Fig. 16.5 (continued)

When building new roads, deficits of all groups should be avoided. This applies at least to visual shadows as well as to concealed curve beginnings when converting and extending roads.

Table 16.1 Required displacement [m] of the beginning of the crest curve behind the beginning of the curve at the transition from straight line to clothoid to circular curve. (FGSV 2012a, b, c, d)

Crest curve radius	Clothoid parameter A [m]			
H_K [m]	150	200	250	≥ 300
3000	25	50	65	80
4000	15	35	55	75
5000		25	50	70
6000		15	40	60
7000			30	55
8000	No displacement required		20	45
9000			10	40
10.000				30

Table 16.2 Effects of deficits in spatial alignment. (FGSV 2012a)

Deficit	Influence on driving behaviour and road safety
Visual shadow (jumping/diving)	High
Hidden start of curve	High
Elongations	Average
Compressions	Low
Design deficits	Low

16.6.1 Visual Shadows

Visual shadow areas occur if the driver (eye point height $h_a = 1.0$ m) over a distance of at least 75 m cannot see a section of the carriageway in the preceding driving area within a distance of 600 m (Fig. 16.6). Visual shadow depths of more than 0.75 m should be avoided. Such visual shadow depths are particularly critical where the oncoming lane must be used for overtaking. If such sections cannot be avoided, overtaking should be prevented. Visual shadow areas can also be created by jumping and diving, i.e. the short sequence of small crest and sag curves (Fig. 16.7).

16.6.2 Concealed Curve Start

The start of a curve is called concealed if, from a distance of 75 m before the start of the curve in the site plan, the road surface ahead cannot be seen at least up to the point where there is a change of direction of 3.5 gon in the site plan (see Sect. 16.5). This means that the point of the relevant change of direction is in the area of the crest visible to the driver. For large clothoid parameters (A ≥ 300 m) it is sufficient if the clothoid is visible at least on 100 m length (Table 16.1).

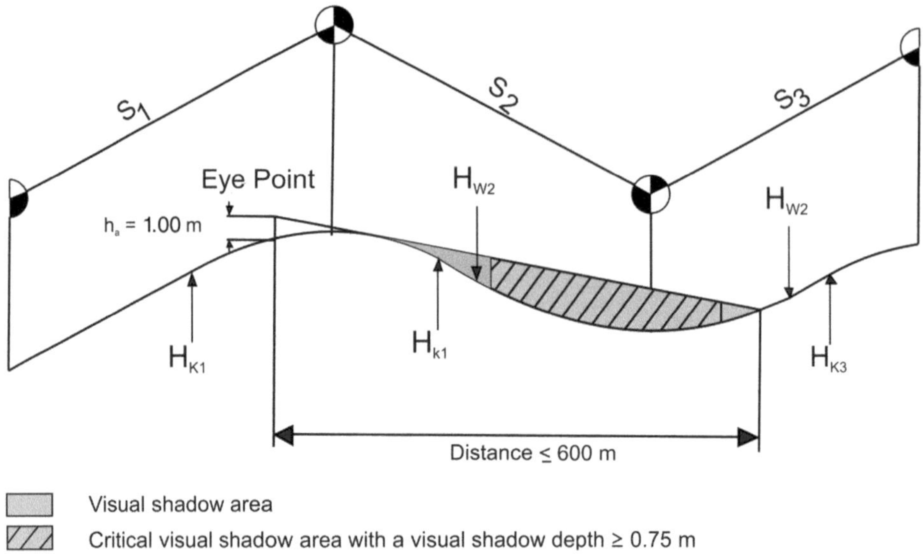

Fig. 16.6 Critical visual shadow area (FGSV 2012a)

Horizontal/Vertical alignment	Perspective image	Note
"Jumping"		- Avoidance of multiple vertical alignment elements with one horizontal alignment element - Use if possible big diameters (>5,000 m) - Longer and deeper viewing shadow with smaller first crest radius - Avoidance of constant longitudinal inclinations between sag and crest curves for changes in longitudinal inclination smaller about 8 % - In the area of the crest vertical curve should for example checked with the help of perspective images the "diving" and "jumping" at areas with constant longitudinal inclination in the vertical alignment
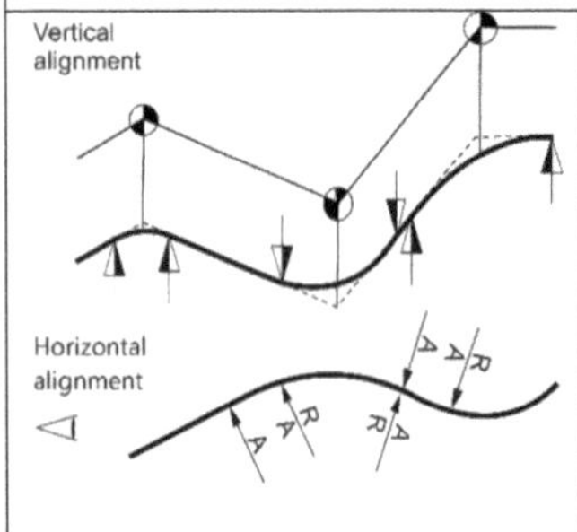		
"Diving"		
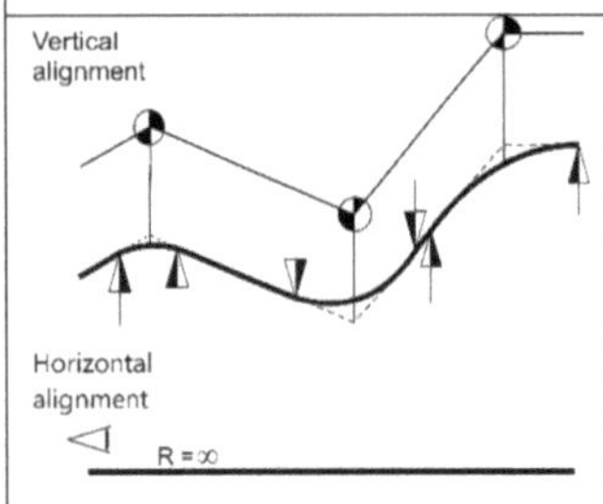		

Fig. 16.7 Visual shadows with jumping and diving

16.6.3 Delongation and Compression

In a sag curve, the radius of a circular curve appears more stretched, which can lead to the fact that the curve is entered too quickly. In a crest curve, the radius of a curve appears more upset, which can lead to a (safety-related) slower entry into the curve. The ratio of the site plan radius to the radius of the elevation (R:H) determines the magnitude of the elongation/compression (Fig. 16.8).

16.6.4 Design Deficits

Irregularities in the carriageway layout give the driver an unbalanced alignment (Figs. 16.9 and 16.10).

Horizontal/Vertical alignment	Perspective image	Evaluation
Expansion by sag		
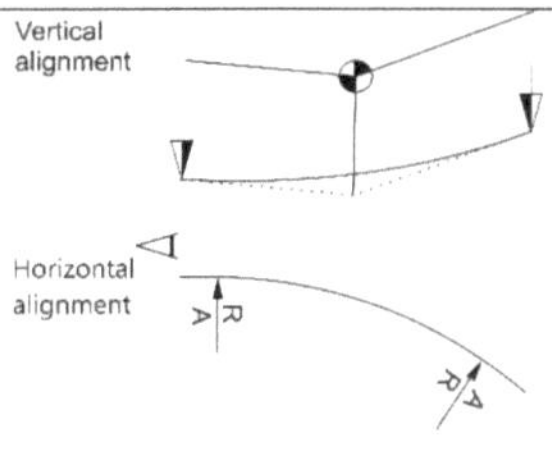		- With R:H > 1:10; safety reduction through higher speed level
Compression by crest		
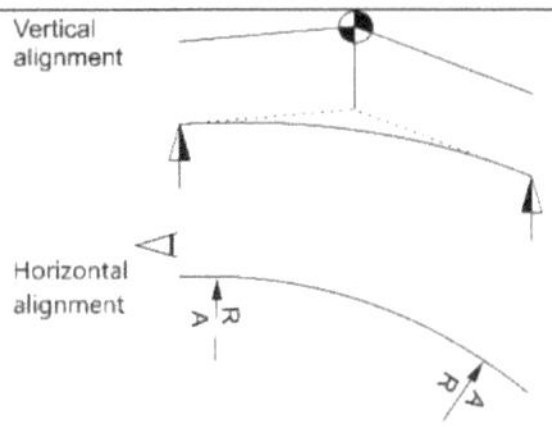		- With R:H ≤ 1:10; increased security through lower speed level - With R:H > 1:10; reduction of the speed damping effect

Fig. 16.8 Elongation and compression of a circular curve with the same radius

Horizontal/Vertical alignment	Perspective image	Evaluation
Kinks in vertical alignment		- Avoidance of straight lines with constant longitudinal inclination - Avoidance of sags with low tangent lengths in vertical alignment between long straight lines with constant longitudinal inclination - Avoidance of minimum crest and sag curves in vertical alignment at changes in longitudinal inclination about > 10 % - Avoidance of small curve radii with small directional changes in horizontal alignment
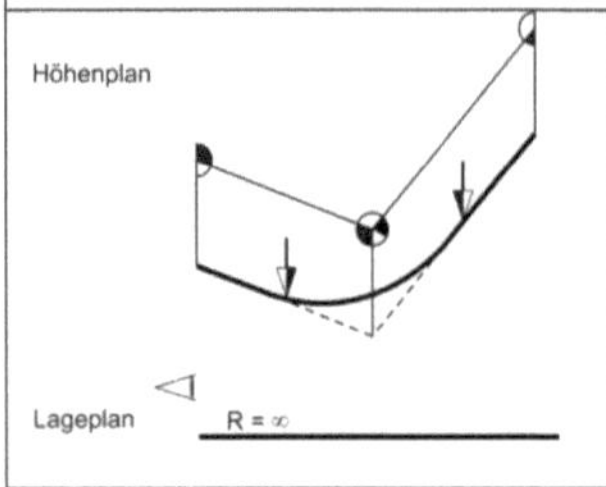		
Kinks in horizontal alignment		
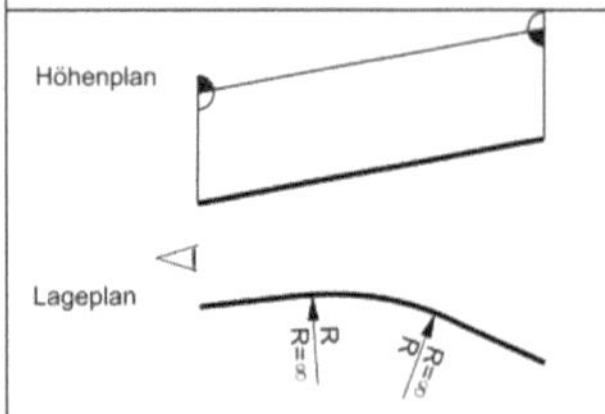		
Kinks in vertical and horizontal alignment		
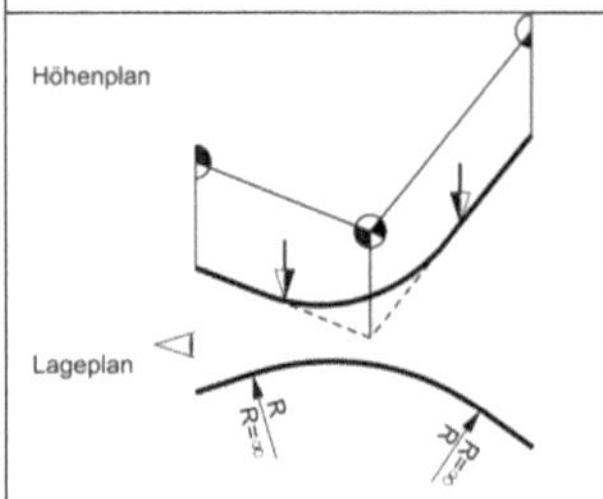		

Fig. 16.9 Kinks as design deficits

Horizontal/Vertical alignment	Perspective image	Evaluation
"Flat, straight board" effect		- Avoidance of straight lines with constant longitudinal inclination between two rectified crests or sags - Bridges are in the continuous line flow in vertical alignment and also as far as possible in horizontal alignment to integrate
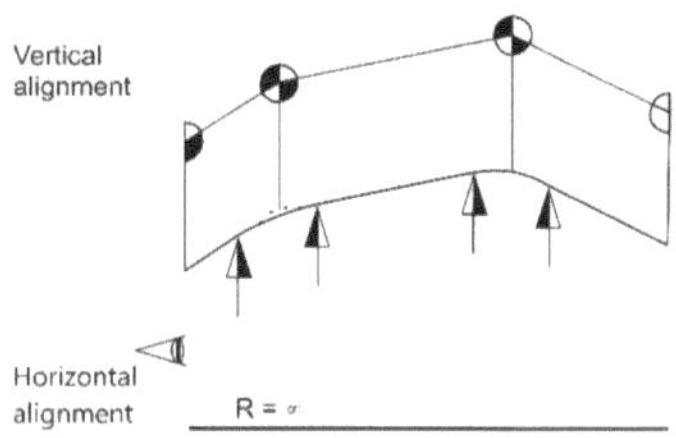		
"Bulge" effect		
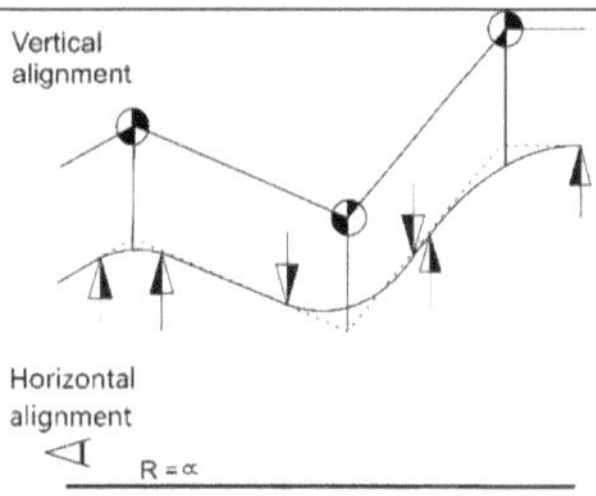		

Fig. 16.10 Flattening and bulging as design deficits

17 Visualization of Roads

Summary

Checking of the spatial alignment is carried out in three stages according to the instructions for visualisation of drafts for roads outside urban areas—HViSt (FGSV Advice on visualization of rural road designs. Cologne: FGSV-Verlag, 2008d).

First of all, it is checked whether the road section can be divided into standard room elements. If a section does not correspond to a standard room element, it is first checked whether a sufficient overlapping of the turning points in the site and elevation plan and thus a standard room element can be achieved by appropriate rescheduling.

If this is not possible, it must be checked whether there are hidden curve beginnings (Sect. 16.6.2) and whether the entire carriageway is visible. If the check is negative, the deficiencies must be remedied by replanning in the site and/or elevation plan.

Finally, the route is checked for design deficits using calculated perspective images. These should also be avoided. However, they can be accepted if, due to constraining points and for economic reasons, a change of alignment does not appear to be sensible (Fig. 17.1).

© Springer Fachmedien Wiesbaden GmbH, part of Springer Nature 2023

T. Richter, *Road Planning - Freeways and Country Roads*,
https://doi.org/10.1007/978-3-658-35189-2_17

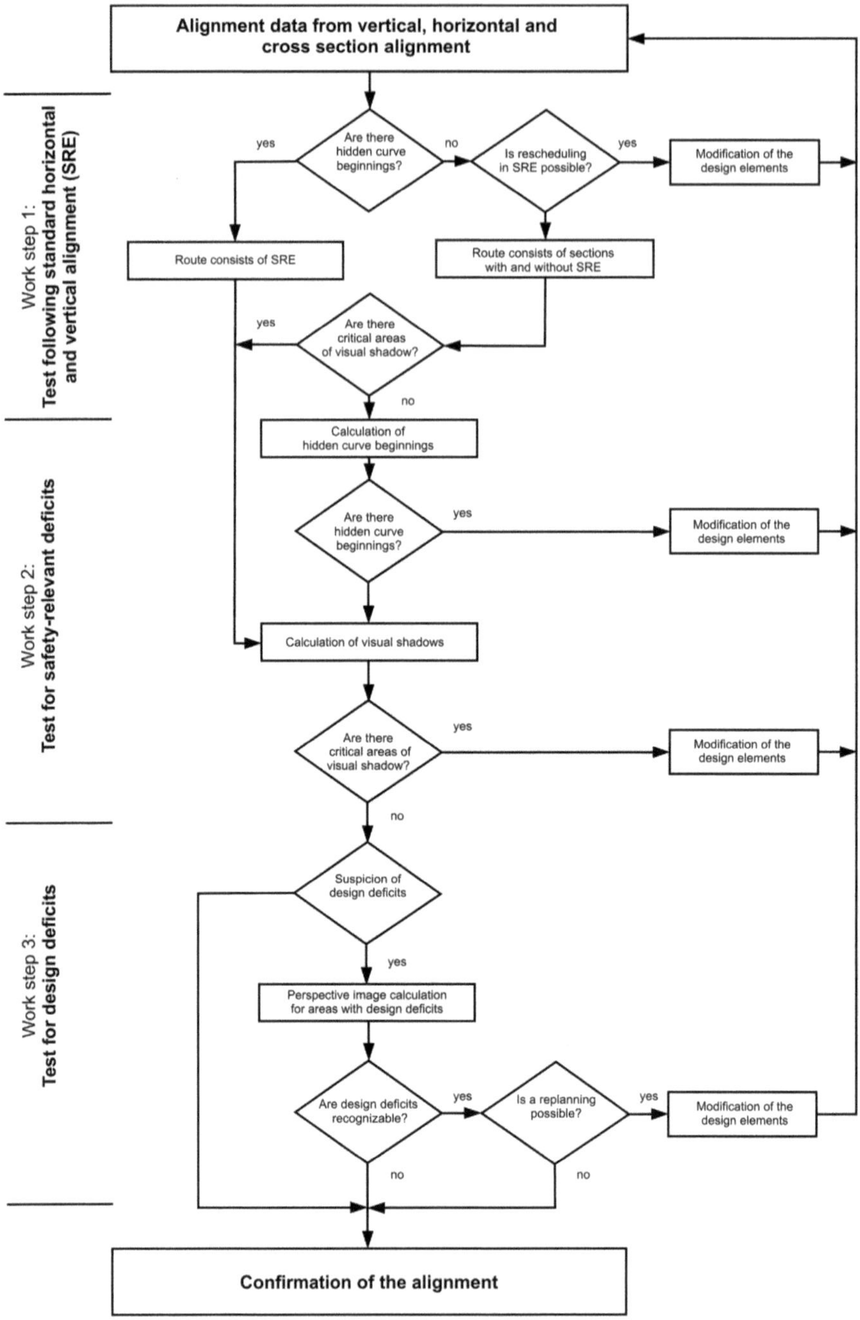

Fig. 17.1 Test sequence. (FGSV 2008d)

Roadside Furniture and Additional Facilities 18

Summary

The roadside furniture and additional facilities of motorways and rural roads are integral parts of planning, design and operation because they have a decisive influence on driving behaviour, traffic safety and traffic quality. Therefore, the roadside furniture, including the intended planting, must be taken into account at all design stages.

The most important legal bases for traffic flow are the Road Traffic Regulations (StVO) (BMVBS Straßenverkehrsordnung. Dortmund: Verkehrsblatt-Verlag, 2013) and the Administrative Regulation on StVO (VwV-StVO) (BMVBS erwaltungsvorschrift zur Straßenverkehrsordnung (VwV-StVO). Dortmund: Verkehrsblatt-Verlag, 2014). The StVO is concise and contains the most important legal provisions. Supplementary notes and comments are contained in the VwV-StVO, to which traffic engineers also contributed. When changing national regulations, international agreements must be taken into account, in particular the "Convention on Road Signalling, opened for signature in Vienna on 8 November 1968" (Vienna World Convention).

The StVO is divided into three parts:

- General traffic regulations (§§ 1–35),
- Signs and traffic facilities (§§ 36–43),
- Implementing, penalty and final provisions (§§ 44–53).

Section I only indirectly influences the design and roadside furniture of roads. Section III assigns the material responsibility for the execution of the StVO to the road traffic authorities (road construction authority designs and builds, road traffic authority orders which traffic signs and markings are to be applied). Section II has the greatest significance for the design of traffic installations. The basis of the order by the traffic authority is a signposting and marking plan (also: operational plan), which the road construction authority must draw up and submit.

© Springer Fachmedien Wiesbaden GmbH, part of Springer Nature 2023

T. Richter, *Road Planning - Freeways and Country Roads*,
https://doi.org/10.1007/978-3-658-35189-2_18

Essential parts of the equipment of traffic facilities are:

- Traffic signs (StVO §§ 39–42),
- Direction signs,
- Road markings,
- Traffic facilities (StVO § 43),
- Vertical guidance and protection devices (delineators, safety barriers of steel or concrete, gates, barrier posts),
- Traffic lights.

18.1 Traffic Signs

Only the traffic signs mentioned in the StVO have official character (supplements possibly according to BMVI decree in the traffic journal). Also dimensions, colours, symbols, writing and the applicability of over- and undersizes (e.g. for cyclists) are defined in the StVO. The installation and type of application of traffic signs in individual cases is only partly regulated in the VwV-StVO. There is an international tendency to replace writing with uniform symbols and justification of danger signs and regulation signs with additional signs. A distinction is made between

- Traffic signs that are part of the construction, even if the road is technically perfect, and
- Traffic signs indicating defects that can be remedied by technical measures.

Positive and negative right-of-way signs are always required, for example at at-grade intersections. The following general rules apply for setting up traffic signs:

- No accumulation of traffic signs and little text, so that the message of all traffic signs is recognized and understood in time,
- Less than 3 Danger signs on the same post or directly next to or above each other,
- Less than 2 Regulation sign on the same post or directly next to or above each other,
- Certain signs (e.g. sign 350: pedestrian crossing, sign 201: St. Andrew's cross, signs 278 to 282: end of route bans) must always be set up alone,
- Traffic signs must be installed and maintained in such a way that they can be seen at all times (i.e. day and night) from a moving vehicle,
- Signs are generally fully reflective or (inside) illuminated,
- Lateral distance outside built-up areas 1.50 m,
- Clear distance lower edge usually 2.00 m (over cycle paths 2.20 m, at sign gantries 4.50 m, on directional islands 0.60 m (not deeper because of dirt), on motorways 1.00 m),
- On important roads, set up on both sides of the road or carriageway or overhead (according to VwV-StVO binding for motorways)

18.1.1 Danger Sign

Danger signs should be placed

- Only where the danger cannot be detected or cannot be detected in time and therefore the safety of traffic is impaired,
- If possible 150 to 250 m in front of the danger zone,
- On motorways 400 m in front of the danger zone (possibly repeat at 200-m distance) and
- Within built-up areas usually directly in front of the danger zone.

18.1.2 Regulation Sign

Regulation signs (§ 41 StVO) contain orders that can have a mandatory, prohibited or mixed character. The sign "Sonderweg für Radfahrer" (special path for cyclists), for example, gives the cyclist the order to use this path, while it prohibits other road users from using it. Road markings (regulation signs in the form of road markings) are also part of the regulation signs.

The sign 274: "Permitted maximum speed"is an important part of the design and will become more important in the future with regard to the goal of the self-explanatory road. In the RAL (FGSV 2012a), a planning speed is assigned to the design classes. The long-term goal should be to define the planning speed also as a permissible maximum speed in the StVO and to make it mandatory by means of the permanently available marking. Broken lines at the edge and not in the middle of the carriageway would then mean, for example, a permissible maximum speed of 70 km/h. So far, only the binding maximum permissible speeds of 100 km/h on country roads and 70 km/h at signalled junctions apply.

The regulation in § 45 (9) StVO "Traffic signs and traffic facilities are only to be arranged where this is absolutely necessary due to the special circumstances. ... Speed restrictions according to part 1d in § 45 (9) may in particular be imposed on flowing traffic only if, due to the special local conditions, there is a danger situation that considerably exceeds the general risk of impairment of the legal interests mentioned in the above paragraphs" (BMVBS 2013). Thus, for example, in many cases, the sensible ordering of overtaking bans in areas with low visibility or the ordering of permissible V = 70 km/h in non-signaled, at-grade intersections is viewed critically by some traffic authorities, because no danger is seen that significantly exceeds the general risk.

18.1.3 Guidance Signs

Guidance signs (§ 42 StVO) are intended to facilitate traffic by providing information. They may also contain instructions. For the design of intersections, the directional signs for signposting (§ 42 (8) StVO) are particularly important.

18.2 Signposting

Even though vehicles are increasingly equipped with navigation systems, direction signage for timely orientation and to avoid abrupt driving maneuvers is an important part of road design. On motorways in particular, signposting is becoming increasingly dynamic in order to indicate traffic jams or routes to major events in good time.

Signposts are required at all grade-separated and at-grade intersection approaches, whereby as few concise destination details as possible are to be used. Information on size, content and layout is contained in the guidelines for signposting on and off motorways (FGSV 2000a, b).

The signposting shall start at a sufficient distance before the intersection, be repeated if necessary and—where necessary—be supplemented by information on how to get into lane.

The following basic rules apply to the direction signposting (FGSV 2000a, b):

- The uniformity rule requires uniform and unambiguous signage in structure and content throughout the network.
- The rule of perceptibility requires that signpost information be perceptible in good time by day and night.
- The (very important) readability rule limits the number of destinations to five per direction, four per colour group and three in coloured inserts.
- The destination selection rule requires the selection of long-distance and short-distance destinations and the supplementation of destination plans (based on the destination directories of the BMVI) by destination spiders in the area of urban regions.
- The flip-over rule requires that the signposting when flipped over in the direction of travel corresponds to the real conditions (long-distance destinations at the top, short-distance destinations at the bottom, upward-pointing arrow for straight-line orientation).
- The (very important) continuity rule requires that once a destination is included in the signposting, it must be repeated in every subsequent signposting until the destination is reached.
- The arrow rule states that a curved arrow is of an anticipatory character, that a diagonal arrow marks the beginning of the lane change area, and that a horizontal transverse arrow is to be placed so that a turn is to be made in front of it or immediately behind it.
- The colour rule determines the ranking between the blue (motorway), the yellow (other roads outside built-up areas) and the white (inner-city roads) signposting system.

18.3 Road Markings

Road markings are indispensable for safe traffic guidance, especially on (unlit) roads outside built-up areas, on roads with more than one lane per direction and at intersections. Good visibility (day and night), skid resistance, durability and evenness are important requirements for road markings. The legal meaning of some road markings is regulated in the StVO. In addition to the VwV-StVO, the guidelines for road markings (FGSV 1993; FGSV 1980/1989/1995) contain dimensions and instructions for use.

Particularly on country roads, in the sense of the self-explanatory road, the markings are an important distinguishing feature of the individual design classes and should therefore also support appropriate driving behaviour in the respective design classes.

The colour of a standard marking is white, temporary markings are yellow, e.g. for construction sites. On unlit roads, the marking should be retroreflective. On motorways and on roads of EKL 1 and EKL 2 in general and on roads of EKL 3 in case of high traffic volumes and on sections with high demands on visibility in darkness and wetness, markings with increased night visibility should be used. Markings separating driving directions should also be clearly profiled so that the driver notices if the vehicle is driven over.

Lane markings (sign 295) and broken lines (sign 340) are applied centrally between adjacent lanes, except on concrete roads with longitudinal joints. Narrow lines are 0.15 m wide on motorways and 0.12 m wide on other roads. Wide lines are twice as wide as the associated narrow lines, but at least 0.25 m. The central traffic stripe on single-carriageway roads of EKL 1 is 1.00 m wide, so that a 0.76 m wide area remains between the lane marking (0.12 m per narrow stripe), which is marked "traffic green" (RAL 6024) in a non-retroreflective manner.

The double line (0.12 m per narrow line) for separating the driving directions on roads of EKL 2 is 0.50 m wide, including the lane markings. This results in a remaining area of 0.26 m between the lane markings that is not marked or otherwise indicated.

Streets of EKL 4 are marked by two broken lines, each 0.50 m from the edge of the carriageway. A central broken line is not used. Since broken lines may be crossed if necessary, the entire width of the carriageway can legally be used to meet large vehicles. The line-to-gap ratio is 1:1 (1.00 m line/1.00 m gap).

18.4 Traffic Guidance Systems

Vertical guidance systems complement and support the horizontal guidance systems (road markings) in the plane of the carriageway in their effect. All devices arranged perpendicular to the plane of the carriageway serve as vertical guidance devices for the optical guidance of the vehicles (delineator posts, guide panels, direction signs in curves). Vertical guidance devices also indicate danger points that may arise due to constrictions in the traffic area or unexpected changes in the alignment of the road.

18.4.1 Delineator (Sign 620)

Delineator posts indicate the lateral boundary of the traffic area and the alignment of the road. They have a day and a night sign, which consists of round reflex signs on the left side of the road. Delineator posts are placed in straight and non-inclined road sections at intervals of 50.0 m. In curves and hilltops, the distances are reduced.

Delineators have a predetermined breaking point at the foot. Delineators are also used, which straighten up again after pushing. At simple intersections without widenings and installations and at intersections of service roads, the delineator, behind which a turn is to be made, receives yellow reflex signs.

18.4.2 Guiding Boards and Guide Marks

Guidance panels and guidance signs in accordance with StVO § 43 Part 3 number 3b are installed if an improvement of the optical guidance is necessary due to a narrowing or change of direction of the roadway. If guidance panels are used to indicate the direction, e.g. in a curve, they are called directional signs.

18.5 Vehicle Restraint Systems (Safety Barriers)

In principle, the side space must be free of obstacles, so that leaving the road does not lead to serious consequences of an accident (forgiving road environment). If such a design is not possible (e.g. bridges, high embankments, parallel to other traffic routes), safety barriers are required.

Safety barriers are technical devices that serve to catch or divert vehicles that have left the carriageway in order to mitigate the consequences of an accident (FGSV 2009).

Safety barriers include protective devices (longitudinal safety barriers at the edge of the road or on the central reserve/side dividing lane), impact absorbers and the corresponding initial, final and transition structures. Safety devices in Germany are usually made of steel (steel safety barriers) or concrete (concrete safety barriers). Safety barriers are required if the minimum distances of the obstacles from the edge of the carriageway according to Table 18.1 are not reached (applies to the same height of the obstacle and the edge of the roadway, other height differences are specified in the RPS [FGSV 2009]).

Four hazard levels are distinguished (Table 18.2). Hazard levels 1 and 2 are about the endangerment of third parties, whose protection has a higher priority in the RPS. This is proven by the larger critical distances of obstacles. Levels 3 and 4 describe the danger to the passengers of a vehicle.

For example, trees on country roads with a permissible maximum speed of 80 to 100 km/h may only be planted at a distance of 7.5 from the edge of the road. Planting trees closer to the edge of the carriageway and then protecting them or the vehicle passengers with safety barriers contradicts the principle of RPS, as obstacle-free side areas are significantly safer than sections of road with safety barriers.

Tubular steel posts with an external diameter $\leq$ 76.1 mm (aluminium 76.0 mm) and wall thickness $\leq$ 2.9 mm (aluminium 3.0 mm) as well as lighting poles and poles of traffic light systems at intersections are not treated as obstacles in the sense of the RPS.

Table 18.1 Critical distances of obstacles. (FGSV 2009)

Permissible maximum speed	Critical distance AE—hazard levels 1 and 2	Critical distance A—hazard level 3 and 4
> 100 km/h as well as on motorways and roads similar to motorways ≤ 100 km/h	20 m	12 m
80–100 km/h	11.5 m	7.5 m
60–70 km/h	7.5 m	4.5 m

Table 18.2 Hazard levels. (FGSV 2009)

Hazard level	Examples
1. Special danger to third parties at a distance AE	– Explosive chemical plants – Intensive use of common areas – Adjacent high-speed railway lines with permitted V > 160 km/h – Buildings at risk of collapse
2. Danger to third parties at a distance AE	– Adjacent heavily frequented footpaths and cycle paths – Adjacent railway lines with more than 30 trains/24 h – Adjacent roads with a DTV > 500 passenger cars/24 h
3. Particular danger to passengers at distance A	– Non-deformable obstacles perpendicular to the road (e.g. bridge abutments) – Non-deformable single point obstacles (e.g. trees with trunk diameter > 8 cm)
4. Danger to passenger at distance A	– Single point obstacles that are still deformable but can no longer be knocked over or sheared off (tubular steel posts with an outer diameter > 76.1 mm (aluminium 76.0 mm) and wall thickness > 2.9 mm (aluminium 3.0 mm)) – Crossing trenches – Rising slopes (inclination >1:3), but only if they are rock or stone slopes. Troughs usually guarantee sufficient rounding → usually negligible – Falling slopes with a height > 3 m and an inclination >1:3 – Waters with a depth > 1 m and white water

The requirements for safety barriers are regulated throughout Europe in DIN EN 1317. The safety barriers must be able to

- Intercept vehicles of different weights without destroying the safety barriers,
- Divert vehicles of different heights without overturning,
- Convert movement energies with a low proportion of reflection by sufficient elasticity,
- Enable unavoidable reflections with small deflection angles (transverse jerk) and
- Avoid driving over and rolling of vehicles around the longitudinal axis.

The main tasks of safety barriers are

- To protect uninvolved parties (e.g. other road users) directly or by avoiding reflection and
- To protect carriageway leaving drivers with the following levels of importance (protection against oncoming traffic, protection against fixed installations, protection against reflection, protection against falling)

Table 18.3 Determination of containment levels according to the RPS. (FGSV 2009)

Hazard level	Permissible maximum speed	ADT (passenger cars)	Increased probability of leaving the carriageway[a]	ADT (trucks)	Containment level
1	> 50 km/h		Yes	> 3000 Veh./24 h	H4b
				≤ 3000 Veh./24 h	H2
			No	> 3000 Veh./24 h	H2
				≤ 3000 Veh./24 h	H1
	≤ 50 km/h				None
2	> 100 km/h			> 3000 Veh./24 h	H2
				≤ 3000 Veh./24 h	H1
	80–100 km/h	> 3000 Veh./24 h		> 500 Veh./24 h	H1
				≤ 500 Veh./24 h	N2
		≤ 3000 Veh./24 h	Yes	> 500 Veh./24 h	H1
				≤ 500 Veh./24 h	N2
			No		None
	60–70 km/h	> 3000 Veh./24 h	Yes	> 500 Veh./24 h	H1
				≤ 500 Veh./24 h	N2
			No		None
		≤ 3000 Veh./24 h			None
	< 60 km/h				None
3	> 100 km/h				H1
	60–100 km/h				N2
	< 60 km/h				None
4	> 100 km/h				N2
	80–100 km/h	> 3000 Veh./24 h			N2
		≤ 3000 Veh./24 h			None
	60–70 km/h	> 3000 Veh./24 h	Yes		N2
			No		None
		≤ 3000 Veh./24 h			None
	< 60 km/h				None

[a]Radius relation outside Fig. 12.4, several curves with small radii in succession (smaller than about 1.2 times the value of the recommended lower radius limit according to Fig. 12.5 (RPS still refers to RAS-L and is therefore only conditionally applicable), accident frequency sections)

In the RPS (FGSV 2009) and in DIN EN 1317, the requirements are defined as the containment levels (breakthrough resistance) (Table 18.3), the impact classes (resiliency) (Table 18.4) and the impact severity levels (injury severity) (Table 18.5). The containment level depends on the hazard level, the permissible maximum speeds, the traffic volume of passenger cars and trucks and the probability of leaving the carriageway.

The classes of the lateral buckling (Table 18.4) depend on the distance between the front edge of the safety barrier and the obstacle. The smaller the distance, the more rigid the safety barrier must be. However, a rigid safety guard usually means a high load on the

Table 18.4 Efficiency classes, data in [m]. (FGSV 2009)

Classes of the lateral buckling	W1	W2	W3	W4	W5	W6	W7	W8
Levels of the lateral buckling	$W \leq 0.6$	$W \leq 0.8$	$W \leq 1.0$	$W \leq 1.3$	$W \leq 1.7$	$W \leq 2.1$	$W \leq 2.5$	$W \leq 3.5$

Table 18.5 Impact severity levels or passenger load levels. (FGSV 2009)

Impact severity level	Characteristic values according to DIN EN 1317–2	
A	ASI[a] ≤ 1.0	THIV[b] ≤ 33 km/h PHD[c] ≤ 20 g
B	$1.0 < \text{ASI} \leq 1.4$	
C	$1.4 < \text{ASI} \leq 1.9$	

[a] Acceleration Severity Index (acceleration on impact)
[b] Theoretical Head Impact Velocity
[c] Post Head Deceleration (maximum acceleration after head impact)

vehicle passengers, so that the largest possible distances should be aimed for. The RPS do not provide any direct specifications for the level of impact severity, but the lowest possible impact severity level (A) should be aimed for. In particularly dangerous places (e.g. edge of high bridges), the highest impact severity level (C) can be accepted for the safe restraint of the heaviest vehicles.

As the driving off the carriageway on level ground can be relatively harmless, the principle of using safety barriers only applies where they mitigate greater dangers than collision and reflection and where the safety barrier themselves do not cause greater dangers. Furthermore, avoidable obstacles must not be created in front of which safety barriers must then be placed.

It is controversial which type of protective devices can best fulfil the formulated requirements. In the Federal Republic of Germany, due to the spectacular collision tests from the 1960s, steel safety barriers were used almost exclusively for many years. In the course of time, these were optimised by impact absorbers on the double-T-beam posts, which are particularly dangerous for motorcyclists who have come off the road, or by underride protection and supplementary ropes in the railings of bridges for fall protection.

Since the mid-1980s, concrete safety barriers have also become established on the basis of positive foreign tests. Today, for reasons of competition, it is necessary to consider both types of construction as basically equivalent and to put them out to tender on a system-neutral basis (Fig. 18.1).

Other protection devices are also used abroad, such as

- Rope safety devices (e.g. in Denmark and Switzerland),
- Box profiles (e.g. in the Netherlands) and
- Trapping fences (e.g. in Austria).

In the case of steel safety barriers, new systems such as Super-Rail or megarail have been developed in recent years in addition to the traditional single/double (distance) safety

ESP 2.0 with underride protection		DB 80 AS-A	
N2 profile A		N2/H2	
W4 W_N=1.3 m		W4/W7 W_N=1.3/2.3 m	
Impact intensity A		Impact intensity B	

Fig. 18.1 Examples of steel safety barrier with underride guard and concrete safety barrier. (Photos: Straßen.NRW and deltabloc)

barriers due to the higher requirements of the RPS and DIN EN 1317 (containment level and lateral buckling).

A distinction is made between double-sided or single-sided concrete safety barriers and single-sided concrete safety barriers with floor filling (raised beds). Concrete safety barriers are rigid protective devices that, due to their design, do not deform or only slightly deform in the event of a collision with a motor vehicle. The impact energy must therefore be converted (instead of by deflection) mainly by the friction of the wheels and the body against the safety barrier and by body deformation.

The standard distance of the safety barrier from the edge of the carriageway is 0.50 m. This distance can be reduced to create larger effective areas (if possible only in exceptional cases) and increased to create sufficient visibility. In front of safety barriers, the height of kerbs should be limited to 0.07 m in order to avoid rollover. When arranging safety barriers, visibility must be taken into account. Outside crest curves, safety barriers with a maximum height of 0.95 m allow the visibility beam to pass over the safety barriers. Otherwise the central reservation may have to be widened, other safety barriers selected or the maximum permitted speed limited (if necessary only in wet conditions).

18.6 Immission Control Devices

The Traffic Noise Protection Ordinance (16th BImSchV) forms the legal basis for protective measures in accordance with the principles of noise prevention. Noise reports are to be carried out in accordance with Annex 1 of the 16th BImSchV or in accordance with the "Guidelines for Noise Protection on Roads" (RLS-90) (FGSV 1990). These guidelines are currently being revised so that a new edition can be expected at the end of 2016.

Exposure to air pollutants can be estimated in accordance with the "Guidelines for determining air quality on roads with no or loose perimeter development" (RLuS) (FGSV 2012d).

Measurement methods can only take into account a load that is actually present at the moment. The results depend on the respective traffic and environmental conditions and are also influenced by the measuring method. In order to obtain a result that is always validated by these situation- and time-dependent coincidences, road traffic noise is determined in a calculation procedure according to the guidelines for noise protection on roads (FGSV 1990).

According to the RLS, the strength of the sound emission from a road or lane (described by the emission level $L_{m,E}$) is calculated from the traffic volume, the proportion of trucks, the maximum permitted speed, the type of road surface and the gradient. The height of the sound level at an immission point also depends on the distance between the immission and emission point (sound source) and on the average height of the beam from the source to the immission point above the ground. It can also be amplified by reflections (e.g. at the front of houses or retaining walls) or reduced by shielding (e.g. by noise barriers, walls, buildings, elevations or by the low level of the road). The influence of road wetness is not considered in the calculation methods.

For comparison with maximum immission values, L_r is the rating level. It is equal to the averaging level, which is increased at signalled intersections and T-junctions by a mathematical addition to take into account the additional disturbances caused by starting and braking of vehicles. The rating level of traffic noise is calculated separately for day and night:

- $L_{r,T}$ for the time from 6:00 to 22:00 hours and
- $L_{r,N}$ for the time between 10:00 and 6:00.

The calculations are based on average daily traffic volumes (ADT) and truck shares averaged over all days of the year.

The results of the calculations according to RLS-90 are compared with the relevant immission limit values (IGW), which are laid down in the 16th Federal Immission Control Ordinance (16th BImSchV) (BMVBW 1990). In order to protect the neighbourhood from harmful environmental effects caused by traffic noise, it must be ensured during the construction or substantial modification of a road that the rating level does not exceed one of the following immission limit values (Table 18.6). Basically, a distinction must be made between noise protection measures during the construction and planning of new roads (noise prevention, mandatory) and noise protection measures on existing roads (noise remediation, voluntary).

Noise protection on roads comprises planning, construction and traffic law measures that must be coordinated and included in the respective planning process. The following order of precedence must be observed in noise prevention (noise protection during the construction of new roads or significant alterations to roads):

- Within the framework of planning, noise impacts on areas in need of protection are to be avoided as far as possible by means of a compatible allocation of land (distance).
- If limit value exceedances cannot be avoided, remedial action must be taken by means of structural protection measures on the road (active noise protection).

Table 18.6 Immission limit values for noise remediation and prevention. (BMVBW 1990)

	Noise remediation		Noise prevention	
	Day [dB(A)]	Night [dB(A)]	Day [dB(A)]	Night [dB(A)]
Hospitals, schools, health residences and retirement homes	70	60	57	47
Pure and general residential areas and small housing estates	70	60	59	49
City centers, village areas, mixed areas	72	62	64	54
Industrial estates	75	65	69	59

- If such measures on the road are not considered, are not sufficient or are much more expensive (if the costs of active noise protection measures on the road are disproportionate to the protection purpose), expenses for necessary noise protection measures on the affected building structure (passive noise protection) must be provided for.

There is also the possibility of reducing the noise emission of the roads by using low-noise surfaces. Here, for example, low-noise stone mastic asphalts and open-pored (single-layer or double-layer) asphalts come into question. In the case of open-pored asphalts, however, the higher costs and more complex maintenance must be taken into consideration.

18.6.1 Noise Protection Measures

18.6.1.1 Modal Shift

By shifting traffic or concentrating traffic on efficient roads, the burden of traffic immissions in residential areas and other sensitive areas can be limited.

18.6.1.2 Planting

The sound propagation can also be influenced and reduced by planting to a small extent. The physical effect of sound insulation planting depends on the degree of foliage over the entire height (thus possibly depending on the season) and on the staggering of the planting in height and depth.

In order to achieve a noticeable or audible level reduction compared to free sound propagation, dense plantings with a depth of more than 50.0 m and sufficient height are necessary. For each 10.0 m depth of planting, a sound reduction of approximately 0.5 dB(A) can be achieved, but not more than 10 dB(A). Individual rows of trees or hedges are almost completely soundproof. Reflections in the treetops can even lead to an increase in level. For the residents, however, a subjective positive effect can still be achieved, as the visual contact to the noise source is interrupted.

18.6.1.3 Soundproof Facilities

Noise barriers are placed between the point of emission and the point of immission in order to impede the propagation of sound. Noise protection measures on the road (active measures) are

- Noise barriers/embankment dams (inclination usually 1:1.5 with 1.0 m wide crown),
- Noise protection walls (if necessary angled to improve the protective effect),
- Cut and trough layers,
- High altitudes and
- Partial and full coverage.

Noise protection installations should be integrated as unobtrusively as possible into the townscape and landscape. This is best achieved with natural and near-natural materials, shapes and colours. Therefore, a near-natural solution should always be given priority in noise protection installations. Thus, noise protection embankment dam, if necessary with an attached wall or retaining wall, should be given preference over steep walls and noise protection walls. The ZTV-Lsw (FGSV 2006b) as well as the "Recommendations for the design of noise protection installations on roads" (FGSV 2005a) contain information on the design and integration into the environment.

Tunnel sections or closed covers for noise protection purposes should only be used if, for example, there are urban planning reasons for doing so, due to the particular problems involved (lighting, ventilation, access restrictions, critical intersections, costs) (Figs. 18.2 and 18.3).

Fig. 18.2 Noise barrier. (Source: Lower Saxony State Authority for Road and Transport)

Fig. 18.3 Noise barrier. (Source: Lower Saxony State Authority for Road and Transport)

Noise barriers are distinguished on the basis of their reflection properties:

Reflective:	$\Delta LA,\alpha,Str. < 4$ dB,
Absorbing:	$\Delta LA,\alpha,Str. \geq 4$ dB < 8 dB and
Highly absorbent:	$\Delta LA,\alpha,Str. \geq 8$ dB.

The space or area required by the different noise screens varies greatly. Noise barriers are mainly used out of town due to their large surface area. Noise barriers and steep walls require a much smaller area and can therefore also be used well in or in front of built-up areas.

18.6.1.4 Encapsulation of the Road

The encapsulation of the road by tunnels, covers or enclosures can achieve level reductions of between 20 and 40 dB (A) if properly executed. However, the investment and operating costs as well as the energy requirements for these measures are very high.

18.6.1.5 Building

Road traffic noise can also be reduced by taking advantage of favourable urban planning and construction conditions. Single or semi-detached houses and also groups of houses up to 50-m length (open construction) do not prevent the noise from passing through the gaps between buildings and therefore do not allow quiet zones to develop. A closed roadside development, on the other hand, acts as a noise protection and effectively protects the areas and buildings behind it from noise, depending on their height and length. Continuous openings, such as open courtyard passages, should be avoided if possible.

18.7 Other Roadside Furniture Elements

18.7.1 Wild Animals Protection Fences

To protect traffic from wild animals crossing the carriageway, protective fences can be erected along individual stretches of road (especially on motorways and on country roads of EKL 1 and EKL 2, where frequent crossing of wild animals is to be expected). For this purpose, the Federal Ministry of Transportation has established Wildlife Protection Fence Guidelines (BMVBW 1985), which differentiate between protective fences against red deer or fallow deer and roe deer or wild boar. Protective fences are a contribution to increasing traffic safety and to avoiding obstacles. However, the road safety obligation alone is already fulfilled with the erection of sign 142 "Deer pass". Another type of wild animal protection fences is fly-over constructions for bats, which are erected as close as possible to the edge of the road. Here, the stop sight distances must be taken into account.

18.7.2 Anti-Glare Fences

Anti-glare fences are generally used to reduce mutual glare when traffic routes run parallel (e.g. road and railway). They can also be used on dual carriageways in the event of a particularly high frequency of accidents in the dark and particularly unfavourable topographical conditions, which, in conjunction with a sag curve, lead to severe glare. Special attention must be paid to the visibility conditions. In straight gradient areas and on crest curves, safety barriers without technical attachments are generally sufficient as glare protection. Furthermore, headlamp technology is developing further, so that in the future, less and less glare can be expected.

18.7.3 Lighting

Motorways and country roads are generally unlit.

18.7.4 Planting

In principle, woodland plantations (trees and shrubs) along roads have been part of the landscape for decades and are subject to the Federal Nature Conservation Act (BNatSchG), which requires the preservation of this landscape. Interference with the landscape must therefore be avoided, minimised or compensated for. Where compensation is not possible, replacement must be provided. Woody planting is also useful to support the spatial alignment, as it can clarify the course of a road.

On the other hand, trees with a trunk diameter of 0.08 m and more are obstacles in the side room, which lead to many accidents with serious consequences. They should therefore only be planted in places where stray vehicles cannot reach them or where the consequences of an accident are minimal. The required distance on roads from EKL 1 to EKL 3 is usually so large (see Sect. 18.5) that trees cannot be integrated into the road cross section. However, trees can be planted where safety barriers have to be installed for other reasons. In this case, the effective area (usually this requires a distance of about 3.00 m from the edge of the carriageway) and the stop sight distances must be taken into account.

In shrub plantings, the holding widths must also be observed and it must be ensured by regular cutting that the trunk diameters do not exceed 0.08 m.

18.7.5 Service Stations

The installation of service stations is based on the recommendations for service stations on roads (FGSV 2011a). They differ between managed and unmanaged service stations. They must be installed on both sides of motorways in sufficient number and size. Rest stops can also be useful on rural roads of EKL 1. They should always be located with entrances and exits at a sufficient distance from intersections and in areas where the alignment is as straight as possible.

Drainage 19

19.1 Safety Aspects in the Drainage of Traffic Areas

Observations of accidents have shown that the risk of accidents on wet roads, expressed by the ratio of the proportion of accidents on wet roads to the share of time with wet roads, is significantly higher than the corresponding ratio on dry road surfaces.

As causes of accidents in this context, the following effects are responsible for this increased risk:

- Spray formation, due to swirled up water on wet roads, the following traffic is hindered in its view.
- The frictional connection between the tyre and the road surface is significantly less in wet conditions than in dry conditions. In borderline cases, aquaplaning occurs. Many drivers are not aware of the significant difference.
- The reflective properties of the road surface change in wet conditions. Particularly at night, light conditions deteriorate in wet conditions due to the transition from diffuse to specular reflection.

The road condition in wet conditions is metrologically described by the parameter "water film thickness". The following parameters can be mentioned as weather-related influencing parameters on the water film thickness:

- Rain intensity (influences the rainwater inflow),
- Rain energy (influences the flow-off process) and
- Wind (influences the flow-off process).

The road design, however, can only influence the carriageway-specific influencing variables on the water film thickness. These are:

- The surface properties of the roadway. In this context, for example, the roughness of the road surface, grooves on the road surface, the evenness of the road surface or the storage capacity of the road surface (e.g. open-pored asphalt).

© Springer Fachmedien Wiesbaden GmbH, part of Springer Nature 2023

T. Richter, *Road Planning - Freeways and Country Roads*,
https://doi.org/10.1007/978-3-658-35189-2_19

- The inclination of the road surface. Especially slopes between 0 and 3% have a significant influence on the water film thickness. In addition, an increase in the gradient only leads to a slight reduction in the water film thickness.
- The runoff path of the precipitation water. Extending the runoff path of rainwater from the roadway leads to a degressive increase in the thickness of the water film. The length of the runoff path of the water can be calculated using the length of the fall line.

19.2 Drainage Measures

In the design of roads, road drainage is to be ensured by the specification of a minimum crossfall of q = 2.5%. This crossfall must be ensured both in straight lines and in arcs with large radii. Since this condition cannot be met in areas of torsion, a minimum ramp gradient is specified in order to keep the area in which the crossfall is less than 2.5% as short as possible. In order to ensure the shortest possible flow path of the water from the roadway in this area with weak drainage, a minimum longitudinal inclination in accordance with Sects. 13.3 and 14.3 must be ensured in the entire torsion area.

If a sufficient longitudinal inclination in the superelevation area cannot be achieved due to special constraint positions and if the superelevation area cannot be moved, the superelevation can be designed as an diagonal superelevation (see Sect. 14.3). With sufficiently large radii (see Sect. 14.2), a crossfall to the outside of the curve can avoid one or two superelevation areas. Further information on the drainage of roads is contained in the Guidelines for the Construction of Roads (RAS), Part: Drainage (FGSV 2005b).

On multi-lane motorways, the water film thickness can reach critical values despite compliance with the design limits. In this case, transverse box gutters or diagonal superelevations can help. Further information is provided in the following (Lippold et al. 2011).

The water running off the road surface is usually collected in drainage hollows on out-of-town roads. As a rule, these connect directly to the terrain or verges at the foot of embankments or cut-and-cover embankments. Their depth must be at least 20 cm and must not exceed $0.2 \cdot b$.

The simplest form of the drainage hollow is the lawn drainage hollow. Only small amounts of surface water can be collected and transferred in it. It is also used for the evaporation and infiltration of street runoff water and thus also has a purifying effect on the surface water.

If larger quantities of water occur due to the size of the area to be drained or if culverts and discharge points require greater depths, road ditches are arranged. They have the same function as drainage hollows, but generally have a greater hydraulic capacity. However, since ditches pose a much higher safety risk if a vehicle leaves the road, the shallow drainage hollow is preferred to the deeper ditch where drainage conditions permit (Fig. 19.1).

For a variety of reasons (groundwater recharge, protection of streams, economic efficiency, etc.), rainwater running off the road should seep away where local conditions and

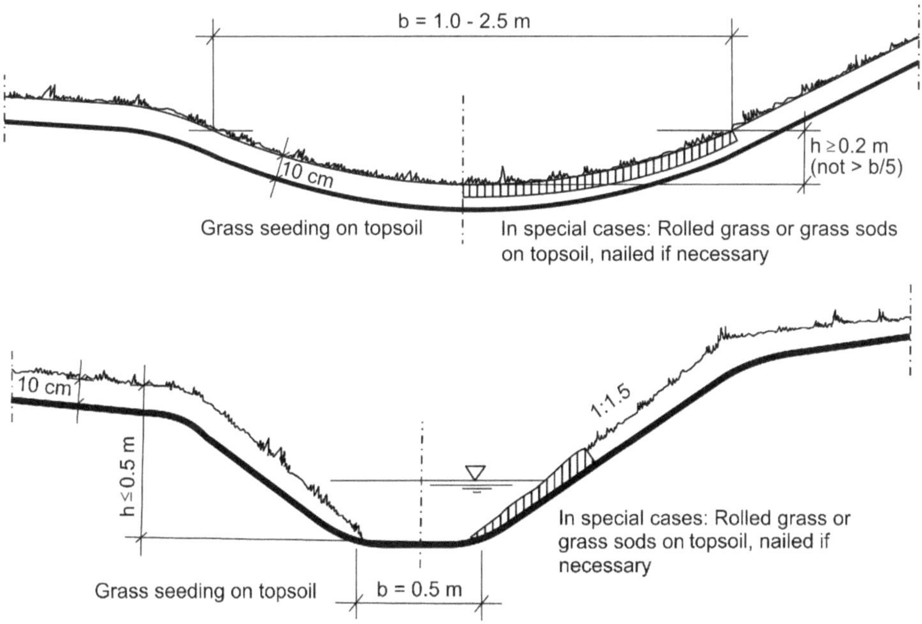

Fig. 19.1 Road drainage hollow and ditch

the subsoil permit. If this is not already done on the flow path over the embankments and in the lawn drainage hollows, the infiltration should be enabled by constructional measures.

The simplest measure to ensure infiltration is the arrangement of infiltration troughs or infiltration ditches. They differ from lawn troughs and ditches by the installation of ground sills, which are erected at right angles to the troughs/ditches at least 20 cm high. The distance between the sills results from the storage space and depends on the longitudinal inclination (Fig. 19.2).

Infiltration basins are structural installations for the infiltration of collected road runoff water. Experience shows that these basins tend to be self-compacting due to the suspended solids carried along by the collected water. The solids gradually settle in the infiltration zone and thus gradually seal the subsoil. For this reason, infiltration tanks are usually constructed with an upstream settling tank or integrated settling zone. Furthermore, the infiltration basins should be maintained at regular intervals (Fig. 19.3).

Rainwater retention basins are used to protect the streams from hydraulic overload due to water runoff peaks. The reduction of the water runoff in a unit of time is achieved by throttling the runoff and storing the road water runoff in a reservoir to be dimensioned for this purpose. Depending on local conditions (subsoil, groundwater level), rainwater retention basins can be designed as wet or dry basins (Fig. 19.4).

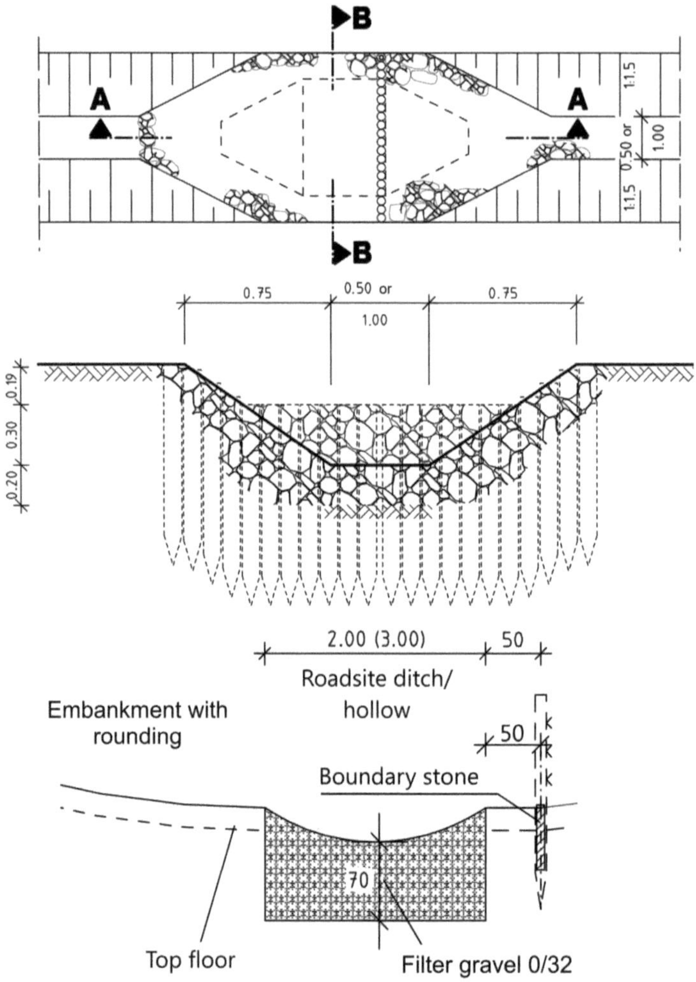

Fig. 19.2 Principle sketch of an infiltration hollow

19.3 Road Construction in Water Catchment Areas

The public water supply on the one hand and the construction and operation of roads on the other hand are important tasks of general interest that have a negative impact on each other. Therefore, roads in water catchment areas should be avoided. Where this is not possible, water must be protected from contamination and adverse effects for the public good. The parties responsible for water supply and road construction must therefore coordinate their planning, construction and operation, especially in water protection areas. Useable water supply is limited and drinking water can only be obtained at locations determined by nature. We obtain water mainly from groundwater, dams and lakes. Water can also be obtained from medicinal springs.

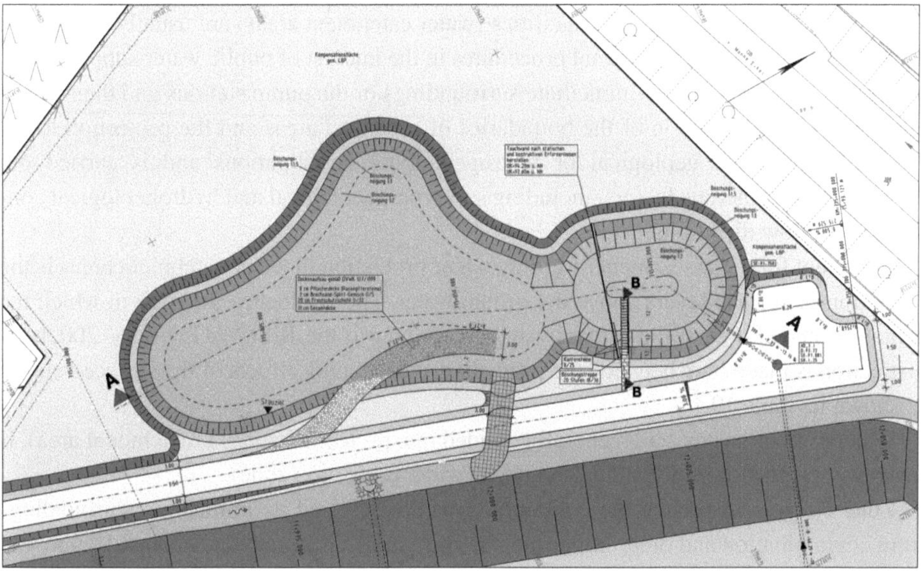

Fig. 19.3 Infiltration basin with upstream sedimentation tank. (Source: NLStbV rGB OS)

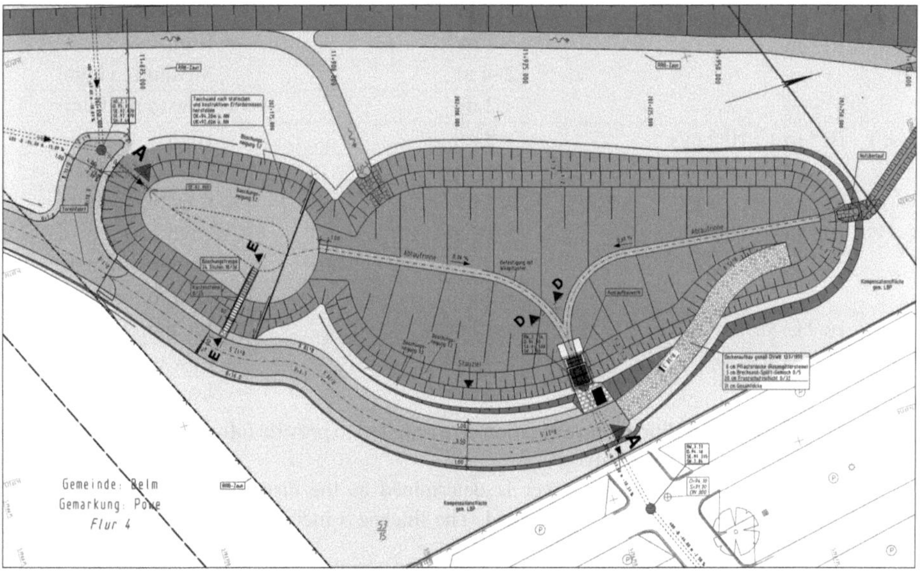

Fig. 19.4 Rainwater retention basin. (Source: NLStbV rGB OS)

Substances hazardous to water are solid, liquid and gaseous substances that are capable of permanently and adversely affecting the physical, chemical or biological properties of water. Liquids are particularly hazardous to water because they spread relatively quickly and can reach surface waters or ground water. Solid substances can enter water bodies in a similar way if, for example, they are dissolved by precipitation water.

Basins of public water supply facilities (water catchment areas) are usually designated as water protection areas in formal procedures in the interest of public water supply. Water protection areas include the immediate surroundings of the pump stations and their catchment areas. Determination of the boundaries of protected areas and the protection levels requires appropriate geological or hydrogeological investigations and is carried out depending on numerous factors, including soil types, geological and hydrogeological conditions and flow directions.

Important for the assessment of groundwater protection in water catchment areas is the formation of surface layers above the groundwater. They influence the time in which the contaminants can reach the groundwater. According to the RiStWAG (FGSV 2002b), a distinction is made here between the protective and cleaning effects of the surface courses as shown in Table 19.1.

Water protection areas are generally divided into protection zones I (catchment area), II (narrower protection zone) and III A and B (wider protection zone).

Zone I serves to protect the collection system itself and its immediate surroundings from contamination and other impairments. The extension amounts to at least 10,0 m, but

Table 19.1 Protective effect of the surface layers above the groundwater. (FGSV 2002b)

Permeability[a]	Depth[b]	Protection[c]
$k_f < 1 \cdot 10^{-6}$ m/s	> 4 m	High
	2–4 m	Medium to high
	< 2 m	Low to medium
$k_f < 1 \cdot 10^{-4}$ to $1 \cdot 10^{-6}$ m/s	> 8 m	High
	4–8 m	Medium to high
	< 4 m	Low to medium
$k_f < 5 \cdot 10^{-4}$ to $1 \cdot 10^{-4}$ m/s	> 10 m	High
	5–10 m	Medium to high
	< 5 m	Low to medium
$k_f < 1 \cdot 10^{-3}$ to $5 \cdot 10^{-4}$ m/s	> 15 m	Medium to high
	5–15 m	Low to medium
	< 5 m	Low

[a]For solid rocks, special additional investigations are required to provide information on the inhomogeneities with regard to permeability

[b]The thickness of the groundwater cover is determined as the distance from the groundwater (MHGW) to the lower lying edge of the road. The thickness includes the near-surface disturbed soil layer

[c]The protective effect is to be assigned to one of the three categories "high", "medium" or "low" on the basis of the local geological and hydrogeological conditions. To determine the protective effect, it may be necessary in individual cases to consult other hydrogeological parameters, e.g. storage conditions, tectonic inventory (fissures, crevices, flows), confined or free groundwater surface, flow conditions in the unsaturated zone

to such an extent that organic fertilization can be permitted in zone II. Inadmissible in protection zone I are, among others, driving and pedestrian traffic or agricultural use.

Zone II extends from the boundary of zone I to a line from which groundwater takes about 50 days to reach the catchment area. This value of 50 days results from the fact that the death of micro-organisms and bacteria introduced into the groundwater occurs within approximately 50 days. With regard to the traffic facilities, construction sites, building material stores, roads or railway lines, among others, are considered to be unsustainable.

Zone III extends from the outer boundary of zone II to the boundary of the catchment area (= outer boundary of the water protection area). If the catchment area extends further than 2,0 km from the collecting system, a breakdown into III A and III B may be appropriate. The zone III A reaches then from the zone II to about 2,0 km from the version, the zone III B up to the border of the water protection area. Accordingly, stricter protection requirements apply to III A than to III B. In zone III, infiltration of road runoff water, storage of water-polluting substances and sewage treatment plants are not permitted.

In road planning, special attention must be paid to water protection areas or water priority areas in the regional planning programmes as early as the preliminary investigation for line determination and the planning intention must be coordinated with the responsible authorities and the water supply companies or operators of drinking water reservoirs concerned. When planning roads in water protection areas, the more favourable design elements specified in the planning guidelines must always be used with regard to traffic safety. In doing so, cuts in water protection areas are to be avoided as far as possible.

Traffic control measures can further reduce the risk of accidents and the associated impact on water bodies. Such traffic control measures include speed limits, bans on overtaking and bans on vehicles with water-polluting cargo. Other measures that contribute to increasing traffic safety, e.g. hard shoulders, additional lanes for slow traffic or safety barriers, also serve the purpose of water pollution control because they can reduce the risk of accidents and the consequences of accidents.

References

Bakaba, J. (10 2005). Streckentypisierung und Geschwindigkeiten - Ein Ansatz zur Differenzierung des Geschwindigkeitsverhaltens unbeeinflusster Fahrer auf zweistreifigen Außerortsstraßen. *Straßenverkehrstechnik* , S. 508-516.

BMVBS. (2008). PlafeR 07 - Planfeststellungsrichtlinien 2007 Richtlinien für die Planfeststellung nach dem Bundesfernstraßengesetz. *Verkehrsblatt* , S. 5.

BMVBS. (2012). *Straßenverkehrs-Zulassungs-Ordnung.* Dortmund: Verkehrsblatt-Verlag.

BMVBS. (2013). *Straßenverkehrsordnung.* Dortmund: Verkehrsblatt-Verlag.

BMVBS. (2014). *Verwaltungsvorschrift zur Straßenverkehrsordnung (VwV-StVO).* Dortmund: Verkehrsblatt-Verlag.

BMVBW. (1985). *Richtlinien fuer Wildschutzzäune an Bundesfernstraßen.* Dortmund: Verkehrsblatt-Verlag.

BMVBW. (1990). *Sechzehnte Verordnung zur Durchführung des Bundes-Immissionsschutzgesetzes (Verkehrslärmschutzverordnung - 16. BImSchV).* Dortmund: Verkehrsblattverlag.

Brannolte, U., Barth, H.-B., Junkers, W., Liu, Y. S., Sighorsson, H., & Stein, J. (1993). *Sicherheitsbewertung von Querschnitten außerörtlicher Straßen.* Bremerhaven: Wirtschaftsverlag NW.

Brilon, W., & Betz, C. (2008). *Entwurf und Bemessung von Autobahnknotenpunkten unter Berücksichtigung der Wechselwirkungen zwischen den Elementen* (Bd. 983). (BMVBS, Hrsg.) Bremerhaven: Wirtschaftsverlag NW.

Brilon, W., & Breßler, A. (2001). *Verkehrssicherheit und Verkehrsablauf an Steigungsstrecken - Kriterien für Zusatzfahrstreifen.* Bremerhaven: Wirtschaftsverlag NW.

Brilon, W., & Geistefeld, J. (2010). *Überprüfung der Bemessungswerte des HBS für Autobahnabschnitte außerhalb der Knotenpunkte.* Bremerhaven: Wirtschaftsverlag NW.

Brilon, W., Geistefeldt, J., Lippold, C., & Kuczora, V. (2007). *Autobahnen und Autobahnknotenpunkte mit vierstreifigen Richtungsfahrbahnen – Gestaltung und Bemessung.* Bremerhaven: Wirtschaftsverlag NW.

Bundesanstalt für Straßenwesen. (2012). www.bast.de . Abgerufen am 2. Mai 2012 von http://www.bast.de/cln_030/nn_510152/DE/Statistik/statistik-node.html?__nnn=true

BVU, Intraplan, IVV, Planco. (2014). *Verflechtungsprognose 2030.*

Durth, W. (1974). *Ein Beitrag zur Erweiterung des Models für Fahrer, Fahrzeug und Straße in der Straßenplanung; Schriftenreihe Straßenbau und Straßenverkehrstechnik, Heft 163.* Bonn Bad-Godesberg: Typo-Druck & Verlags-GmbH.

Durth, W., & Levin, C. (1991). *Differenzierte Kuppenbemessung.* (BMVBS, Hrsg.) Bonn: Typo Druck Verlag.

© Springer Fachmedien Wiesbaden GmbH, part of Springer Nature 2023

T. Richter, *Road Planning - Freeways and Country Roads*,
https://doi.org/10.1007/978-3-658-35189-2

Durth, W., Bald, S., & Wolff, N. (1988). *Wirksamkeit von trassierungstechnischen Ausgleichsmaßnahmen bei Unter- oder Überschreitung von Trassierungsgrenzwerten* (Bd. 520). (BMVBS, Hrsg.) Bremerhaven: Wirtschaftsverlag NW.

FGSV. (2001a). *Bemessungsfahrzeuge und Schleppkurven zur Überprüfung der Befahrbarkeit von Verkehrsflächen.* Köln: FGSV Verlag.

FGSV. (2002a). *Beurteilung und Abwägung in der Verkehrsplanung mit Hilfe des formalisierten Abwägungs- und Rangordnungsverfahrens (FAR).* Köln: FGSV Verlag.

FGSV. (2003). *Empfehlungen für die Einbindung von Straßen in die Landschaft (ESLa).* Köln: FGSV Verlag.

FGSV. (2005a). *Empfehlungen für die Gestaltung von Lärmschutzanlagen an Straßen.* Köln: FGSV-Verlag.

FGSV. (2010a). *Empfehlungen für Radverkehrsanlagen (ERA).* Köln: FGSV Verlag.

FGSV. (2011a). *Empfehlungen für Rastanlagen (ERS).* FGSV-Verlag.

FGSV. (1997). *Empfehlungen für Wirtschaftlichkeitsuntersuchungen an Straßen (EWS).* Köln: FGSV Verlag.

FGSV. (2006a). *Empfehlungen zum Schutz von Unfällen mit Aufprall auf Bäume (ESAB).* Köln: 2006.

FGSV. (2015a). *Handbuch für die Bemessung von Straßenverkehrsanlagen (HBS).* Köln: FGSV Verlag.

FGSV. (2011b). *Hinweise für barrierefreie Verkehrsanlagen H BVA.* Köln: FGSV Verlag.

FGSV. (2010b). *Hinweise zu Einsatzbereichen von Verfahren zur Entscheidungsfindung in der Verkehrsplanung.* Köln: FGSV Verlag.

FGSV. (2015b). *Hinweise zu Turbokreisverkehren.* Köln: FGSV-Verlag.

FGSV. (2008d). *Richtlinien für die Anlage von Autobahnen (RAA).* Köln: FGSV Verlag.

FGSV. (2001b). *Leitfaden für Verkehrsplanungen.* Köln: FGSV Verlag.

FGSV. (2008b). *Merkblatt für die Anlage von Kreisverkehren.* Köln: FGSV Verlag.

FGSV. (2008c). *Merkblatt zur Anlage von Querungshilfen für Tiere und zur Vernetzung von Lebensräumen an Straßen (M AQ).* Köln: FGSV Verlag.

FGSV. (2001c). *Merkblatt zur Umweltverträglichkeitsstudie in der Straßenplanung (MUVS).* Köln: FGSV Verlag.

FGSV. (2002b). *Richtlinien für bautechnische Maßnahmen an Straßen in Wasserschutzgebieten RiStWag.* Köln: FGSV Verlag.

FGSV. (1990). *Richtlinien für den Lärmschutz an Straßen (RLS).* Köln: FGSV-Verlag.

FGSV. (2008a). *Hinweise zur Visualisierung von Entwürfen für außerörtliche Straßen.* Köln: FGSV-Verlag.

FGSV. (2012a). *Richtlinien für die Anlage von Landstraßen (RAL).* Köln: FGSV Verlag.

FGSV. (1995). *Richtlinien für die Anlage von Straßen (RAS), Teil: Linienführung (RAS-L).* Köln: FGSV Verlag.

FGSV. (2005b). *Richtlinien für die Anlage von Straßen (RAS), Teil: Entwässerung (RAS-Ew).* Köln: FGSV-Verlag.

FGSV. (2012b). *Richtlinien für die einheitliche Entwurfsgestaltung von Entwurfsunterlagen (RE).* Köln: FGSV Verlag.

FGSV. (2011c). *Richtlinien für die Erstellung von Umweltverträglichkeitsstudien im Straßenbau (RUVS).* Köln: FGSV-Verlag.

FGSV. (2012c). *Richtlinien für die landschaftspflegerische Begleitplanung im Straßenbau - mit Musterkarten LBP (RLBP).* Köln: FGSV Verlag.

FGSV. (1993). *Richtlinien für die Markierung von Straßen (RMS), Teil 1.* Köln: FGSV-Verlag.

FGSV. (1980/1989/1995). *Richtlinien für die Markierung von Straßen (RMS), Teil 2.* Köln: FGSV-Verlag.

FGSV. (2000a). *Richtlinien für die wegweisende Beschilderung an Autobahnen.* Köln: VGSV Verlag.

FGSV. (2000b). *Richtlinien für die wegweisende Beschilderung außerhalb von Autobahnen.* Köln: FGSV-Verlag.

FGSV. (2008e). *Richtlinien für integrierte Netzgestaltung (RIN).* Köln: FGSV Verlag.

FGSV. (2010c). *Richtlinien für Lichtsignalanlagen - Lichtzeichenanlagen für den Straßenverkehr.* Köln: FGSV Verlag.

FGSV. (2009). *Richtlinien für passiven Schutz an Straßen durch Fahrzeug-Rückhaltesysteme* . Köln: FGSV Verlag.

FGSV. (2012d). *Richtlinien zur Ermittlung der Luftqualität an Straßen ohne oder mit lockerer Randbebauung (RLuS).* Köln: FGSV-Verlag.

FGSV. (2006b). *Zusätzliche Technische Vertragsbedingungen und Richtlinien für die Ausführung von Lärmschutzwänden an Straßen (ZTV-Lsw).* Köln: FGSV-Verlag.

Hiersche, E.-U., Lamm, R., & Dieterle, K. ,. (1984). *Auswirkungen von Ausbaumaßnahmen gemäß RAS-L auf die Verkehrssicherheit zweispuriger Straßen.* Bonn: Typo Druck Verlag.

Hoffmann, S., Irzik, M., & Friedrich, B. (2006). *Verkehrsablauf an Autobahnausfahrten des Typs A 2 der RAL-K-2* (Bd. 946). (BMVBS, Hrsg.) Bremerhaven: Wirtschaftsverlag NW.

Hoffmann, S., Irzik, M., Friedrich, B., & Matschke, I. (2008). *Zweistreifige Einfahrten an dreistreifigen Richtungsfahrbahnen - Verkehrssicherheit und Verkehrsablauf* (Bd. 992). (BMVBS, Hrsg.) Bremerhaven: Wirtschaftsverlag NW.

Ihs, A. (2006). *The effect of road equipment on traffic safety, accessibility and comfort – A driving simulator study regarding the effect of centre line and guide posts respectively.* VTI Rapport, Swedish National Road and Transport Research Institute, Linköping/Schweden.

ITP München, BVU Freiburg. (2007). *Prognose der deutschlandweiten Verkehrsverflechtungen 2025.*

Kasper, H., Schürba, W., & Lorenz, H. (1954). *Die Klotoide als Trassierungselement.* Bonn: Ferd. Dümmler Verlag.

Kempfe, B., Schlag, B., & Weller, G. (11 2005). Streckencharakteristik und Fahrfehler. *Straßenverkehrstechnik* , S. 564-571.

Kölle, M., & Schnüll, R. (1999). *Sicherheitseigenschaften außerörtlicher Knotenpunkte.* Bremerhaven: Wirtschaftsverlag NW.

Köppel, G., & Bock, H. (1979). *Fahrgeschwindigkeit in Abhängigkeit von der Kurvigkeit* (Bd. 269). (BMVBS, Hrsg.) Bonn: Typo Druck Verlag.

Krebs, H., & Klöckner, J. (1977). *Untersuchungen über Unfallraten in Abhängigkeit von Straßen- und Verkehrsbedingungen außerhalb geschlossener Ortschaften.* Bonn: Typo Druck Verlag.

Krebs, H.-G. (8 1973). Einfluss des Kurvenschneidens auf Bogengeschwindigkeiten und Trassierungsgrundsätze. (Kirschbaumverlag, Hrsg.) *Straße und Autobahn* .

Leutzbach, W., & Zoellmer, J. (1989). *Zusammenhang zwischen der Verkehrssicherheit und den Elementen des Straßenentwurfs.* Bonn: Typo Druck Verlag.

Leutzbach, W., Handschmann, W., & Schuler, M. (1982a). *Einfluss von Verkehrsbeschränkungen und Aufbauformen auf die Sicherheit an plangleichen Knotenpunkten ohne Lichtsignalanlage außerhalb bebauter Gebiete.* Bonn-Bad Godesberg: Typo Druck Verlags.

Leutzbach, W., Maier, W., Handschmann, W., & Schuler, M. (1982b). *Einfluß von Verkehrsbeschränkungen und Ausbauformen auf die Sicherheit an plangleichen Knotenpunkten ohne Lichtsignalanlage außerhalb bebauter Gebiete.* Bonn: Typo Druck Verlag.

Lippold, C. (1997). *Weiterentwicklung ausgewählter Entwurfsgrundlagen von Landstraßen.* Dissertation, Technische Hochschule Darmstadt, Bauingenieurwesen.

Lippold, C., Krüger, H.-P., Schulz, R., & Scheuchenpflug, R. (2007). *Orientierungssichtweite – Definition und Beurteilung* (Bd. 977). (BMVBS, Hrsg.) Bremerhaven: Wirtschaftsverlag NW.

Lippold, C., Lehmann, T., Vetters, A., Ressel, W., & Benner, A. (2011). *Entwurfshinweise zum sicheren sechsstreifigen Ausbau von Autobahnen* (Bd. 1057). (BMVBS, Hrsg.) Bremerhaven: Wirtschaftsverlag NW.

Lippold, C., Schulz, R., Krüger, H.-P., Scheuchenpflug, R., & Lorenz, C. (2009). *Einfluss der Straßenbepflanzung auf Fahrverhalten und Verkehrssicherheit* (Bd. 1018). (BMVBS, Hrsg.) Bremerhaven: Wirtschaftsverlag NW.

Maier, R., & Berger, R. (2012). *Schmale zweibahnig vierstreifige Landstraßen (RQ 21).* Bremerhaven: Wirtschaftsverlag NW.

Maier, R., Berger, R., & Schüller, H. (2013). *Bewertungsmodell für die Verkehrssicherheit von Landstraßen* (Bd. V 226). (B. f. Straßenwesen, Hrsg.) Bremen: Fachverlag NW.

Meewes, V., & Eckstein, V. (2002). *Knotenpunktgrundformen Verkehrsregelung Zufahrten* (Bd. Mitteilungen Nr. 40). Köln: Institut für Straßenverkehr Köln.

Palm, I., & Schmidt, G. (1999). *Querschnittsbreiten einbahniger Außerortsstraßen und Verkehrssicherheit.* Bremerhaven: Wirtschaftsverlag NW.

PTV/TCI. (2009). *Gesamtverkehrsprognose 2025 für die Länder Berlin und Brandenburg.* Berlin.

Richter, T. (1993). *Entwurfsstandards für Knotenpunkte an Ortsumgehungen.* Bonn-Bad Godesberg: Typo Druck Verlag.

Richter, T., & Hüsken, B. (1996). *Einsatzkriterien für Kreisverkehrsplätze außerhalb bebauter Gebiete* (Bd. 757). (BMVBS, Hrsg.) Bonn: Typo Druck Verlag.

Richter, T., & Ruhl, S. (2014). *Untersuchung von Maßnahmen zur Prävention von Überholunfällen auf einbahnigen Landstraßen.* (U. d. GDV, Hrsg.) Berlin.

Richter, T., & Schnüll, R. (1994). *Sicherheitsvergleich der Knotenpunktgrundformen Kreuzung und Rechtsversatz an Straßen außerhalb bebauter Gebiete.* Bonn-Bad Godesberg: Typo Druck Verlag.

Spacek, P. (2 1999). Fahrverhalten und Unfallgeschehen. *Straßenverkehrstechnik* , 68-75.

Spacek, P. (1998). *Fahrverhalten und Unfallgeschehen in Kurven – Fahrverhalten in Kurvenbereichen* (Bd. Forschngsbericht 410). Zürich: VSS .

Spacek, P., & Belopitov, I. (1999). *Geschwindigkeit in Kurven* (Bd. Forschungsbericht 420). Zürich: VSS.

Steierwald, G., & Buck, M. (1992). *Geschwindigkeitsverhalten auf einbahnigen Außerortsstraßen* (Bd. 621). (BMVBS, Hrsg.) Bonn: Typo Druck Verlag.

Steinauer, B., Schuckließ, W., & Becher, T. (2008). *Übergreifende Bewertung von Streckenabschnitten und Knotenpunkten auf Landstraßen* (Bd. 1004). (BMVBS, Hrsg.) Bremerhaven: Wirtschaftsverlag NW.

Vieten, M., Dohmen, R., Dürhager, U., & Legge, K. (2010). *Quantifizierung der Sicherheitswirkungen verschiedener Bau- Gestaltungs- und Betriebsformen auf Landstraßen.* Bremerhaven: Wirtschaftsverlag NW.

Vieten, M., Dohmen, R., Kesting, T., Dürhager, U., & Funke-Akbiyik, R. (2011). *Bewertung von Ortsumgehungen aus Sicht der Verkehrssicherheit.* Bremerhaven: Wirtschaftsverlag NW.

Weber, R., & Löhe, U. (2003). *Verkehrssicherheit und Verkehrsablauf auf b2+1 Strecken mit allgemeinen Verkehr.* Bremerhaven: Wirtschaftsverlag NW.

Weise, G., & Steyer, R. (1999). *Gestaltungskriterien für Landstraßenkurven* (Bd. V63). (B. f. Straßenwesen, Hrsg.) Bremerhaven: Wirtschaftsverlag NW.

Weise, G., Dietze, M., Ebersbach, D., & Kuczora, V. (2002). *Entwicklung eines praktikablen Verfahrens zur Berücksichtigung der räumlichen Linienführung von Außerortsstraßen* (Bd. 849). (BMVBS, Hrsg.) Wilhelmshaven: NW Verlag.

Weiser, F., Sillus, A., Bondzio, L., & Lippold, C. (2006). *Verkehrssicherheit und Verkehrsablauf an planfreien Knotenpunkten mit Fahrstreifensubtraktion.* Bremerhaven: Wirtschaftsverlag NW.

Zierke, B. (2010). *Sichere Gestaltung von Landstraßen durch definierte Straßentypen.* Aachen: Shaker Verlag GmbH.

Zimmermann, M., Loeben, W. H., Dietze, M., & Lippold, C. (2007a). *Kontrolle und Bewertung der räumlichen Linienführung von Außerortsstraßen auf der Grundlage quantitativer Parameter* (Bd. 960). (BMVBS, Hrsg.) Bremerhaven: Wirtschaftsverlag NW.

Zimmermann, M., Roos, R., & Cindric-Middendorf, D. (2013). *Festlegung der Einsatzbereiche für die Rampentypen gemäß RAA unter Berücksichtigung der Zusammenhänge zwischen Verkehrsstärke und Geschwindigkeit sowie anhand der Verkehrssicherheit* (Bd. 1085). (BMVBS, Hrsg.) Wilhelmshaven: NW Verlag.
Zimmermann, M., Roos, R., Beeh, J., & Schulz, S. (2014). *Sicherheitstechnische Überprüfung von Elementen plangleicher Knotenpunkte an Landstraßen.* FE-Vorhaben 02.0316/2009/GGB im Auftrag des BMVBS, Karlsruhe.
Zimmermann, M., von Loeben, W., Roos, R., Dietze, M., & Lippold, C. (2007b). *Entwicklung eines Verfahrens zur Kontrolle und Bewertung der räumlichen Linienführung von Außerortsstraßen auf der Grundlage quantitativer Parameter* (Bd. 960). (B. u. Bundesministerium für Verker, Hrsg.). Wilhelmshaven: NW Verlag

Index

© Springer Fachmedien Wiesbaden GmbH, part of Springer Nature 2023

T. Richter, *Road Planning - Freeways and Country Roads*,
https://doi.org/10.1007/978-3-658-35189-2

D

E

U

V

W

Z

MIX
Papier aus verantwortungsvollen Quellen
Paper from responsible sources
FSC® C105338

If you have any concerns about our products,
you can contact us on
ProductSafety@springernature.com

In case Publisher is established outside the EU,
the EU authorized representative is:
Springer Nature Customer Service Center GmbH
Europaplatz 3, 69115 Heidelberg, Germany

Printed by Libri Plureos GmbH
in Hamburg, Germany